Building
Bioinformatics Solutions

Building Bioinformatics Solutions

with Perl, R, and SQL

Conrad Bessant
Darren Oakley
Ian Shadforth

Second Edition

OXFORD
UNIVERSITY PRESS

Building Bioinformatics Solutions. Second Edition. Conrad Bessant, Darren Oakley and Ian Shadforth.
© Conrad Bessant, Darren Oakley, and Ian Shadforth 2014. Published 2014 by Oxford University Press.

OXFORD
UNIVERSITY PRESS

Great Clarendon Street, Oxford, OX2 6DP,
United Kingdom

Oxford University Press is a department of the University of Oxford.
It furthers the University's objective of excellence in research, scholarship,
and education by publishing worldwide. Oxford is a registered trade mark of
Oxford University Press in the UK and in certain other countries

First Edition published in 2009
Second Edition published in 2014

Published in the United States of America by Oxford University Press
198 Madison Avenue, New York, NY 10016, United States of America

British Library Cataloguing in Publication Data
Data available

Library of Congress Control Number: 2013943290

ISBN 978–0–19–965856–5

Acknowledgements

We would like to take this opportunity to acknowledge the army of very smart people who have contributed to the open source software and publicly available databases used herein. Without these tools, bioinformatics would never have advanced to the state it is in today and this book would probably not exist.

We would also like to acknowledge our colleagues who have contributed to this book by providing advice, ideas, and material. Particular thanks are due to Michael Cauchi, Jenny Cham, Elena Chatzimichali, Jun Fan, Dan Klose, Fady Mohareb, and Will Stott. Thanks also to the team at Oxford University Press who have supported *Building Bioinformatics Solutions* since the beginning and have brought this second edition to the shelves.

Finally, we are indebted to our friends and family who once again provided their invaluable support as we inevitably spent longer than we should have done locked away in our offices.

Preface to the Second Edition

Bioinformatics is a fast moving subject, so while updating this book for the second edition it has been interesting to consider which aspects of the field have changed and which have remained the same. As the commissioning of a second edition indicates, bioinformatics skills continue to be in high demand, as the acquisition of complex multivariate data sets becomes the norm across many areas of biology. Although many new tools have emerged since the first edition of this book was published, and many existing tools have been further developed, there is still a need for the development of new tools, and the creation of bespoke software that brings existing tools together, increasingly across the different 'omics' disciplines.

So the need to build bioinformatics solutions remains. The general approach to building these solutions also remains the same, typically involving a database element, programming, some quantitative analysis, and possibly the creation of a web front end. We have therefore kept this structure within the book. The core tools around which we wrote the first edition—MySQL, Perl, and R—are all still popular within the bioinformatics community. The choice of programming language was a tough one, with Python and Ruby both gaining ground in the bioinformatics community. Ultimately, however, we decided that Perl was still the right choice for this book as it remains the single most widely used language in bioinformatics, and has the benefit of a thriving ecosystem that gives a head start when developing most types of bioinformatics applications. However, in recognition of the increased diversity of programming languages used in the community, we have added new material explaining how to get started in Python, Ruby, and Java to help you experiment with these languages so you can make up your own mind which best suits your needs. Having said that, we should note that all these languages are underpinned by the same fundamental programming concepts, so time spent learning Perl will give you a head start in any of the other languages.

Something else we have witnessed over recent years is a gradual increase in the size and complexity of bioinformatics projects. More complex projects call for more systematic coding practices, such as those proposed by the discipline of software engineering. We have added a new chapter dedicated to this subject, covering version control, documentation, user-centred design, and unit testing. These techniques can help streamline the development of your software—especially when working as part of a team—and make it more reliable, easier to use, and easier to maintain and extend.

The technical area that has changed the most since the first edition is web development. Perl CGI's dominant position in bioinformatics web development has been usurped by web frameworks that are far more powerful while, at the same, make the development and debugging of web applications much easier. At the same time, the arrival of HTML5—the most significant update to the core web standard for over a decade—together with increasingly sophisticated JavaScript libraries such as jQuery, has ushered in a new era of more engaging web applications. Chapter 5 has been substantially revised to take account of these latest developments.

We sincerely hope that this book provides the overview of bioinformatics software development that you are looking for, and that it gives you the knowledge and confidence to go ahead and build your own bioinformatics solutions. We very much welcome your comments and questions—the best way to find us is via `www.bixsolutions.net`.

Conrad, Darren, and Ian
November 2013

Contents

CHAPTER 1

Introduction

1.1 From data to knowledge: the aim of bioinformatics

The term 'bioinformatics' can mean different things to different people. For the purpose of this book, we adopt the broad definition that bioinformatics is the process of extracting novel biological knowledge from bioanalytical data. Such knowledge has immense value because it may be used to better understand biology, to combat disease, and to mitigate environmental catastrophes, but what we actually have is lots of data—genome sequences, protein structures, metabolomic profiles, and more.

Bioinformatics bridges this gap between data and knowledge. It is neither tied to a particular type of experimental data nor to a particular biological application. Indeed, the data may come from any of today's plethora of bioanalytical methods, such as high throughput sequencing, nuclear magnetic resonance, or mass spectrometry, and the knowledge sought can be as varied as the identity of a new disease biomarker, a phylogenetic tree, or a system-wide understanding of a particular biological process. Similarly, the tasks involved in bioinformatics range from simply organizing data for future use, through to sophisticated analysis, visualization, and sharing of that data and the results derived from it. Bioinformatics is therefore a truly interdisciplinary subject, requiring an understanding of at least some biology, analytical science, mathematics, statistics, and information technology.

There are, of course, many generally available bioinformatics tools that can be used to analyse data with a view to extracting new knowledge, and many of these tools are free of charge. However, due to the high complexity and bespoke nature of biological data sets, it is often necessary to produce in-house software to organize, analyse, and visualize data. In this book, we introduce some of the main tools and general approaches employed to produce such software.

The book is primarily aimed at readers with a background in the life sciences who have some bioinformatics knowledge, but little or no experience in the development of software and databases. A typical reader will most likely have used online databases such as GenBank or Ensembl (if not, these are introduced briefly later in this chapter), and experienced the power of tools such as BLAST (Basic Local Alignment Search Tool) or analysis platforms such as Galaxy, and now want to take the next step and develop their own bioinformatics tools, either for their

Building Bioinformatics Solutions. Second Edition. Conrad Bessant, Darren Oakley and Ian Shadforth.
© Conrad Bessant, Darren Oakley, and Ian Shadforth 2014. Published 2014 by Oxford University Press.

own use or for sharing with their collaborators or with the life science community at large. Our aim in writing this book was to fill the gap between texts that introduce the field of bioinformatics, such as *Introduction to Bioinformatics* (Lesk, 2008), and software development books, such as those published by O'Reilly (e.g. Laurie & Laurie, 2013; Christiansen *et al.*, 2012; Tahaghoghi & Williams, 2009). We therefore cover computing material from a fairly elementary level (for example, Appendix A explains how to use a command-line interface), while the biological background necessary to understand the applications is assumed.

1.2   Using this book

The book has been written with the intention of being read linearly from beginning to end, with a structure that generally mirrors a typical bioinformatics project. First, it is necessary to assemble the data, be it from a laboratory or from online resources, typically into a structured database using a relational database management system (RDBMS) such as MySQL (Chapter 2). Then, due to the nature of bioinformatics data sets, some programming is usually required to automate various data manipulation tasks—Perl (covered in Chapter 3) is the tool of choice for this among bioinformaticians. Often, advanced numerical data analysis is then required to extract useful information from the data gathered, which is where R comes in (Chapter 4). In Chapter 5, we bring everything together and the real power of integrating Perl, R, and MySQL becomes apparent as we show how to combine these ingredients with HTML5 and web frameworks to make complex bioinformatics tools available via the web. Finally, Chapter 6 introduces some good software engineering practice—essential when working on larger multi-developer projects—and introduces alternative programming languages that are purported to have some technical advantages over Perl. Having said all that, Chapters 2, 3, and 4 have been written to be reasonably standalone, so if you have an urgent need to go straight to a particular chapter, that is possible. Where necessary, cross-references are provided to help locate explanations of concepts that appeared earlier in the book. Similarly, if you are already experienced in one of the three main tools covered, you should be able to skip that chapter without any problems.

One thing that this book is definitely not is a reference manual. The index should help you find a relevant passage in the book when you need it, but don't expect to see every single Perl function, R package, or Apache configuration option listed there. A simple reason for this is that there just is not enough space, but more importantly these tools have excellent online documentation and we find such documentation to be a much more convenient format for reference materials because it is easy to search and frequently updated. This book is a companion to those reference materials, and a starting point in terms of knowing where the reference materials are and how to use them efficiently.

1.2.1   About the coverage of this book

Deciding what to include and what not to include in this book was not easy. Whole books have been written about MySQL, Perl, and R, not to mention all the various types of bioinformatics applications to which they can be applied. It is, therefore, impossible to claim that this book provides exhaustive coverage of all the subjects covered. Selecting the topics to cover was very much like writing a tourist guide to a particular country—it is not feasible to cover everything of interest, so information needs to be carefully selected so that it helps newcomers orientate themselves, outlines the practicalities of survival, covers a few highlights that give a flavour of what is possible, and tells you where you can go to find out more. This is the approach we have taken in this book.

1.2.2 Choice of tools

The toolkit for building bioinformatics solutions that we put forward in this book is loosely based on the so-called LAMP toolkit, which has been widely used by developers in all sorts of domains, not just bioinformatics, for many years. LAMP is an acronym for four popular open source software packages: Linux (the operating system), Apache (web server), MySQL (database), and Perl[1] (for programming). Together, these packages make a powerful combination for gathering, storing, and serving up data over the Internet. To this mix we add R, which brings with it the sophisticated data analysis and visualization capabilities frequently needed in bioinformatics applications. A key factor in the appeal of this suite of tools is that they are open source and freely available. To be frank, the technical benefit of these tools being open source is minimal for the typical user. Although one is able to view and edit the source code of the tools one is using, the chances are that one will never need to do so and, if one did, the complexity of these packages is such that making worthwhile modifications would be very time consuming. Even the fact that the packages can be obtained free of charge is of little direct relevance if one has a software budget. What really makes the use of open source tools appealing is that they are ubiquitous and this widespread uptake has two very important consequences. First, there are a lot of people around who know how to use them, so getting help should not be a problem. Second, if someone wants to produce a piece of software or an add-on, they are most likely to do it with these tools to allow maximum exposure. This latter benefit is really the most important, as it means that there is, for example, a constant stream of add-on modules for Perl and packages for R that perform common bioinformatics tasks. We refer to the collection of community activities and third-party add-ons for a particular tool as the tool's *ecosystem*. Choosing to use a development tool that has a well-established and dynamic ecosystem saves a huge amount of development effort and allows one to concentrate on the science, which is, after all, the main priority in bioinformatics.

1.2.3 Choice of operating system

Throughout this book we have worked to ensure that everything is operating system independent as far as possible. All examples have been tested on Windows

---

1 Depending on who you ask, the P might actually stand for a similar language called PHP, or even just programming in general. It doesn't really matter.

8, Mac OS 10.8.3, and Ubuntu Linux 12.04. This is not just a ploy to maximize sales of the book, but a reflection of the fact that, in real-world bioinformatics, all three operating systems are commonly used. In general, Linux is the operating system of choice in bioinformatics, because it is well suited to running servers and is very scalable, allowing solutions to be developed for anything from individual desktop PCs through to multi-processor supercomputers. However, the familiarity, usability, and availability of Windows and Mac OS ensure they are also widely used, particularly by those starting out in bioinformatics, and among the biologists using the software that we produce. Indeed at the time of writing, some of the most popular freely available software tools for proteomics run exclusively on Windows.

Where there are differences between operating systems, typically when installing software, we have provided instructions for all three. Clearly, we have not been able to test the material in this book on every available Linux distribution, or on versions of Windows or Mac OS that were released after the book went to press. If you have any problems getting the examples in the book to work, we recommend that you head over to the book's website (`www.bixsolutions.net`) in search of a solution.

1.2.4   www.bixsolutions.net

To help you as you work through this book, we maintain a companion website, `www.bixsolutions.net`, where you can find the main example programs and data from the book, as well as up-to-date lists of recommended reading. The site also has a discussion forum, which is monitored by the authors, who are happy to help if you have any difficulties while working through the examples. Obviously, we have checked the examples very carefully and believed them to work correctly when the book went to press, but they may not work forever. This is because the tools and, indeed, the operating systems, used in this book are constantly evolving—new versions are released frequently, so functions may be deprecated, database schemas may change, and new features may become available. As we become aware of any changes that affect material in the book we will post updates or workarounds on the website.

For readers familiar with code repositories, we also maintain a GitHub repository (`github.com/dazoakley/bbs-v2`) containing the main example programs from the book. For those new to code repositories, this topic is covered in Chapter 6.

1.3   Principal applications of bioinformatics

Since its origins in genomic sequence analysis, bioinformatics has spread across the whole of molecular biology, in its broadest sense. There are already many fine texts explaining the various applications of bioinformatics (e.g. Lesk, 2008), so we don't replicate such material here. However, we include below a brief outline of the main areas of contemporary bioinformatics with particular emphasis on the application to those areas of the tools introduced in this book.

1.3.1 Sequence analysis

Sequence analysis is a massive field, covering all manner of analysis of textual sequences representing genomes (DNA) and proteins (sequences of amino acids). Applications within genomics are wide ranging and include sequence assembly, prediction of coding regions (i.e. genes), determination of genomic structure, research into the purpose of non-coding DNA, translation of DNA into protein sequences, comparison of sequences to infer evolutionary relationships, rates of evolution, the study of variation between individuals, and prediction of gene function and regulation. DNA sequence analysis is also used in the design of experiments that employ techniques such as PCR and microarray technology.

Protein sequence analysis has a similarly wide range of applications, including protein structure prediction, inference of protein function based on sequence similarity, determination of protein similarity for building protein families and understanding protein evolution, and identification of structural or functional subsequences (*motifs*). Protein sequences are also used when designing and interpreting the results from proteomic (protein expression) experiments.

From a technical point of view, a distinguishing feature of sequence analysis applications is the frequent use of relatively large data sets such as whole genome sequences (for example, the human genome, which consists of about 3,000,000,000 bp), or multiple genomes when studying variation or metagenomics. As well as having to deal with the sequences themselves, it is also necessary to handle *annotations*, without which the sequence data is meaningless. Such annotations include things like the species from which a selected sequence originated, its location in the genome, and any functions or gene products that have been associated with it. A RDBMS, such as MySQL, is essential to store such data in a useable way. Another distinguishing feature of sequence analysis is the need to efficiently process textual data, essentially long strings of As, Cs, Gs, and Ts, or the 20-letter alphabet used to represent amino acids. Typical tasks include looking for motifs within much larger sequences and looking for similarities between sequences. Handling large amounts of textual data is something at which Perl excels, as we will see in Chapter 3.

1.3.2 Transcriptomics

With the invention of DNA microarrays it became possible, for the first time, to monitor the expression level of all known genes in an organism (e.g. approximately 22,000 genes for humans) in a single analysis. Such genome-wide expression analysis has become known as *transcriptomics*. This yields large sets of highly multivariate quantitative data, as studies tend to involve multiple samples so that differential expression or behaviour over time can be observed.

More recently, the high throughput sequencing-based RNA-seq method has become a popular alternative for transcriptomic studies. A key advantage of RNA-seq is that it sequences cDNA generated from the transcribed mRNA, whereas the fundamental principle of microarray analysis is the hybridization of cDNA to pre-fabricated DNA probes. RNA-seq is therefore a more open technique, particularly useful for non-model organisms or studies where multiple species or sequence variation are of particular interest.

Just storing expression data and associated information about that data (the so-called *metadata*, which describes the sample and experimental conditions) can be an issue, but the major challenge lies in the processing and statistical analysis of gene expression data. In microarray experiments, the raw data is in the form of an image that needs to be processed and cross-referenced with metadata describing the array design (i.e. which spot relates to which gene) to yield numerical expression values. The raw data in RNA-seq studies consists of millions of short sequence reads, which are typically mapped to a reference genome for identification purposes, and the abundance of which are used to achieve quantitation.

Once the expression values have been extracted, assigned to genes, and tabulated, we can begin statistical analysis to identify biologically important features, such as differentially expressed genes in a comparative study or co-regulated genes in a temporal study. Due to the large data sets and sometimes sub-optimal experimental designs (the number of genes monitored usually exceeds the number of samples analysed) the statistical methods required can be complex. This is where R is extremely valuable, as the common processing algorithms and even some of the more exotic methods have already been implemented for us in R packages, such as Bioconductor. Chapter 4 will get you started with R and, having worked through it, you should be in a position to start working with packages like Bioconductor.

1.3.3   Proteomics

Gene expression studies are essential in understanding gene regulation and related phenomena, but to gain a deeper insight into how a biological system functions it is arguably more useful to look at the expression of the functional molecules themselves: the proteins. Proteomics is the science of identifying and, where possible, quantifying proteins in a sample. At the time of writing, the majority of proteomics methods are based on mass spectrometry (MS), coupled with prior protein digestion and separation steps to help ensure that peptides are delivered to the mass spectrometer individually. Due to the complexity and variety of proteomic protocols, handling the data from these experiments can be a major challenge. Some laboratories separate proteins using two-dimensional gel electrophoresis, after which image analysis is employed to identify and define gel spots, prior to excision and MS analysis. Today it is more common to use liquid chromatography (LC) instead of gels as this supports a much higher throughput, resulting in large data sets comprising several thousand spectra per sample.

Whichever separation technique is used, data analysis is required to identify the peptide represented by each mass spectrum. The most common techniques are peptide mass fingerprinting (PMF) or, if peptides have been subjected to secondary fragmentation in a subsequent MS stage, a search against simulated spectra derived from a database of known protein sequences for the species being studied. Identified peptides then need to be assigned to proteins, which is not easy due to the relatively short length of each peptide and the many proteins that could potentially be present in a sample—in human tissue this could be several hundred thousand if splice variants and post-translational modifications are taken into account. Due to the large data sets and the experimental metadata needed to make use of this, a well-designed relational database (perhaps implemented in MySQL) is essential to organize the information, and the increasingly high throughput nature of proteomics necessitates analysis pipelines to be built (often in Perl) if data analysis is to keep pace with data acquisition. Having said that, certain steps of the processing have been shown to be amenable to large-scale, non-RDBMS (NoSQL) methods, which we touch on briefly in Chapter 2.

Having identified proteins from all the samples in a given study, statistical analysis then needs to be performed across samples to extract the biologically significant information from the acquired experimental data. Some proteomics protocols are only qualitative—they aim to determine which proteins are present in a sample, but not how much of each protein is present. However, a raft of quantitative protocols is now in use, which can reveal not just the identities of proteins, but also their abundance. Statistical techniques similar to those used for gene expression analysis are clearly applicable to this data and, as these protocols are adopted more widely, relevant R packages are emerging.

1.3.4 Metabolomics

Metabolomics deals with the identification and quantification of small molecules in biological samples. Analysis of metabolites is one of the most well-established bioanalytical techniques that we come across in bioinformatics. The primary technologies used—nuclear magnetic resonance (NMR) and mass spectrometry following gas chromatography (GC) or liquid chromatography (LC)—have been around considerably longer than high throughput sequencing, microarrays, or proteomic MS. Indeed, it is even possible to buy pocket-sized devices for personal monitoring of medically important metabolites, such as glucose and cholesterol. However, metabolomics brings a new emphasis to high throughput analysis and the desire to quantify as many analytes as possible in each sample. The desire for such a global view of the metabolome is being driven mainly by the search for diagnostic biomarkers and the growth of systems biology. As in proteomics, increasing data volumes are encouraging people to put together relational databases and software pipelines for metabolomic data analysis. As metabolite analysis is well established, so are the core algorithms for dealing with the data from such analysis—such methods often come under the banner of *chemometrics*, about which some very good introductory texts have been written (e.g. Brereton, 2007; Otto, 2007). R is a perfect environment for such analysis, as demonstrated by some of the examples in Chapter 4.

1.3.5 Systems biology

Traditionally, bioanalytical science, and even the bioinformatics that supports it, is broken down into the areas outlined previously—genomics, transcriptomics, proteomics, and metabolomics. In terms of how organisms function, these delineations are artificial because in reality the genes, proteins, and metabolites are free to interact. Systems biology recognizes this and, in bioinformatics terms, comprises integration of both data storage and analysis to permit system-wide analysis and modelling of living organisms without being constrained to a particular class of molecule. This is likely to be a theme of bioinformatics for some time to come, because the potential outcome of such work—cell and tissue models with application in areas such as toxicology—is so valuable, but the challenge of achieving this is immense. The core elements of database design, programming, and numerical data analysis covered in this book are all important in systems biology, and much of the effort in systems biology is focused on the design of databases for meaningful storage of heterogeneous data and novel methods for data analysis and visualization.

1.3.6 Literature mining

PubMed, which contains bibliographic information on journals primarily associated with biomedicine (see Section 1.5.1), is growing by around 3,000 citations every day. Keeping up with the work described in all these papers by reading them is clearly impossible, and that is before we consider the backlog of at least 20 million papers already out there. There is, therefore, a lot of interest in literature mining methods that help to facilitate the high throughput machine reading of papers to extract salient information, and advanced ways of searching through and annotating papers to assist human reading. Perl's text handling capabilities make it ideal for this type of work, as does its ability to automate querying of bibliographic databases and retrieval of papers from websites. There can also be a need to visualize the results of literature mining, which can be handled by Perl or R. For example, if we wrote a Perl program to extract protein–protein interactions from text, it would be convenient to display the resulting network of interactions graphically, as this is a representation with which biologists are familiar.

1.3.7 Structural biology

Structural biology is the study of the physical architecture of biological molecules—particularly proteins. Research typically focuses on topics such as the relationship between structure and function, structural similarity between proteins, simulation of interaction between proteins and other molecules, and the relationship between protein sequences and their structure (particularly the process of protein folding). Some of these topics can be tackled using the tools introduced in this book, but due to the mathematical complexity of molecular simulations, it is common for researchers to use existing modelling tools (some of them commercial products) or to produce bespoke software in lower level languages, such as C. Such tools are sometimes pipelined using Perl, with a program being written to retrieve a structure from a repository of protein structures, such as PDB (see Section 1.5.1), pass it to a modelling program, and deposit the final result of the modelled experiment into a local database.

1.4 Building bioinformatics solutions

The premise of this book is that, regardless of the particular type of data being analysed, or the scientific purpose behind the analysis, the tools and general approaches used to solve a bioinformatics problem are often the same. This is because most bioinformatics projects share a similar aim—to bring together data (be it public or proprietary) with analysis tools (be they existing or novel) to generate new biological knowledge. We refer to this as building a bioinformatics solution because it best describes what we are doing—putting things together to solve a problem.

A schematic representation of this concept is shown in Fig. 1.1. Regardless of the particular type of data being analysed or the scientific aim behind the analysis, the general structure of a bioinformatics solution tends to follow this pattern, although not all components are required in all applications. For example, it may be possible to analyse novel data collected locally without recourse to any public databases. Similarly, we might be able to rely entirely on our own in-house analysis software, instead of sourcing tools from the public domain. Conversely, we may not have access to any novel data or novel analysis routines, and instead focus entirely on the analysis of public data using publicly available tools. The way in which results are output from our system may also vary. The figure shows two potential routes, which may be used together. These are deposition of the results in a local database, and presentation of the results to one or more users via a web browser. Unless the software is truly single user—intended only for use by the developer—some kind of user interface is required, and a web-based interface is convenient because web interfaces are familiar to users, are relatively easy to implement, and should be platform independent. The web obviously allows you to make your software available to remote users around the world, but even if you are developing a solution for local use in a single group or organization, a web interface is often the best way to go.

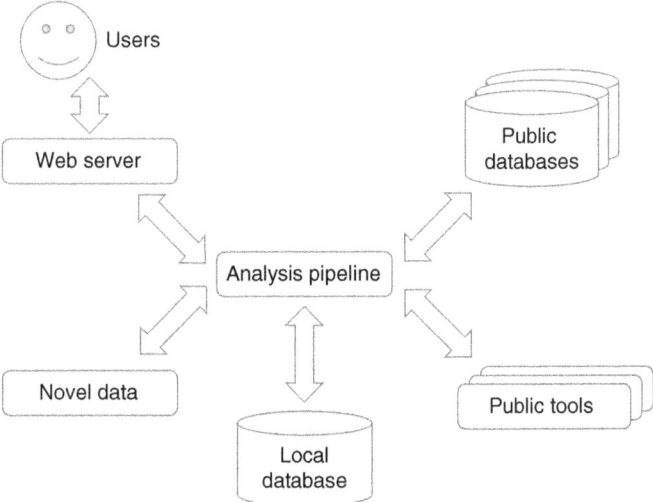

Fig. 1.1 A generic bioinformatics solution showing typical components that may be used.

So, the key components of a system like the one shown in the figure are the analysis pipeline, the local database, and the web interface. The process of building a bioinformatics solution starts with capturing the project requirements, typically by manually carrying out analysis on a small subset of the data being collected and, if appropriate, talking to users about what they want from the software. This should reveal which components are needed and what functionality they should have. These components can then be produced using the tools described in the following chapters, specifically the pipeline (with Perl and R), the local database (with MySQL), and the interface (with HTML5, a web framework and Apache). For completeness, a brief overview of other potential components—the public tools and databases—is provided in the next section.

1.5   Publicly available bioinformatics resources

We are very fortunate to have a substantial body of high quality, freely available bioinformatics resources accessible via the Internet. Most of the core resources, or services as they are sometimes called, are hosted by two major organizations—the American National Centre for Biotechnology Information (NCBI) and the European Bioinformatics Institute (EBI), which is based in the UK. Resources provided by these organizations are easily accessed via their websites (`www.ncbi.nlm.nih.gov` and `www.ebi.ac.uk`, respectively). These resources can roughly be broken down into databases and analysis tools, although the distinction is becoming ever more blurred as the websites through which databases are accessed begin to include more sophisticated integrated analysis tools. What follows is a brief overview of the key resources at these sites and on the wider Internet.

1.5.1   Publicly available data

There exists a substantial global collection of data covering a wide range of biological areas, from gene sequences to protein structures and medical information. Furthermore, this collection is growing all the time, both in terms of the number of databases and the amount of data in each database. Indeed, many of the databases are growing exponentially.

This impressive collection of data forms the basis for much bioinformatics work, especially among the many researchers who do not have laboratories or laboratory-based collaborators. Even when we are working with novel proprietary data, we often need to analyse this in the context of publicly available data. For example, if we are looking at SNP data, we may want to map this to existing genomic annotations to identify potential consequences of that SNP. In gene expression studies, it is common to use publicly available information about gene function to provide context for the analysis. In proteomics, it is common practice to identify peptides by searching acquired mass spectra against protein sequence databases.

Due to the extent and rapidly changing nature of the database landscape, we are unable to provide a thorough review of all available resources here. Instead, we would refer you to the annual *Database Collection* published in the journal *Nucleic Acids Research* (www.oxfordjournals.org/nar/database/c), which has summaries of many hundreds of databases. However, it is worth taking a little space here to introduce what we consider to be the key core databases at the current time, particularly as some of these are referred to in later chapters.

Genome sequences

The core repositories for nucleotide sequence data are Genbank (at NCBI), ENA (at EBI), and DDBJ (at the Japanese National Institute of Genetics). These three databases comprise the International Nucleotide Sequence Database Collaboration (INSDC), which collectively capture all the public genome sequence data ever collected. There is even some putative—though seemingly dubious—dinosaur DNA in there if you look hard enough (accession number[2] U41319). As part of the collaboration, the contents of these three databases are automatically synchronized daily, so a sequence submitted to one will appear in the other two. All three databases therefore contain the same data. This may seem wasteful, but having the databases in different geographical locations increases data security and ensures that most researchers have the data reasonably close to them, which improves efficiency of access. Also, the three databases are distinct in that they have different user interfaces.

Although the three INSDC databases are exhaustive in terms of cataloguing all available nucleotide data, the organization of the data is fairly rudimentary and, as a result, they are not the easiest databases to browse. Consequently, many *secondary databases* have appeared, in which the primary data from repositories such as those in the INSDC has been compiled and indexed into a form that is highly structured, easy to browse, and well integrated with other resources. An excellent example of this is Ensembl (www.ensembl.org), which provides information about a number of completed eukaryotic genomes. As well as the original sequence data, assembled into whole chromosomes, Ensembl also provides annotations for this data, such as known genes, gene predictions, gene structure, gene products, orthologues, and SNPs. As well as being accessed via the web interface, Ensembl can also be accessed by programs directly via an application programming interface (API), or can even be downloaded in its entirety for working with locally, which can be faster and arguably offer more privacy than connecting to the database via the Internet.

Protein sequences

The most comprehensive central resource for protein sequences and functional annotation is UniProt (www.uniprot.org). The UniProt knowledgebase comprises two main sections: a large database called TrEMBL, which contains protein sequences produced by automatic translation of ENA nucleotide sequences, and a

---

2 An accession number is a unique identifier for a particular record in a database. The record can be found simply by searching for this number.

much smaller database of manually curated protein sequences called Swiss-Prot. Thanks to the extensive manual curation, which includes the addition of cross-references to many other databases, the data in Swiss-Prot is recognized as being of very high quality and it is considered to be the gold standard protein sequence database. As well as individual protein records, the UniProt knowledgebase also contains complete proteomes for an increasing number of organisms. Many other protein databases exist, providing information related to or derived from UniProt sequences. A notable example is InterPro (www.ebi.ac.uk/interpro), which provides information about protein families, domains, and functional sites.

Transcriptomic data

There are two main databases of gene expression data: Array Express (www.ebi.ac.uk/arrayexpress) and the Gene Expression Omnibus—GEO (www.ncbi.nlm.nih.gov/geo). These databases support data from both microarrays and RNA-seq. A key element in establishing these databases was finding agreement within the community for a common way in which to report all the relevant details of a microarray experiment. This was finally achieved by the definition of the MIAME (Minimum Information from A Microarray Experiment) reporting standard, and databases such as GEO and ArrayExpress are fully MIAME compliant. Having this consistent method of representing data is important because it means that we can automatically download expression data complete with all the metadata necessary to interpret it, regardless of its origin. Also, having a common data standard means that people are willing to spend time developing software that supports it. Indeed, we will see in Chapter 4 that there are R packages to deal directly with this type of data.

Protein expression data

Proteomics is following transcriptomics both in terms of uptake in the laboratory and in terms of data repositories. Like transcriptomic data, proteomic data is only useful if metadata is available to provide context to that data, and this issue has largely been addressed through the development of reporting guidelines and data formats by the Proteomics Standards Initiative (PSI). Substantial databases of protein identifications and associated mass spectrometry data have already emerged, most notably PRIDE (www.ebi.ac.uk/pride) and PeptideAtlas (www.peptideatlas.org).

Metabolomic data

Repositories of metabolomics data sets have taken some time to appear due to the difficulty in coherently capturing data from the many diverse protocols employed. In contrast to proteomics and transcriptomics, many metabolomics studies employ profiling methods where phenotypic differences are inferred directly from analytical data without first identifying all (or sometimes any) of the molecular entities represented by that data. This is, in part, because the analytical signatures of many metabolites are not known so they simply cannot be identified. However, the situation is improving thanks to the development of databases that catalogue small molecules and their analytical signatures. These include the

Human Metabolome Database (`www.hmdb.ca`), Metlin (`metlin.scripps.edu`), PubChem (`pubchem.ncbi.nlm.nih.gov`), and ChemSpider (`www.chemspider.com`). These databases are often used as libraries to identify compounds from mass spectra acquired during metabolomics experiments. In terms of the data and findings resulting from such experiments, MetaboLights (`www.ebi.ac.uk/metabolights`) leads the way at the time of writing.

Molecular structures

The pre-eminent protein structure database is the Protein Data Bank—PDB (`www.pdb.org`). This contains protein structures determined primarily using X-ray diffraction or NMR. The information about each structure is quite exhaustive, including information about who determined the structure and how they did it, biochemical information about the protein and, of course, the structure itself, in formats that can be viewed within PDB's web interface, and that allow the structure to be downloaded and analysed locally. Other protein structure databases, many derived from PDB, are available, as well as databases of smaller molecules.

Interactions and pathways

The databases mentioned thus far are primarily collections of experimental data from laboratory instruments, albeit augmented with manually generated annotations. Pathway databases are different in that they contain interaction data *derived* from experiments, rather than the experimental data itself. Such databases are becoming increasingly important as systems biology studies become more common. One of the most well-established pathway databases is KEGG (the Kyoto Encyclopedia of Genes and Genomes) Pathway (`www.genome.ad.jp/kegg/pathway.html`). KEGG comprises several databases, with the Pathway database dedicated to data pertaining to molecular interaction networks. Much of this information has been available for some time in books and papers, but having it online in electronic form facilitates easier access, new applications, and better integration with other resources.

Other useful databases in this general field include the Reactome pathway database (`www.reactome.org`), the IntAct database of molecular interactions (`www.ebi.ac.uk/intact`), and BioModels (`www.ebi.ac.uk/biomodels-main`), which is a collection of mathematical models of biological systems.

Literature

When we think of scientific literature, journals come to mind and, indeed, it is here that most scientific literature can be found. Despite the existence of the repositories of experimental data described above, journal articles still play an essential role, acting as the glue that links a lot of this data together and gives it context. It is also where the biological knowledge extracted from experimental data is reported. Today, the vast majority of journal articles are available online, although not all are available free of charge (open access). A large body of bibliographic information about papers, including abstracts, is available in the PubMed database (`www.pubmed.gov`).

Journal articles are not the only source of textual information. One notable repository is OMIM (`www.ncbi.nlm.nih.gov/omim`), which contains textual descriptions of all known human genetic disorders. These descriptions include links to supporting data in many of the repositories described above. Similarly, there are links to OMIM from many of these repositories.

Ontologies

In bioinformatics terms, ontologies are essentially lists of terms with strictly defined meanings that have been agreed upon by the scientific community. Where more than one word can be used to describe the same thing, synonyms are listed along with the definition. To avoid any confusion or duplication, each term has a specific accession number associated with it. The most commonly used ontology in bioinformatics is the Gene Ontology—GO (`www.geneontology.org`). It may sound like an ontology is little more than a dictionary, but an important additional feature is that relationships between terms are also captured. For example, in GO the term 'carbohydrate binding' is defined as a subset of the molecular function 'binding', and many specific types of binding, such as 'glucose binding', are linked to that more general 'carbohydrate binding' term.

Ontologies have many uses in bioinformatics. At their simplest, they can be used as *controlled vocabularies*—lists of terms that users are restricted to selecting when entering data. A typical way of enforcing a controlled vocabulary on users is to only allow input via a drop-down box, which contains only the permitted terms. Retrieving and analysing data using automated systems is a lot easier if it is annotated in this way, rather than with free text descriptions. For this reason, many of the standard data formats, such as the MIAME microarray standard, make use of ontologies for their metadata. Ontologies can also be used to facilitate advanced querying of data. For example, searching for documents using a normal keyword search for 'carbohydrate binding' would simply return documents containing that term, whereas a search augmented by GO could additionally return documents that do not contain the term 'carbohydrate binding', but do contain terms that GO defines as being related to this, such as 'glucose binding'. Thanks to such utility, ontologies appear in almost every area of bioinformatics.

GO is currently the most ubiquitous of biological ontologies, as it covers three key areas of interest: biological process, cellular component, and molecular function. However, there are many complementary ontologies, most of which can be found at the Open Biomedical Ontologies Foundry (`www.obofoundry.org`). Although there are various ontology formats, the important thing is that these ontologies are freely available, easily machine readable, and can therefore be incorporated into our own programs with relative ease.

1.5.2   Publicly available analysis tools

As well as data, there is also a wealth of data analysis tools freely available via the web. Some of these are add-ons for specific software such as Perl (Perl modules) and R (R packages), which we will deal with in the later chapters. Many other tools are available in standalone form, either accessible via web front ends, or as programs that you can download and run locally. The most frequently used tools can be found among the services listed at the EBI (`www.ebi.ac.uk/services`) and via the Tools tab on the NCBI resources page (`www.ncbi.nlm.nih.gov/guide/all`). Due to the fundamental importance of sequence analysis, these toolboxes tend to be biased towards sequence analysis, with BLAST and the ClustalW multiple sequence analysis tool arguably being the most well known. However, other tools, such as OMSSA (at NCBI) for proteomic mass spectrometry and DaliLite (at EBI) for pairwise structure comparison, clearly cover other data types.

For occasional analyses, as done by the typical laboratory-based biologist, the web-based interfaces to these tools are very convenient. The good news for those of us seeking to build bioinformatics solutions with higher throughput is that most of these tools can also be accessed programmatically, allowing us to incorporate them into our own software and automate analysis. This can be done either by downloading versions of the tools that can be run locally and incorporated into your programs, or by connecting to a server at the EBI or NCBI, passing across the data of interest and running the tool there. The latter is achieved using an API that allows data and commands to be sent directly to the tool on the server, and results returned. The details of how to do this vary from tool to tool, but comprehensive instructions and examples are available at both the NCBI and EBI toolbox websites, and an example is given in Chapter 3.

Another suite of tools worth knowing about, especially if you are working with sequence data, is the European Molecular Biology Open Software Suite—EMBOSS (`emboss.sourceforge.net`). This open source suite mainly comprises tools for sequence analysis, such as sequence alignment, picking primers, and looking for sequence motifs. The programs that make up EMBOSS are well respected and easy to incorporate into your own software, making them a popular choice for bioinformatics developers.

There are many other freely available tools out there, but because these tools tend to be developed by different academic groups, they can be hard to find. In our experience, the best way to discover new tools is by monitoring relevant journals, although even that is a challenge, as tools might be announced in bioinformatics or domain-specific journals, such as those covering genomics, proteomics, or metabolomics.

1.5.3 Publicly available workflow solutions

As already mentioned, many bioinformatics applications require the connection or integration of various individual tools. Writing Perl programs is arguably the most flexible way to do this, but not everyone has the skills necessary to do that. In response to this need, a number of software platforms have emerged that allow integration of different bioinformatics tools and resources without any need for programming. These platforms act as a host environment for other tools, enabling them to be joined together into complex pipelines that can be saved for future use and shared among the scientific community. Most of these have graphical interfaces that allow workflows to be created by dragging graphical representations of tools into position and connecting them together, and the best of them take care of optimizing the execution of the workflow on whatever computer hardware you have to hand.

Bioinformatics workflow platforms include Taverna (`www.taverna.org.uk`), Knime (`www.knime.org`), and Galaxy (`usegalaxy.org`). At the time of writing, Galaxy is most popular of the three. Indeed if you ask biologists which bioinformatics tools they use, many will say Galaxy ahead of—or even instead of—the tools within Galaxy that actually do the work. The vast majority of tools supported by Galaxy are for analysis of sequence data, but efforts are underway to add support for other data, such as mass spectra, to serve the proteomics and metabolomics communities.

Do workflow platforms such as Galaxy spell the end for programming in bioinformatics? Not at all. These platforms can do a lot of the tedious work of connecting tools and distributing compute jobs, but the development of new tools for data storage, analysis, and visualization is needed as much as ever. Indeed, environments like Galaxy provide an excellent route to generating demand and maximizing uptake of new tools that we produce. Furthermore, whenever it becomes necessary to add a newly created tool or data format to a workflow platform, there is invariably some programming to do.

1.6   Some computing practicalities

Finally, before we get into the substance of this book, we need to mention a few practical issues that apply to all of the following chapters.

1.6.1   Hardware requirements

A question we are often asked by people looking to develop their bioinformatics skills is: 'what kind of computer do I need?' For some people, the expectation is that, with all the talk about large data sets, whole genome analysis, and complex visualizations, high-end computing resources must be required. In practice, however, most bioinformaticians are able to do what they need on a standard desktop computer, and that is definitely the case for this book. Any recently produced PC or Apple Mac, in either desktop or high performance laptop format, connected to the internet, should be fine for running the examples in this book. For the avoidance of doubt, tablets, chromebooks, and similar lightweight devices are not a sensible choice for this type of work.

The alternative to using your own computer is to adopt a client–server approach, as practised in many organizations. In this scenario, a powerful computer is set up as a server and core software is installed on it. In bioinformatics, the server would typically be running a Linux operating system, with MySQL (or an equivalent such as Oracle), Perl, R, and Apache web server installed. You would connect to the server from a separate (client) computer to query databases, execute commands, and run software that you have created. This may sound like a lot of unnecessary hassle, but it has several benefits if multiple users or software developers are involved. First, because it is not necessary to physically sit at the server to use it, it can be used simultaneously by a number of people. This makes it possible to share, for example, a single installation of MySQL or even a specific database. This considerably reduces administrative duties, because the software only needs to be installed and configured once, regardless of the number of users. This approach is therefore very popular in larger organizations, such as universities and companies that are big enough to have dedicated system administrators. The client–server approach also provides separation between the server and the personal computer on which you write program code, check your email, browse the web, and so on. This has several advantages:

• The server can be in a remote location, such as a data centre with better security and an uninterruptible power supply.

• You can run a different operating system on your personal computer to the operating system that is installed on the server.

• You can switch off, reboot, or otherwise abuse your computer without affecting the server.

All this is particularly useful if you are developing bioinformatics tools that need to be accessible to other people and therefore need a high level of reliability.

Cloud computing is a modern manifestation of the client–server approach. The main difference compared to client–server is the fact that in a cloud environment the remote computing resource may be geographically distributed across multiple servers and it may be supplied as a product, paid for according to the amount of storage and processor time that your software uses. This has the attraction that you are not burdened with the upfront costs of purchasing and commissioning large pieces of computer hardware, and brings the benefits of economy of scale through sharing running costs among many other users.

Despite the advantages of working in a server or cloud-based environment, when starting out we recommend you use a single computer for which you have administrator rights. This is what we assume throughout the book. However, if you are in an organization where you wish to, or are forced to, use a server you shouldn't have any major problems with the examples. You will just need to liaise with the server administrator to find out what software is available on your particular server and how to access it.

1.6.2 The command line

In most of the examples throughout this book we interact with the computer via command-line interfaces. In a command-line interface, instructions are issued to the computer simply by typing commands. Any feedback is returned to the screen. This may seem like a step backwards in a world of mice, touch screens, and graphical user interfaces (GUIs), but for some tasks commonly carried out in bioinformatics, such as querying databases or manipulating large data sets, command-line interfaces can actually be much more efficient than GUIs. Furthermore, if we can get the computer to do something by typing in a command, we can incorporate that command into a program as part of an automated process.

For those readers unfamiliar with command-line interfaces, we provide a basic primer in Appendix A. If you are not confident with working at the command line, you should take a look at that before moving on to the following chapters.

1.6.3 Case sensitivity

When using command-line tools or writing software, it is essential to be aware of whether the tool or language you are using distinguishes between capital and lower case letters. For example, would it consider something called 'bioinformatics' to be distinct from 'BIOINFORMATICS' or, more subtly, 'Bioinformatics'? This can lead to all sorts of problems, especially for beginners, because commands that superficially look correct may not work. To make things even more confusing it is not just the tools that differ in their case sensitivity—the behaviour of operating systems differs too. Specifically, Windows ignores case, so a file called 'DNA.TXT' could be referred to on the command line as 'dna.txt'. If you tried to do this on Linux or Mac OS you would receive an error saying that the file could not be found. Indeed, in such case-sensitive operating systems two distinct files 'dna.txt' and 'DNA.TXT' could co-exist in the same directory, although this is hardly a desirable situation.

Our recommendation throughout this book is to assume case sensitivity, regardless of the tool, language, or operating system you are using, but make sure that you avoid using names that would be identical if the case was ignored. This is the approach we have followed in the examples in the chapters that follow. Not only does this avoid confusion, it also helps ensure that any programs you write are portable between operating systems.

1.6.4 Security, firewalls, and administration rights

Ever since computers started being connecting to networks, there have been concerns about unauthorized people connecting to computers and accessing data or installing malicious software. This is clearly a particular concern for those in bioinformatics working with sensitive data, such as novel compounds, or clinical data that needs to be guarded for ethical reasons. There is also the danger of criminals taking over networked computers and using them to send out spam email, which can have serious consequences for organizations as they can end up being *blacklisted*—having all their outgoing mail flagged as spam. For these reasons, organizations have become increasingly strict in limiting what users can do with their computers, and operating systems and server software has become loaded with more security features. Basic security features include the use of password protected user accounts to access a computer or particular data, and a firewall to filter network traffic to and from the computer.

We mention this here because the material in this book goes beyond what many organizations expect you to be doing with your computer. In particular, you will need to install software on your computer if you don't already have it, and in Chapter 5 your will be using your computer as a web server. Installing software requires you to have administrator rights on the computer and running server software may require you to modify firewall settings if you want to access the server from another computer. Neither of these things are inherently dangerous, but getting such access can be tough if your computer belongs to your organization, and is set up and administered by them. If you are working or studying in an organization, our advice is to talk to your IT support people and explain what you want to do. Depending on the organization, it is possible that the software you need may already be installed, either locally or on a server.

Even if you are using your own computer, you may need to spend a little time ensuring you have administrator rights and tweaking firewall settings. The good news is that security issues are only likely to be a problem when installing and setting up the tools used in this book, typically at the start of the chapter. Once you have successfully got MySQL, Perl, R, and so on up and running, you should be able to proceed through the chapter without further hindrance. Also, software installation and configuration is a generic issue, not specific to people using this book or even to people working in bioinformatics, so there is likely be help available on the web for almost every eventuality. Indeed, you can always head over to `www.bixsolutions.net` to report the issue and seek a solution.

References

Brereton, R. G. (2007). *Applied Chemometrics for Scientists*. Wiley: Chichester, UK.

Christiansen, T., Foy, B., Wall, L., & Orwant, J. (2012). *Programming Perl*. O'Reilly: Sebastopol, California, USA.

Laurie, B. & Laurie, P. (2013). *Apache: The Definitive Guide*. O'Reilly: Sebastopol, California, USA.

Lesk, A. M. (2008). *Introduction to Bioinformatics*. Oxford University Press: Oxford, UK.

Otto, M. (2007). *Chemometrics: Statistics and Computer Application in Analytical Chemistry*. Wiley: Chichester, UK.

Tahaghoghi, S. M. M. & Williams, H. E. (2009). *Learning MySQL*. O'Reilly: Sebastopol, California, USA.

CHAPTER 2

Building biological databases with SQL

A database is at its simplest a set of stored information, such as a filing cabinet or a computer's hard disk. Generally pieces of similar or related information are gathered together in the same place, as common sense would dictate and as you probably already do when you create folders and subfolders for information held on your computer. Database concepts provide a way of formalizing the gathering together of this data such that the relationships between pieces of information are consistent. They can therefore be more efficiently used, whether through manual or automated processes, and the structure provides a means by which data consistency may be maintained.

This chapter focuses primarily on a type of database called a *relational database*. Relational databases are powerful because they enforce a great deal of security and consistency on the data within them. The software tools that are used to create and manage relational databases are called *relational database management systems* or RDBMSs. These allow the data contained within a database to be queried in immensely powerful ways, often using very simple commands created using a special programming language called the Structured Query Language (SQL).

Also briefly introduced in this chapter are two other types of database commonly encountered in bioinformatics: flat text files, such as FASTA files, containing sequence information, and Extensible Markup Language (XML) files, which are a key component of most modern data standards, such as the HUPO-PSI standards for proteomic data. Understanding these types of database is easier than for relational databases, so they do not form the bulk of this chapter.

Finally we will also introduce the concept of NoSQL databases, a class of diverse solutions to data storage and access that have arisen in response to the increasing need to access records across very large data sets in very short time frames. There are some instances where the use of such a technology in bioinformatics applications is more appropriate than using an RDBMS.

At first glance the ordering of this chapter may seem strange—the installation of a relational database system follows extensive sections on databases and database design, and database access through SQL is covered last. This is deliberate. The hardest aspects of understanding and dealing with databases occur at the

Building Bioinformatics Solutions. Second Edition. Conrad Bessant, Darren Oakley and Ian Shadforth.
© Conrad Bessant, Darren Oakley, and Ian Shadforth 2014. Published 2014 by Oxford University Press.

design stage, which is also the most important. Installing a RDBMS is straightforward and also unnecessary for good database design. However, a working system is needed to experiment with accessing databases, and hence installation and connection issues are discussed after database design but before database interaction.

2.1   Common database types

2.1.1   Flat text files

As stated above, a database is merely a store of data, one of the simplest forms of which would be to write the data as a set of text files, often termed *flat files*. These flat files could be any text file created in a commonly readable format, such as the `.txt` files created in text editors like Windows Notepad. A collection of text documents on a hard disk is one example of a database of flat files. In order to assist automated access and, indirectly, readability, it helps if some sort of structure is imposed upon the data within a text file. In terms of scientific papers, this order is often along the lines of Introduction, Materials and Methods, Results, Discussion and References, or variants of this type. The structure helps the reader to quickly locate information of interest by navigating first to the relevant section. Subheadings further help this cause. For automated reading, or *parsing*, of data within a file, it helps if the structure is highly consistent. The headings within scientific papers may differ due to a number of factors, such as different journal formats. On the other hand, if we consider a basic implementation of the FASTA flat file format for storing sequence files, presented below, the structure is much simpler but allows for both intuitive human and machine reading.

A simple example of a FASTA format file would be:

```
>ENSP00000630516 | a protein description
SEQUENCEAPPEARSHERE
>ENSP00000295897 | another protein description
THESEQUENCEOFTHISPROTEIN
```

The FASTA file features four structural elements:

1   Information about each protein is introduced by a greater than (>) character

2   The first piece of data to follow the > character is the protein accession number, in this case the Ensembl accession number. This is followed by a bar (|) character.

3   Following the bar, we have the protein description. There is then a newline character that is not directly visible, but results in a new line being started. This is the indication that the protein sequence follows. In terms of human readability, this is clearly indicated by the start of a new line on which text that looks like a sequence is presented, but in machine readability terms, it's the invisible newline character that is used to differentiate between protein description and protein sequence.

4 Another newline character is used to terminate the sequence, followed immediately by one more newline character. Visually this results in a blank line
   separating successive proteins.

The pattern of these structural elements may be repeated until the end of the
file, which may contain any number of proteins.

The important point to note about the above description is that it separates
the data contained in the file from the structure of that data. No information
about what a protein naming convention, description, or sequence actually is,
or what this data looks like, is required to correctly assign data elements into
one of three groups: protein accession numbers, protein descriptions, and protein sequences. Furthermore each piece of data is explicitly related, by its location in the file, to the other two belonging to the same protein. The data
structure is therefore valuable in itself, regardless of any data contained within
the file.

Far more complicated forms of structured flat file exist and are used in everyday
bioinformatics applications. A good example is the Genbank format used to store
sequence data—each individual sequence *record* contains not just the sequence but
many additional *fields* of metadata such as the species name and genomic location
(see `www.ncbi.nlm.nih.gov/Sitemap/samplerecord.html`). Regardless of
the specific format, the principle of flat files remain the same: there is a set of
consistently used structural elements that allows data to be sorted into like types
and also grouped as appropriate, for example by protein, as in the preceding
example.

2.1.2 XML

Extensible Markup Language (XML) is a commonly used file format in bioinformatics applications. It adds a *syntax* to the concept of a structured text file. A
syntax defines the order of language elements such that what is written is generally comprehensible. In the case of the English language, sentences are most
readily understood if the correct grammar and punctuation are used. These elements therefore form the syntax for English. In other languages, such as Perl, or
in this case XML, a strict syntax helps both humans and computers to understand
exactly what is meant.

As with FASTA files, XML files may be written in, and viewed in, a simple text
editor. However, the structures have been designed to be primarily machine readable, not necessarily human readable, and therefore can appear a lot harder to
understand. Having said this, the basic XML syntax is made up of a small number
of structural elements that are easily understood.

XML structure

Each XML file can begin with a declaration of certain information, such as the
type of XML being used, the way characters in the file are encoded, and other information, for example:

```
<?xml version="1.0" encoding="UTF-8"?>
```

You will find this at the top of the XML file before the information-containing body of the file starts.

The generic structural elements in XML are called *names*, *attributes*, and *values*. This structure is used for the declaration too, but note the use of the question marks, indicating that this information is part of the declaration, not the body information. Textual content may also be entered into each element. Generically, these are written in the following way:

```
<name attribute_1="value_1" attribute_2="value_2"... attribute_
n="value_n">Some text about this named element</name>
```

A named set of information is thus introduced using the syntax <name; the attributes and values belonging to this follow until the closing > symbol. To close a named group, the syntax </name> is used. For example, if we were representing the proteins identified in a proteomics experiment, the following could be used:

```
<protein_identified id_number="1" probability="1.00">Protein
identified using mass spectrometry</protein_identified>
```

Named elements may be nested below one another, such that a subset of information belongs to the named element that surrounds it. So, the various names by which an identified protein may be known follow the opening of the protein_identified, but precede its closing statement, as in:

```
<protein_identified id_number="1" probability="1.00">
...
<annotation protein_description="Cerulaplasmin precursor"
ipi_name="IPI00017601" refseq_name="NP_000087"
swissprot_name="P00450" ensembl_name="ENSP00000264613"
trembl_name="Q9UKS4" locus_link_name="1356">
...
</protein_identified>
```

A great innovation of XML, and the reason it has 'extensible' in its name, is that the names of elements and how they can be used in a particular application can be decided by anyone. The only requirement is that this specification is encoded in an XML schema definition (XSD) file and made available somewhere on the Internet. The URL of the XSD to which a particular XML file should conform must be referenced in a special section at the top of that file. Software can therefore validate a given XML file against the XSD to check that elements are being used correctly.

A real-world example

The above examples have been formatted with some bold text such that they are clearer for us to read. However, this formatting is not part of the XML format and the convention may not be used in all situations. For instance, Fig. 2.1 provides a sample of an XML file that was generated by ProteinProphet, a system for identifying proteins from mass spectrometry data. In this file there is little formatting, initially rendering it very difficult to read.

```xml
<?xml version="1.0" encoding="UTF -8"?><xml-stylesheet type="text/xsl" href="regis/sbeams/
archive/edeutsch/HUPOPP12/HUPO12_run31/HsIPI_v2.21/interact-prot.xsl"?>
<protein_summary xmlns=http://regis-web.systemsbiology.net/protXML xmlns:xsi=http://
www.w3.org/2001/XMLSchema -instance" xsi:chemaLocation=http://regis-web.systemsbiology.net/
protXML/tools/bin/TPP/tpp/schema/protXML_v3.xsd
summary_xml="regis/sbeams/archive/edeutsch/HUPOPP12/HUPO12_run31/HxIPI_v2.21/interact-prot.
xml">...

...<protein_group group_number="1" probability="1.00"><protein protein_name="IPI00017601"
n_indistinguishable_proteins="1" probability="1.00" percent_coverage="9.0"
unique_stripped_peptides="KLVYREYTDASFTNRK+IYHSHIDAPKDIASGLIGPLIICKK+LVYREYTDASFTNR+YKK
VVYR+LVYREYTDASFTNRK+KLISVDTEHSNIYLQNGPDR+HYYIGIIETTWDYASDHGEKK+IGGSYKKLVYREYT DASFTNRKER+IYH
SHIDAPK+IGGSYKKLVYREYTDASFTNRK" group_sibling_id="a" total_number_peptides="15" pct_spectrum_
ids="1.18"><annotation protein_description="ceruloplasmin precursor" ipi_name="IPI00017601"
refseq_name="NP_000087" swissprot_name="P00450" ensembl_name="ENSP00000264613"
tremble_name="Q9UKS4" locus_link_name="1356"/><peptide peptide_sequence="KLISVDTEHSNIYLQNGPDR"
charge="2" initial_probability="1.00" nsp_adjusted_probability="1.00" weight="1.00"
is_nondegenerate_evidence="Y" n_enzymatic_termini="2" n_sibling_peptides="8.00"
n_sibling_peptides_bin="6" n_instance="1" is_contributing_evidence="Y"
calc_neutral_pep_mass="2299.4919"></peptide><peptide peptide_sequence="IYHSHIDAPKDIASGLIGPLIIC
KK" charge="2" initial_probability="1.00" nsp_adjusted_probability="1.00" weight="1.00"
is_nondegenerate_evidence="Y" n_enzymatic_termini="2" n_sibling_peptides="8.00"
n_sibling_peptides_bin="6" n_instances="1" is_contributing_evidence="Y" calc_neutral_pep_
mass="2761.1919"></modification_info modified_peptide="IYHSHIDAPKDIASGLIGPLIICKK"
mass="161.138794"/><modification_info></peptide>
<mod_aminoacid_mass position="23" mass="161.138794"/></modification_info></peptide>
```

Fig. 2.1 A sample of an XML file generated by a protein identification system, ProteinProphet.

This is pretty much incomprehensible to the untrained eye, and even if the reader understood the XML structure, as we do now, this would not be a preferred manner in which to view it. Rather XML files are often viewed through an interface, such as a web browser, which itself refers to a second document, a stylesheet (XSL document), for instructions as to how to display data contained within the XML structure. This XML file, when viewed with its stylesheet, is shown in Fig. 2.2.

To highlight each element as it appears in the body of the example, this is reproduced as Fig. 2.3 with names appearing in bold, attributes in grey, and values in normal text.

All of the information presented in Fig. 2.3 relates to the `protein_group` with a `group_number` of 1, because it falls in between the opening of this group (`<pro-tein_group...n...>`) and the closing tag (`</protein_group>`). Later in the file, another `protein_group` may be started with a `group_number` of 2. This would be a different protein and hence have a different set of alternative protein names nested within it as annotations. In this way, information that is naturally related is linked together within the XML structure, very much as information is linked within the tables of a relational database, as we shall see in the next section.

2.1.3 Relational databases

Relational databases take the concepts of order and structure of data one step further. This is achieved through compartmentalizing data into boxes of related elements and then linking these boxes such that pieces of data in one box may be accessed alongside related information in another box. The essence of a relational

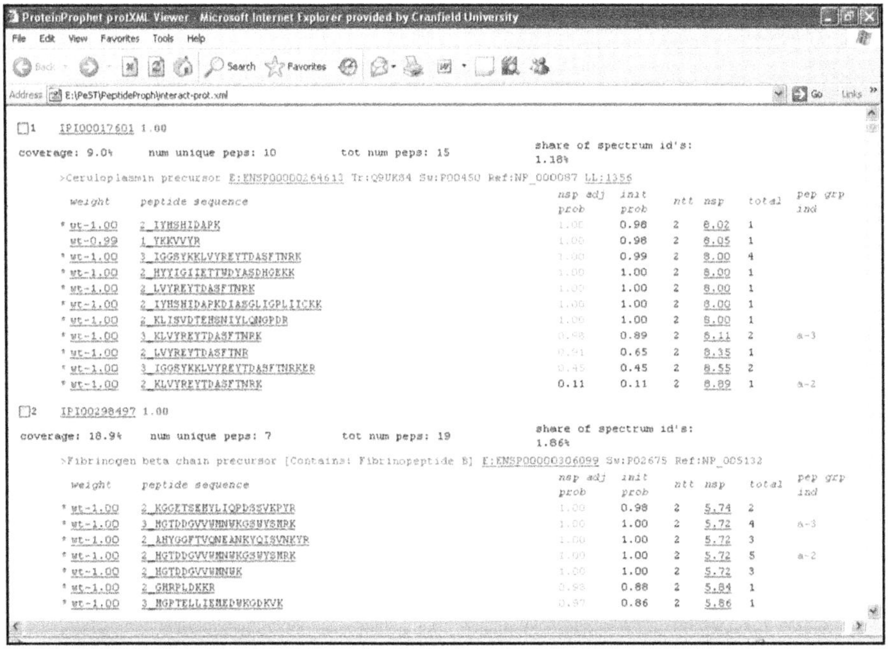

Fig. 2.2 The ProteinProphet XML file, shown in Figure 2.1, viewed in Internet Explorer with the help of a stylesheet.

```xml
...<protein_group group_number="1" probability="1.00"><protein protein_name="IPI00017601"
n_indistinguishable_proteins="1" probability="1.00" percent_coverage="9.0"
unique_stripped_peptides="KLVYREYTDASFTNRK+IYHSHIDAPKDIASGLIGPLIICKK+LVYREYTDASFTNR+YKK
VVYR+LVYREYTDASFTNRK+KLISVDTEHSNIYLQNGPDR+HYYIGIIETTWDYASDHGEKK+IGGSYKKLVYREYT
DASFTNRKER+IYHSHIDAPK+IGGSYKKLVYREYTDASFTNRK"group_sibling_id="a" total_number_peptides="15"
pct_spectrum_ids="1.18"><annotation protein_description="ceruloplasmin precursor"
ipi_name="IPI00017601" refseq_name="NP_000087" swissprot_name="P00450"
ensembl_name="ENSP00000264613" tremble_name="Q9UKS4" locus_link_name="1356"/><peptide
peptide_sequence="KLISVDTEHSNIYLQNGPDR" charge="2" initial_probability="1.00"
nsp_adjusted_probability="1.00" weight="1.00" is_nondegenerate_evidence="Y"
n_enzymatic_termini="2" n_sibling_peptides="8.00" n_sibling_peptides_bin="6" n_instance="1"
is_contributing_evidence="Y" calc_neutral_pep_mass="2299.4919"></peptide><peptide
peptide_sequence="IYHSHIDAPKDIASGLIGPLIICKK" charge="2" initial_probability="1.00"
nsp_adjusted_probability="1.00" weight="1.00" is_nondegenerate_evidence="Y"
n_enzymatic_termini="2" n_sibling_peptides="8.00" n_sibling_peptides_bin="6" n_instances="1"
is_contributing_evidence="Y" calc_neutral_pep_mass="2761.1919"><modification_info
modified_peptide="IYHSHIDAPKDIASGLIGPLIICKK"><mod_aminoacid_mass position="23" mass="161.138794"/>
</modification_info></peptide></annotation></peptide></protein></protein_group>
```

Fig. 2.3 The same sample of XML as provided in Figure 2.1 featuring, in the lower portion, highlighted structural elements. Here name elements appear in bold and attributes appear in grey, with their values remaining in the normal font. For completeness name group terminators have been added at the bottom of this section.

database, and much of their power, comes from the design of these boxes and their relationships to one another.

As a physical example of a relational database, we might consider a library indexing system. We can think of a library as a store of information with all of the books in one, very big, box. The books are often stored in a number of smaller, subject-specific boxes and then indexed alphabetically by author name. To find any book in the library for which you know the author and subject area, you can look in the correct subject area and work your way through the books until you hit the correct author and then work through all of their works until you find the one you want. However, if what you want is to find all of the books in the library written by a specific author, regardless of subject, you may be in for a long search using this method—you would have to search through all the subject areas in the library to be sure of finding all those written by the author of interest. To help in this case, we may create another box, such as a filing cabinet, in which to store the author, book title, and key subject information in another way, as a series of cards linking each author to all the books they have written. Now, if you know the name of an author, you can look in this box, work through the list until you find the one of interest, and written next to the name should be the list of books, each one assigned the correct subject area. It would then be possible to fairly quickly locate all the books in the library that had the same author.

To recap, the above example features two boxes of naturally related data items. The first of these, the library, contains sets of books grouped by subject and then by lead author, and the second contains key cards listing below every author's name their complete set of published works and the subject areas in which they could be found. In database terms, these boxes are called *tables*. Each table has a number of pieces of information stored within it; subject, author name, title, and book contents in the case of the library, and author name, book title, and subject in the case of the filing cabinet. Most importantly, each box is also linked to the other by three pieces of information: author, title, and subject area.

In these ways the two boxes of information are related. In database terms, the tables are now related by three of their fields. This situation can be rendered pictorially, as shown in Fig. 2.4.

One further important concept highlighted by the above example is that to describe the process of accessing data and to represent the information that is stored, we have not had to refer at all to a specific author, title, or publisher. Compare this to the examples of flat text files and XML formats presented above. In each of these the example contains the stored data. In the case of Fig. 2.4 we

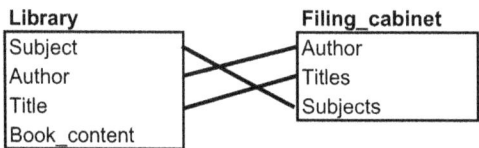

Fig. 2.4 Schematic representation of the contents of a library linked to author information stored in a filing cabinet. The library and filing cabinet are represented as two distinct *tables*.

have a representation of the data, but no data is shown. The representation of the database forms an abstract layer describing the underlying data, and therefore we can discuss each item of data generically without referring to a specific example.

When dealing with database design and access we are therefore able to talk about access and manipulation generically, a very powerful concept. Why? To go back to the library example, the books in each section are assumed to be ordered alphabetically by author name. This makes a lot of sense, but what if we want to locate a book by title? If we don't know the author then a rather long search may be required. We can get around this by having a set of index cards that link titles to authors, but this duplicates data which leaves us prone to errors caused by mismatches between the titles and authors printed on the covers of the books themselves and those on the index cards. In this case, because the data and the representation of the data are inextricably linked, we have to choose one ordering mechanism and use additional means if we wish to use an alternative order. With a relational database, however, we can order the information contained in a table by any attribute we wish. It's like having a library that will automatically move all the books around to suit the type of search we are doing, whether searching by subject area, by author, or by title.

As we have seen, the fundamentals of a relational database are not complicated. However, there is a lot of complicated jargon and even mathematics associated with formal database design that can be very off-putting for beginners. In the next section we present a natural approach to database design that, for most applications, allows new databases to be created without the need to wade through this complexity.

2.2   Relational database design—the 'natural' approach

The aim of designing a database is to produce a functional map of the database structure, which we call the database *schema*. The steps to producing a fairly robust, useable database schema without worrying about the complexities of formal database design are as follows:

1   Gather together a list of pieces of data to be contained within the database.

2   Group these so that they fit naturally together.

3   Assign consistent and descriptive short names to each piece of data.

4   Define the type of each piece of data: number, text, binary, etc.

5   Check for atomicity—can any data item be broken down further?

6   Index your database.

7   Link the tables of your database through relationships.

Each of these steps is described in detail through a worked example in the following sections. As will be seen, the above steps are ordered, but it may be necessary to repeat certain of them as part of an iterative design process. It may be useful for you to attempt to design a database of your own whilst working through the examples given here, or you might prefer to work through the examples and then re-read the section with your own database design in mind. Either way, to get the most out of this chapter you should carefully consider each step as it happens: why are you doing it and what has it achieved? In this way your appreciation of the operations will be enhanced and the easier you will find it to apply these concepts in any situation.

2.2.1 Steps 1–3: gather, group, and name the data

The first important point to keep in mind when designing a database is to think about what you want to get out of the database, not what you want to put into it. There are a number of reasons for this, the most compelling of which is that you are designing a database for a purpose and it is that purpose that should define its form and content. It's unlikely that you are designing a database only to store your data—you, and perhaps other people, will want to access it too.

Having said that, we have to start from somewhere and often the easiest point to start from is indeed to consider just what is available. For the rest of this chapter we will concentrate on a specific example to illustrate database design and implementation. In this example, we are interested in building a repository of information about PCR experiments that have been carried out in a particular organization. The benefits of such a repository to the organization are to make the results of the PCR more readily available, to avoid duplication of effort, and to facilitate sharing of best practice between experimentalists. The example was inspired by an MSc student project—if you want to find out more we refer you to the relevant thesis (Simecek, 2007) which is available via the web (`dspace.lib.cranfield.ac.uk/handle/1826/1773`).

Data to be captured from a PCR experiment

Rather than considering the experimental process as a whole, we first break the process down into potential sources of information for the database. Here we may have five sources of information from an experiment:

1 The PCR kit used.

2 Experimental parameters (annealing temperature, cycle times, etc.).

3 Primers used.

4 The scientist performing the experiment.

5 Results of the experiment.

You may disagree with the above, wish to combine or regroup the sources, or split them down further, but in essence you will achieve the same result—a set of headings under which it is simpler to list the individual pieces of information that are to be stored than would be the case if you were to consider the whole experiment in one go. The result of assigning information under these five headings is shown in Table 2.1.

It is at this point in the design process that you can start asking yourself questions. The first might be 'Is this complete?' This would be a fair question, but

Table 2.1 Example items of information from a PCR experiment grouped under source headings

PCR Kit	Experimental parameters	Primers used	Scientist	Results
Manufacturer	Denature temperature	Primer 1 sequence	Name	Gel image
Kit name	Denature time	Primer 2 sequence	Title	
Order number	Annealing temperature	Primer concentration	Department	
Supplier plus their address	Annealing time	Primer design software used	Telephone number	
Cost	Elongate temperature		Email	
Buffer	Elongate time			
Buffer concentration	Number of cycles			
Enzyme	Completion temperature			
Enzyme concentration	Completion time			
Nucleotide mix				
Nucleotide concentration				

tends to lure the unwary database designer down a never-ending quest for completeness that is not necessary to meet the application for which the database is intended.

There are some items that are undoubtedly missing from Table 2.1 that you may wish to include. One might be the purpose of the experiment. There may be other experimental parameters that you wish to capture, or variations in protocol that could be used. The results column is currently very sparse—would readings summarizing the information content of the gel resulting from the PCR be useful? You might also think that some of the pieces of information are not particularly useful and we might even be misguided to put them in. For instance, if the laboratory follows a set protocol for the majority of PCR experiments—might it not be safer (and more efficient) to point the user to the standard protocol sheet and hence eliminate the need to store, and potentially enter incorrectly, most of the pieces of information in the first two columns?

So, to get a better feel for what the database should be able to store, we need to consider how we intend to use the database. For the purposes of this example, it is to provide a record of protocols and kits that have been used in an attempt to amplify sequences that have proven tricky to analyse using standard methods. The benefit of storing the results from these experiments is that the database will provide a reference point across the laboratory for scientists to quickly determine if the sequence they are working with has been considered before and, if so, what the best protocol to use is.

With this in mind, it can be seen that perhaps the only extra pieces of information required would be the sequences that are to be amplified and some sort of judgement as to how successful each protocol was in amplifying each sequence. Perhaps we may also wish to add in some more data such as the time and date of the experiment and also an experiment identification number so that we can use this as a quick way of referring to the protocol and its outcome.

Refining Table 2.1 to include these additional pieces of information results in Table 2.2. At this stage there don't appear to be any pieces of information that should be removed from these lists.

A secondary use of the database might be to research whether there are certain trends in the data that might indicate suitable starting points for amplifying novel sequences. Are there any further pieces of information that would be needed for this?

Once you are satisfied that the pieces of information you have collected are sufficient to fulfil the purpose of the database, the next stage is to turn these lists into a database design.

Schema design and normalization

Database design can be broadly approached in two ways. The officially 'correct' way to design a database is to start with a single database table containing all of

Table 2.2 A refinement of the items of information to be stored from a PCR experiment, once again grouped under source headings. Additional fields are highlighted in *italics*

PCR Kit	Experimental parameters	Sequence and primer information	Scientist	Results
Manufacturer	*Experiment identifier*	*Sequence to be amplified*	Name	Gel image
Kit name	*Date*	*Purpose of experiment*	Title	*Assessment of method*
Order number	*Time*	Primer 1 sequence	Department	
Supplier plus their address	Denature temperature	Primer 2 sequence	Telephone number	
Cost	Denature time	Primer concentration	Email	
Buffer	Annealing temperature	Primer design software used		
Buffer concentration	Annealing time	*Primer supplier*		
Enzyme	Elongate temperature	*Cost*		
Enzyme concentration	Elongate time			
Nucleotide mix	Number of cycles			
Nucleotide concentration	Completion temperature			
	Completion time			

the pieces of information that you are seeking to store in the database and then use a process called *normalization* to break this table into a set of tables linked by common pieces of information.

A more natural approach to database design can be followed that takes advantage of a lot of the thought that we put into collecting the information above. This is a method that we have evolved and tested many times with feedback indicating that it is more intuitive than the formal method, and also allows a quicker understanding of the steps involved in the normalization process. For these reasons this 'natural design' approach is presented in this chapter. However, if you are going to be doing database work regularly, then you should definitely investigate other design approaches as each has its own strengths and weaknesses and, therefore, suitability to different tasks. There are many papers and books dedicated to the subject: Codd first introduced the concept of normalization in 1970 (Codd, 1970), followed by numerous other papers. For complete guides to database design you may like to consider the popular *Database Design for Mere Mortals* (Hernandez, 2003), or, by the same publisher, *Database Solutions: A Step-by-Step Approach to Building Databases* (Connolly & Begg, 2003). If you would prefer a MySQL specific guide to both design and the fundamentals of the MySQL system, try *Beginning MySQL Database Design and Optimization* (Stephens & Russell, 2004).

If you talk with database administrators in your organization, they may wish to ascertain the level of normalization of your proposed schema. For most purposes a schema designed to the '3rd Normal' specification will provide a good balance of functionality and robustness. The approach outlined in this chapter tends to result in a schema that is 3rd Normal and, therefore, should suit your needs.

Here we will cover how to present a schema and some of the more standard terminology used in database design.

Presenting the schema

The information in Table 2.2 is fairly clear for this small example, containing names for tables and the types of information within each table. It does not, however, look very much like a database schema. Before continuing to develop the database it is useful to draw it in a slightly different fashion. Specifically, each table should be presented in its own box with the table name above the box and the types of information contained within each appearing within the box, as shown in Fig. 2.5. At this point it is useful to choose simple, but descriptive, names for each of the tables and for each type of information. Such names should be consistent with one another. For instance, in Table 2.2 there is currently one table named Scientist and another named Results. The first of these is singular and the second plural. Although this seems to be intuitive, it can make using the database harder in the future as not only does one have to remember the name of the table, but also if it is plural or singular. A standard convention is to name all tables in the singular, so Results becomes Result.

The types of data contained in each database are referred to as *fields*. Each field should also be named descriptively and consistently. Field names should generally not be long as they may need to be typed often, and so prove frustrating, but

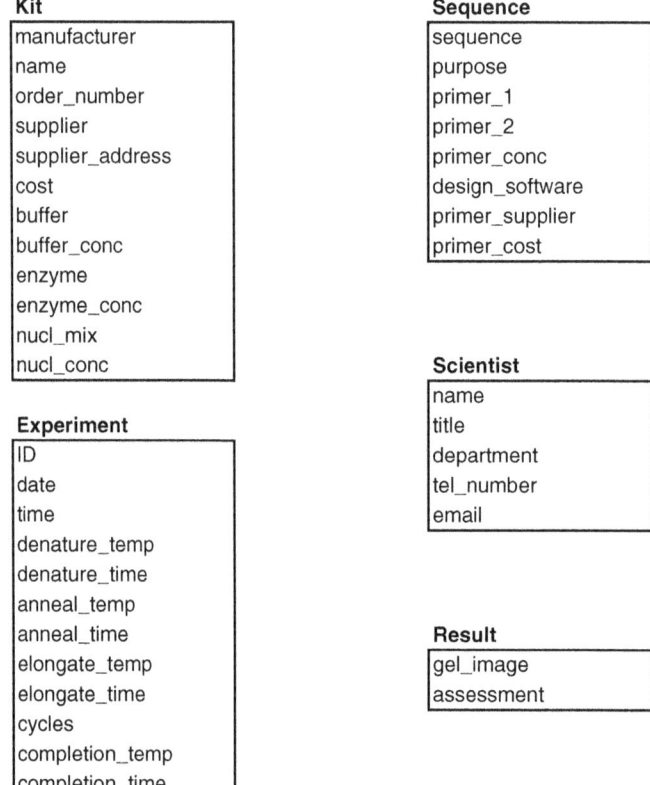

Fig. 2.5 Schema representation of Table 2.2.

they should convey the sort of information contained in the table. For example, naming each field 'a', 'b', 'c', and so on may be consistent and certainly speeds up typing the names, but it will be impossible for anyone, including the database designer, to remember what is actually contained in the database.

It was with these guidelines in mind that the schema shown as Fig. 2.5 was created. The immediate improvement in the clarity of presentation of each table is clear when comparing Fig. 2.5 to Table 2.2. This now looks much closer in form to the simple library schema given in Section 2.1.3.

During the process of providing tables and fields with short, but descriptive, names, a number of interesting issues have come to light. For instance, does the information now contained in Sequence look right? There are eight fields in this table; six of them are related to primers and primer design and the other two are related to the experiment. In this case, the renaming operation has helped to guide the formation of the database—we can now see that the fields sequence and purpose in the table Sequence may be better placed in the table Experiment as they represent the sequence that is to be amplified in this experiment and the reason why we are doing the experiment. Once they are removed, all other information in the table relates to primers and so the table might be better named Primer. These changes are shown in a revised schema later in the chapter (Fig. 2.6).

We now have the outline of a schema. It's not yet complete, there are a number of further operations to perform before it is, but much of the conceptual hard work has been done. The following sections describe the processes by which this loose set of disconnected tables may be made more robust, to help ensure data integrity, and also linked together into a truly relational database.

2.2.2 Step 4: data types

The above processes have been fairly intuitive—gathering information, grouping it, and naming each piece. We now need to assign a specific data type to each piece of data. Specifying data types is important because it helps to maintain the integrity of the database, as well as helping to minimize its size and maximize the speed with which we can get data in and out.

There is a set of standard data types defined within the ANSI (American National Standards Institute) standards for SQL. Most or all of these are used within RDBMSs such as MySQL, Oracle, and Access, but are often given different names in each RDBMS. For this reason, there may be some problems when designing a database schema for one particular RDBMS and implementing it on another. It is, therefore, always worth checking that the data types you have used are compatible between the two and, if not, how they should be modified.

As this book assumes that you will be working with open source software, the data types discussed in this section are for MySQL. Most of these will be the same, or there will be equivalents, in other systems, but the online help for each RDBMS will identify any differences.

Numeric data types

A number of the most commonly used numeric data types are given in Table 2.3. These can be split into two types: integer (whole number) types and floating point types (which can include numbers with fractional parts after a decimal point). Once you have decided whether your data requires a decimal point or not, the choice of the exact type will depend on how large you expect your stored numbers to be and to what level of precision you require the number to be stored. For instance, if you are only storing integers between 1 and 10 in a field, then the MySQL data type TINYINT would be adequate.

To store currency data, two decimal places may always be required and therefore the data type could be set to DECIMAL (n,d) where n specifies the total precision and d the number of digits that follow the decimal point, in this case $d=2$ and n defines the maximum amount that can be stored in the column. (It is worth noting here that MySQL versions since 4.1 store DECIMAL fields as a text string, not as a numeric type, which can have implications when programming.)

Finally, to store the results of calculations that may result in real numbers, FLOAT may be used. In general it is not necessary to define a precision for FLOAT, and indeed your schema will be more transferable if you don't, but the option is there in MySQL.

Text data types

Once again, there are a number of choices when deciding how to store textual information. The most commonly used of these are shown in Table 2.4. If the text

Table 2.3 Standard SQL and MySQL-specific numeric data types. Adapted from the MySQL Reference Manual

SQL Type	MySQL Type	Minimum Value (Signed)	Maximum Value (Signed)	Notes
SMALLINT INTEGER	TINYINT SMALLINT MEDIUMINT INT BIGINT	−128 −32768 −8388608 −2147483648 −9.22337E + 18	127 32767 8388607 2147483647 9.22337E + 18	An unsigned attribute is also available in MySQL. The range of this may be calculated by adding the numeric parts of the signed range (e.g. 0 to 255 (=128 + 127) for TINYINT)
DECIMAL	DEC or DECIMAL or DECIMAL (n, d)?	−1E + 38	1E + 38 − 1	Fixed precision: Decimal numbers are expected to match the precision defined (e.g. currency data)
FLOAT	FLOAT or FLOAT (M, D)	−1.79E + 38	1.79E + 38	Floating point number with optional (non-standard) user definable precision
DOUBLE PRECISION	DOUBLE or REAL	−3.4E + 38	3.4E + 38	15 digits of floating precision

Table 2.4 Standard SQL and MySQL-specific text data types. Adapted from the MySQL Reference Manual

SQL Type	MySQL Type	Capacity	Notes
CHAR(n)	CHAR	n = 0 to 255	n sets the stored size of the string
VARCHAR(n)	VARCHAR	n = 0 to 255	n sets the maximum length of the string
	TINYTEXT	2^8 bytes	Up to 255 characters
TEXT	TEXT	2^{16} bytes	Up to 65,535 characters
	MEDIUMTEXT or LONG or LONG VARCHAR	2^{24} bytes	Up to 16,777,215 characters
	LONGTEXT	2^{32} bytes	Up to 4,294,967,295 characters
	TINYBLOB	2^8 bytes	256 bytes
BLOB	BLOB	2^{16} bytes	~65 KB
	MEDIUMBLOB	2^{24} bytes	~16.7 MB
	LONGBLOB	2^{32} bytes	~4.3 GB

itself is to be stored in the database in a readable form, then a text type is used. CHAR(n) sets aside storage space for a character string of length n, no more and no less. VARCHAR(n) defines a more flexible data type that allows any length of string to be entered to a maximum length of n, with storage only being used for approximately the length of string entered. Always using a VARCHAR(255) may seem like an attractive option for storing any text information (as long as it is shorter than 256 characters). However, if the data entered is likely to be of uniform length, such as serial numbers or certain accession numbers, then a CHAR of suitable length will generally be more efficient in terms of storage space.

For longer text strings, the variants on the TEXT type may be used. These may seem just like longer VARCHAR strings and do behave rather like them, but care should be taken when grouping and sorting using TEXT values as MySQL defaults to using only the first 1024 characters[1] for such operations. So, if two TEXT strings start the same, but diverge after 1024 characters, they will be treated as being equal for sorting and grouping, which may not yield the desired result.

Binary Large OBjects (BLOBs) are treated as binary strings (as opposed to character strings). Files such as PDFs, Microsoft Word documents, image data, and so on could be stored within the database as BLOBs, ordered, and searched by comparing their binary strings. As with TEXT, such comparisons are limited by default to the first 1024 bytes of the string.

Choosing and representing the data types for our example

Back in Fig. 2.5 we captured all the fields in our database—each of which needs to be assigned a data type. Fig. 2.6 presents the data types assigned to each field in a tabular format. Consider the field types that have been suggested. Do they make sense? Can you see any problems with any of them or restrictions that they may impose? Some of these will be discussed below. Importantly, it is fine if you disagree with a number of these assignations. They are not perfect for all eventualities, but when you design your own database you will need to carefully consider the application for which it is to be used and pick data types accordingly.

As mentioned above, VARCHAR(255) has been suggested for most of the text fields in this example. An exception to this is the order_number field in Kit which assumes that the order numbers assigned by the department are of length 16 and hence there is no need to use a variable string length.

All temperatures and times have been assumed to be integer based, with time in minutes. This assumption may be correct or a finer scale of temperature or time could be required. A FLOAT data type would be considered under these circumstances.

DATE and TIME are special data types that have not been discussed up to now. MySQL has a number of ways of representing these, which can be found in the MySQL Reference Manual (dev.mysql.com/doc/#manual), however these are the most common, with DATE representing yyyy-mm-dd and TIME hours:mins:secs.

Kit cost and Primer primer_cost are both represented as decimals with two decimal places and a maximum number of six figures, allowing a maximum cost of 9999.99 to be represented in each of these fields. The unit of currency is not specified or stored in these fields.

---

1 The comparison is actually restricted to the first 1024 bytes, so it will compare fewer than 1024 characters if you are using a character set that requires more than one byte to encode each character.

Kit

Field	Type
manufacturer	VARCHAR(255)
name	VARCHAR(255)
order_number	CHAR(16)
supplier	VARCHAR(255)
supplier_address	TEXT
cost	DECIMAL(6,2)
buffer	VARCHAR(255)
buffer_conc	FLOAT
enzyme	VARCHAR(255)
enzyme_conc	FLOAT
nucl_mix	VARCHAR(255)
nucl_conc	FLOAT

Primer

Field	Type
primer_1	VARCHAR(255)
primer_2	VARCHAR(255)
primer_conc	FLOAT
design_software	VARCHAR(255)
primer_supplier	VARCHAR(255)
primer_cost	DECIMAL(6,2)

Scientist

Field	Type
name	VARCHAR(255)
title	VARCHAR(255)
department	VARCHAR(255)
tel_number	VARCHAR(255)
email	VARCHAR(255)

Experiment

Field	Type
ID	INT
sequence	VARCHAR(255)
purpose	TEXT
date	DATE
time	TIME
denature_temp	INT
denature_time	INT
anneal_temp	INT
anneal_time	INT
elongate_temp	INT
elongate_time	INT
cycles	INT
completion_temp	INT
completion_time	INT

Result

gel_image	MEDIUMBLOB
assessment	TEXT

Fig. 2.6 Fields and suggested data types by table, for PCR database example.

As gel images can be large, these have been assigned a MEDIUMBLOB type that will allow storage of files just over 16 MB. Alternatively, a link to the location of the image on a hard drive or server could have been placed here (in the form of a text string) to save space within the database whilst still providing easy access to the image. This may often be a better solution than storing the image itself, as few database operations will be usefully employed on an image.

Finally the assessment of the experimental result has been assigned a TEXT type allowing a longer description, of up to around 65,500 characters, to be stored here. This might be excessive and perhaps a TINYTEXT type would be more appropriate.

One point of note where the use of VARCHAR(255) might seem a little odd is the tel_number field in Scientist, which intuitively might have been an INT. However, often telephone numbers contain other characters, such as +, spaces, and brackets which would not be compatible with a numeric type. This problem arises mainly because the telephone number is not atomic, which brings us neatly on to the next topic.

2.2.3   Step 5: atomicity of data

The term *atomicity* may at first appear overly abstract, but all it means is that each piece of information in a field should be as small as it can be—that is it should only contain data about one item. If we consider the example of the phone number above, this seems to be just one data item, a phone number. However, it may contain three, or more, distinct pieces of information, such as country code, area code, and the number itself. As discussed above, the ways of representing such pieces of information generally require that characters other than numbers be entered. This in itself can introduce inconsistency and error into the database, and therefore any way of avoiding this will be desirable.

One way of doing this is to split the data up into its smaller parts and thereby create a field for country code, one for area code, and one for the number. All of these will contain only integer values and so the type INT can be used. This may appear to make the database more complicated—we have just replaced one field by three—but these steps serve in the longer term to make the database more robust and in many cases more useful. For instance, it is now possible to search within the database for scientists with offices in specific countries using the country code.

Another example of this is in the `supplier_address` field in `Kit`. An address generally consists of a number of parts, including: number/name of building, street name, town/city, county/state, post code/zip, and country. `supplier_address` is therefore not atomic. To correct this, six other fields need to be created and the `supplier_address` field removed.

The field `name` in `Scientist` should be treated similarly. The revised schema is presented as Fig. 2.7.

For each field in your proposed database, ensure that it cannot be split into smaller parts and you will have achieved atomicity. The full advantages of this will be seen in the next section, but for the time being it is worth thinking of this as a useful way of making each piece of data in the system as simple as it can be—simple things are always easier to deal with.

2.2.4   Steps 6 and 7: indexing and linking tables

We now have an atomic database containing all of the information that we think we need to know about the PCR experiments being performed. However, this database is not yet relational—there are no links between tables and therefore the information contained within them is not linked. This format works for a paper example as we think intuitively that if we talked about two different experiments then we could imagine two separate pages containing the data, one for each experiment. This is very much a spreadsheet view of the data. Much of the power of a relational database comes from being able to search through the data contained to spot trends, order by different variables—such as scientist or manufacturer to identify systematic errors—and generally to search the data as an interlinked whole, not as a series of discrete items. For this reason we will need to add some fields into the database that allow the tables to be linked to one another.

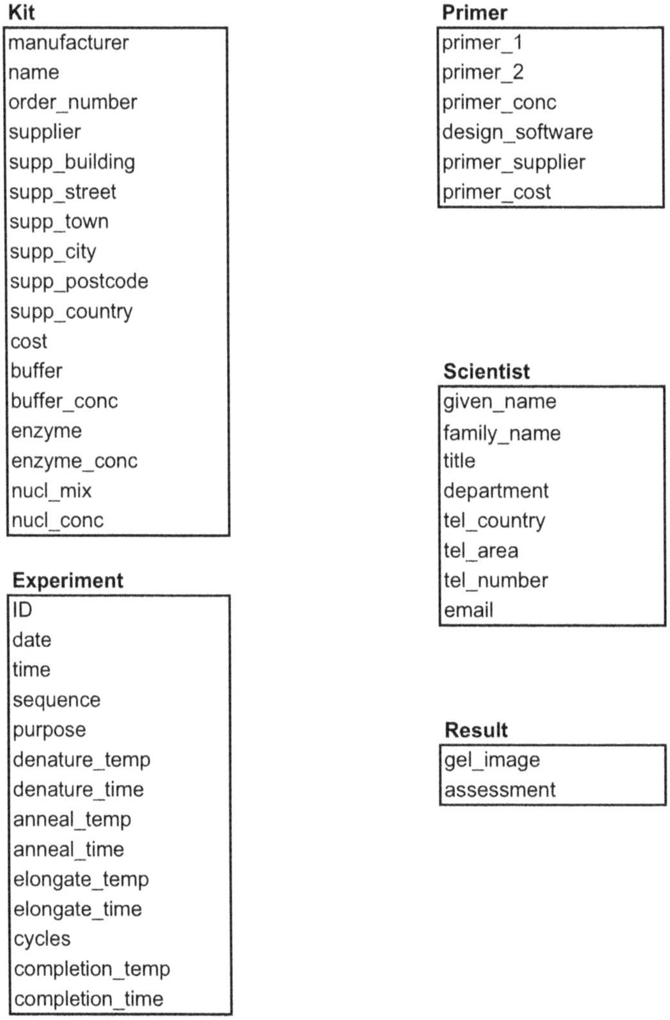

Kit

manufacturer
name
order_number
supplier
supp_building
supp_street
supp_town
supp_city
supp_postcode
supp_country
cost
buffer
buffer_conc
enzyme
enzyme_conc
nucl_mix
nucl_conc

Experiment

ID
date
time
sequence
purpose
denature_temp
denature_time
anneal_temp
anneal_time
elongate_temp
elongate_time
cycles
completion_temp
completion_time

Primer

primer_1
primer_2
primer_conc
design_software
primer_supplier
primer_cost

Scientist

given_name
family_name
title
department
tel_country
tel_area
tel_number
email

Result

gel_image
assessment

Fig. 2.7 Atomic schema for the PCR database example.

Before we attempt this, we need to first ensure that our tables can be efficiently searched individually. For this they should be indexed. As in a printed dictionary, what indexing does is provide a structure (and order) to the data such that it can be searched and the correct information retrieved. In a dictionary the sorting is by alphabetical ordering of the first few letters in each word and the correct information for each word is retrieved because each word is generally unique. This is sometimes not true, and it is good to think about the difficulties that can be caused when two words are spelt the same but have different meanings—it would be more convenient if the datum, or meaning of the word, could be unambiguously accessed by a truly unique key, or word. This is what we are attempting to do by indexing our database.

An index therefore allows us to unambiguously select any row of data from any of our tables, with each table having its own index *key*. Such a key can consist of one or more of the pieces of data within the table. A simple example of this is in the table `Experiment`. In this table, each experiment has been given an identification number (`ID`). If this identification number were simply an integer, each one greater than the last identification number, it would be sufficient on its own to uniquely identify any experiment in the table.

A number of database designers will recommend that each table within the database has a unique identifier based on an incremental number. This may result in certain performance increases under some circumstances, but we disagree with this approach on two fronts. The most important of these is that it complicates the database by introducing unnatural fields, with nothing to do with the data, into each table. This makes the database harder to think about, design, and query. The second is that it breaks one of the formal rules when designing databases, which is that all information in a table should be directly related to the key of that table—if we introduce an arbitrary running number into the key, we break this relationship.

So, how can keys be created other than by using numeric identifiers? In just the same way as any object is identified everyday, through distinct characteristics. As an example of this, let's consider the table `Scientist`. Here a candidate for a primary key immediately presents itself: the scientist's name. In this table an index key could be created on just the `family_name`. But many family names are common and hence we might add in the given name of the scientist. In this way we could build a *compound key* to the table consisting of both the given name and the surname. If we could guarantee that no two scientists were going to have the same combination of given name and family name, then this would be fine, but it is not true—lots of people share the same name. At this point someone might suggest adding in an employee number and referring to the scientists using this. Doing so has some uses, especially if this database were to be connected to their organization's human resources database. However, this is unlikely, and most people don't know their employee number so getting these might be difficult. It is also just as unintuitive as using the incremental number method. There is also a better option already in this table—the e-mail address. By definition, this will be unique to each scientist, provided, of course that they have an e-mail account. Most scientists now do have these and, if not, they are easily obtained, even if they never use it!

The `Primer` table does not seem to have any natural key in its present form. Eventually this table will also need to be linked to the table `Experiment`. Linking two tables is achieved when two tables share at least one field. In this case neither contains a field present in the other and we therefore need to choose at least one field from one table to place in the other. Here the `ID` seems like a good choice as the primer sequences will be related to the experimental sequence we are seeking to amplify. This would also allow `ID` to be used as the primary key for the `Primer` table. Once this is done, it can be seen that the two tables now both have `ID` as their key. As all the information in both tables is uniquely identified by the same key, logically all of this information should appear in the same table, although we thought earlier that they should be separate. (There is an argument that the information in `Primer` should in fact remain separate because the same primer combination could be used for more than one experiment and hence the same information could be repeated many times in the `Experiment` table, which would be undesirable. The final choice would be determined by end use—does each experiment run in this laboratory generally use different primers or not?)

Similarly, no key within `Result` naturally suggests itself, and these data are also directly related to the experiment. These fields should therefore also be placed within the `Experiment` table. `Experiment` is now a much larger table, as shown in Fig. 2.8.

Conversely, when considering the `Kit` table, an eligible key for the table might be a combination of manufacturer and the kit name (`manufacturer,name`). This works for most of the fields in this table except for those that are related to the supplier—a supplier may provide many makes and versions of kit and hence their details are not uniquely identified by a single kit manufacturer and name. This suggests that the supplier details should be brought out of `Kit` and placed in their own table, `Supplier`. As suppliers of similar products should have different names, we will assume that `supplier` will form an adequate key to this new table, as shown in Fig. 2.9.

However, further consideration of the `Kit` table shows that this primary key is also incorrect. If the same kit is ordered twice, then much of the information within the table will have to be repeated, as this will have a different order number. This demonstrates either that (`manufacturer,  name`) is an incorrect primary key, or that this table is still not yet properly designed. The answer is the latter: `order_number` should not be in this table as most of the other

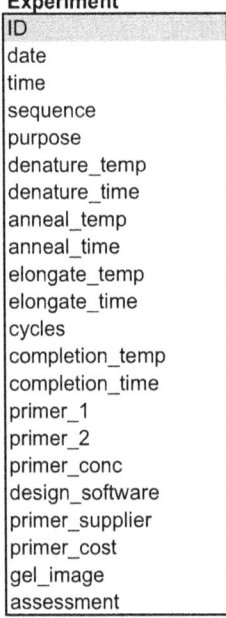

Fig. 2.8 The new `Experiment` table including fields originally within `Primer` and `Result`. The primary key is shaded in light grey.

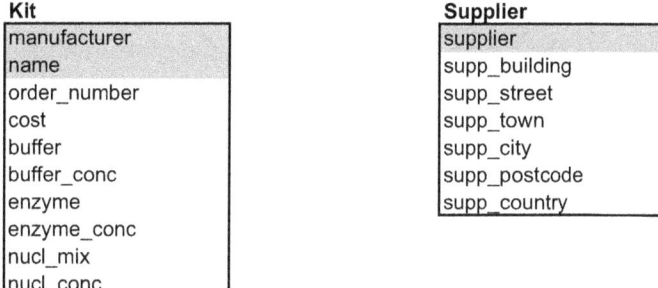

Fig. 2.9 The reduced table `Kit` and the new table `Supplier`. Primary keys are shaded in light grey.

information in the table does not depend on the order number. This should therefore be brought out into another table.

The order number does uniquely link to the manufacturer and kit name and therefore these fields may also feature in the new table and thus serve to link the two tables. Furthermore, the order number is naturally assigned to a supplier and this field can therefore appear in the `Supplier` table. The resultant tables and links are shown in Fig. 2.10.

To link `Experiment` to `Kit`, it may be tempting to put the `ID` field into the `Kit` table, but this would break the constraint that all the information in Kit should be uniquely identified by (`manufacturer`, `name`)—the experiment identification number has nothing to do with this. A better choice of fields by which to link the two tables is to do this indirectly through the `Kit_order` table by placing `order_number` in `Experiment`. This will therefore serve to uniquely identify any kit that is used in an experiment. It's no coincidence that the key to one table forms a good link to another table in this way—they are routes to identify any unique row in their own table and hence can be used for a similar purpose in related tables.

Similarly, `Experiment` may be linked to `Scientist` by putting the key of `Scientist`, email, into `Experiment` as well.

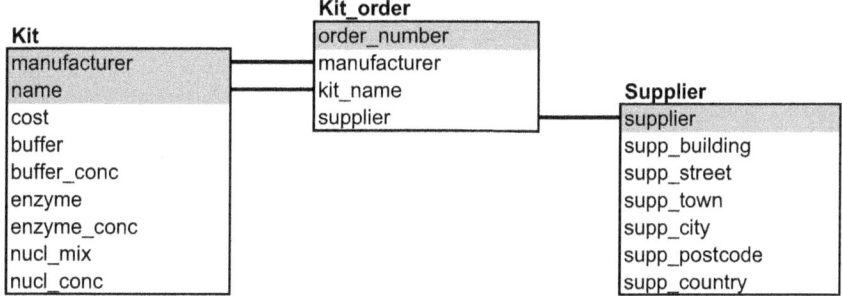

Fig. 2.10 Reconfigured `Kit` table linked through `Kit_order` by `order_number` to `Supplier`. Note that linked fields do not need to have identical names in each table.

Defining relationships between tables

All of the tables in the schema now have keys that may be used to uniquely identify their every row. Each table is also linked to at least one other table. This is the basis of our relational database. This section now looks a little more closely at how the relationships between the tables are formed and at the terminology used to refer to them.

The keys that have been defined above are known as the *primary key* of their respective tables (e.g. the primary key of the `Scientist` table is email). A field in a table that links to another table such that any specific value may not be entered unless the same value already exists in the other table is called a *foreign key*. This helps maintain data integrity, and also implies an order for entering data into the tables. Figure 2.11 shows the complete schema with the primary and foreign keys linked.

Observant readers will notice that subtle changes have been made to two of the field names in this figure. Specifically, the order number added to the `Experiment` table has been called `kit_order_number`, and the `supplier` field in the `Supplier` table has been renamed `supplier_name`. This is because it is not considered good practice to use identical names for linked fields.

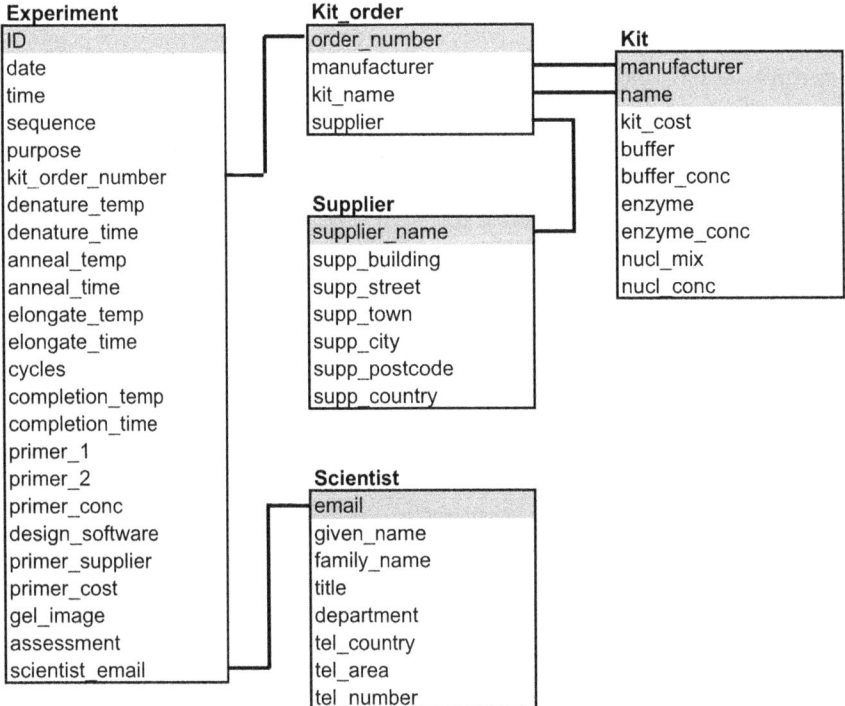

Fig. 2.11 Final schema for the PCR example. Primary keys are shaded in light grey. Links (via shared fields) between tables are also shown. Foreign keys are the fields to which the primary keys are linked.

A note on many-to-many relationships

When creating your database design using the above method, you should end up with a schema that links a field such that for any one instance of a field in one table, there will be zero, one, or many occurrences of that instance in another table. For example, in `Scientist` a scientist's e-mail address will appear once, whereas it may appear zero, one or many times in `Experiment`. In the latter case, this type of relationship may be described as *one-to-many*. If, however, your database design is not yet fully complete, you may find that you have a situation in which many instances of a field in one table might be linked to many instances of that field in another table.

Imagine that we had linked `Scientist` to `Experiment` using the first name of the scientists rather than their unique e-mail addresses. If 'Bob Andrews' had performed six experiments, his contact details would appear once in `Scientist` and 'Bob' would appear six times in `Experiment`. If 'Bob Barrows' was also in the lab group and had performed four experiments, then his contact details would appear once in `Scientist` and 'Bob' would feature an additional four times in `Experiment`. Now, if we had linked the two tables on the first name field, then both instances of 'Bob' in `Scientist` will be linked to all ten instances of 'Bob' in `Experiment`. When performing queries against these tables, this would likely result in incorrect information being returned and also queries taking much longer to return than they would have done otherwise.

As indicated above, this should not happen in a well-designed schema, and it is hard to see why we would have done anything like this contrived example. It can happen though, especially in more complex databases, so it is always worth checking for any relationships that do not feature a primary key on one side to see whether they may cause a similar problem.

2.2.5 Departure from design

Once you have designed your database, you are ready to start implementing it in your preferred RDBMS. As discussed in the introduction to this chapter, it is tempting to dive straight in to typing away at your computer without giving sufficient thought to what it is that you wish to achieve. If you approach all new database design on paper and spend the time and effort to design your database well before reaching for the keyboard, implementing and using your database will be much quicker and a lot simpler.

The next section covers installing a MySQL database server onto your computer and how to use this to first implement your database, then populate it with data, and finally to search through the information stored within.

2.3 Installing and configuring a MySQL server

2.3.1 Download and installation

Getting hold of a copy of MySQL to install on your machine is a simple process, but can vary depending on the operating system that you are using. The specific product you are looking for—at least as a beginner—is the free 'MySQL

Community Edition', as opposed to the paid for alternatives such as 'MySQL Enterprise Edition'.

If you are using Microsoft Windows, simply visit the MySQL website's Developer Zone (`dev.mysql.com/downloads/mysql`) to download and run the MySQL installer, which will guide you through the familiar Windows installer process. During this process, you will be prompted to choose a setup type. For the bulk of the material in this chapter, only the MySQL Server is required, so you could opt for the 'Server only' installation option. However, unless you are particularly tight for storage space, we recommend you choose 'Developer Default' as this includes other things that you will find useful as you get more deeply into MySQL. To make MySQL and associated tools directly accessible via the Windows command line,[2] we need to open a command-line window (see Appendix A) and add the directory in which MySQL is installed to the path using a command like this:

```
set PATH=%PATH%;C:\Program Files\MySQL\MySQL Server 5.6\bin
```

For Mac OS users, we would recommend installing MySQL via a command-line package manager called homebrew (`mxcl.github.io/homebrew/`), but in order to do this you will need to setup and install homebrew:

• First you will need to install Apple's developer tools, known as XCode. If you are on OS Lion or above, you will find this in the App Store as a free download. For older editions of OS, visit `developer.apple.com/downloads` and find the most recent version available to you.

• OS X Lion (or above) users will also need to install the XCode command-line tools. You will find the option to install this within the Preferences pane of the app, under 'Downloads'.

• Next we can install homebrew via the following command in your terminal:

```
ruby -e "$(curl -fsSL https://raw.github.com/mxcl/homebrew/go)"
```

Follow the instructions given to you, if there are any problems please consult the homebrew website, otherwise ask a question on the `www.bixsolutions.net` forum and we will do our best to help you out.

• Finally, now you have homebrew installed, you can install MySQL with the following simple command:

```
brew install mysql
```

This will take a short while to download and install the MySQL software for you, and it will then give you one or two more lines of instructions to follow in order to prepare and start your server.

The reason we recommend this slightly more complicated installation process on OS is threefold; it first enables easy updates of MySQL in the future—with

---

2 In Windows, MySQL installs its own MySQL command-line client, which is perfectly functional; but if you want to follow the examples in this chapter verbatim then you will need to access MySQL from the standard Windows command line.

the simple command: `brew upgrade mysql`; second, it sets up your environment correctly for installing programming libraries that talk to MySQL (in languages such as Perl, Python, and Ruby that we cover later in the book); and finally, now that you have homebrew setup, a wide variety of Unix tools and libraries are at your disposal and easily installed—we shall be taking advantage of this in later chapters.

If you are using Linux, before going to the MySQL website it is best to look into your distribution's package management system as MySQL is nearly always available for easy installation from there (often this can even be only a couple of mouse clicks for a complete install). In addition to the MySQL-server packages you will also want to install the MySQL-devel packages as these will be needed later on for building/installing third-party code libraries when working with Perl and other programming languages. In the unlikely event that you cannot find MySQL within your package manager, head over to the MySQL Develop Zone instead.

During configuration of the MySQL server you will be given the opportunity to set up a password for the *root* user. The root user is effectively the administrator of the database server, with a lot more power than regular users. It is good practice to set a strong root password, as long as it is something you can remember of course! You can also set up accounts for other users at this point. It is also good practice to set up an account for yourself separately from the root account, but we will deal with this later so no action is needed for the moment. For all other configuration options, the default or recommended options should be fine. If you chose one of the fuller installation options, the installer may invite you to open MySQL Workbench once installation is complete. We don't need the Workbench at the moment, as for the time being we will be interacting with MySQL through the command line, but we will come back to it later.

Once MySQL has been installed, you need to start it from your operating system's command-line (see Appendix A if this is unfamiliar). You can start MySQL by typing `mysql` at the command line and hitting the Enter key. However, if you chose to set up a root user and password during the installation of MySQL, you will instead need to specify these when starting MySQL, by typing the command below and entering the password when prompted.

```
mysql -u root -p
```

Once MySQL has started you will see some version information appear followed by the MySQL user prompt:

```
mysql>
```

You have now moved from the operating system command line to the MySQL command line. Almost all of the commands that you enter into MySQL will be entered at this prompt, so assume that you need to have got to this point before attempting any action described below, unless otherwise stated. Once you've had enough of MySQL, you can close it by typing `quit` (or `exit`) and hitting Enter. If you get totally stuck installing MySQL, head over to `www.bixsolutions.net` to seek professional help.

2.3.2   Creating a database and a user account

At this point, you will be logged into the MySQL server as the root user. This is a very powerful position to be in and, although it is unlikely that you will do anything to damage anything on your own system, if you were to log into a shared database server within your organization, then this may be a more important issue. In either case, you should create a different user account by which to access your database.

First, you will need a database on which to work. This may be created using the following command:

```
CREATE DATABASE dbname;
```

Here *dbname* is any suitable name for your database that you would like to use. (Throughout this book we use italics in generic command examples to identify placeholders for parameters than can be passed to the command.) Just as with table names, as discussed above, database names should be descriptive and simple.

Note also the semi-colon (;) that follows the command—this is necessary to tell MySQL that you have finished writing the command. If it is missing you will simply pass on to another line when you hit Enter (the command prompt will change to ->). You can then type in the semi-colon and the command will complete as if it had been written on the one line. This is a useful, although sometimes confusing or annoying, feature as it allows you to split long commands over a number of lines and therefore read them back more clearly. SQL conventions are discussed further in Section 2.5.

You can now create a user account through which to access your database in preference to the root account. This is done by issuing the following command:

```
GRANT ALL ON dbname.* TO 'username'@'localhost' IDENTIFIED BY 'password';
```

For instance, if you had created a database called sandpit (a name often given to systems intended for playing around in while learning) and wanted to create an account with a user name of 'Ian' and a password of 'BBSbook' you would do this by typing:

```
GRANT ALL ON sandpit.* TO 'Ian'@'localhost' IDENTIFIED BY 'BBSbook';
```

Be careful to get the quotation marks in the right place—note in particular that 'Ian' and 'localhost' are separate entities. The * means that all tables within the database will be covered by the GRANT statement. The term localhost refers to the machine from which you will be accessing the account. If you are working on your own machine and have installed your own MySQL server, this will be fine. However, if you are connecting to a different server from your machine, you will need to put either the full DNS name of your machine here, or its IP address (or if you would like your new user to be able to access the database from any other computer—we'll leave you to ponder the security implications of this—you can just replace `'localhost'` with `'%'`). See your database administrator for more information if this is the case for you, as in this instance it is likely they will have to create a user account for you and should then be able to tell you exactly how to access the database server.

Once you hit Enter, MySQL should give a message stating that this operation has completed successfully. It also tells you how long the command took to execute—not very significant for this command, but an indication of how critical performance can become when databases get large. Following this you should exit MySQL and then reconnect using your new account.

```
QUIT;
```

To reconnect using this new account, you would now type:

```
mysql -u Ian -p
```

and then enter the password (BBSbook in this case) when prompted to do so. Note that passwords are case sensitive, even when using MySQL in Windows.

You now have access to your relational database management system (RDBMS), in this case MySQL. There are others available, which we shall discuss briefly in the next section.

2.4 Alternatives to MySQL

MySQL is only one of many RDBMSs available. There are a number of reasons why we have chosen to use it for this book: it is free, widely available, and has very large community usage, which means that should you have a question about how to achieve a specific function, or need help solving a problem, almost undoubtedly there will be someone on the Internet who has had a similar issue and solved it and posted the solution.

However, you may well want to consider, or at least feel that we have not just ignored, other available systems. Some of these are discussed very briefly below, but we encourage you to do your own research in this area with one caveat—comparisons between systems are rife on the Internet, and many are out of date, so it is probably best to take all you read with a judicious pinch of salt.

2.4.1 PostgreSQL

PostgreSQL is another fully featured open source RDBMS. It is freely available from `www.postgresql.org` in versions for Windows, Linux, and Mac OS. Historically, PostgreSQL has been more feature rich than MySQL. However, there comes a point where the feature set is likely to be more than you really need. There are enterprise installations of both of these open source systems that work extremely well under very high volumes of data and large numbers of concurrent users.

2.4.2   Oracle

Oracle is regarded as the industry standard for database management and has the richest feature set of any current database management system. It is available for many operating systems, including Windows, Linux, and Mac OS. Oracle is often used by the biggest companies as, despite the positive experiences with MySQL and PostgreSQL, it has a reputation as the only choice for supplying the stability, performance, and administrative (recovery) capabilities necessary for a large organization. As an enterprise solution, it is also one of the more expensive options. Versions of Oracle are available for free from `www.oracle.com` for you to develop and distribute your own databases, although not for all operating systems. If you know that you are going to need to interface with a company instance of Oracle, a good point to start would be to download the most up to date free version you can, and work with this locally until you are ready to hand your database and any associated tools to your database administrator. MySQL was acquired by Oracle in 2010, but at the time of writing remains a separate product.

2.4.3   MariaDB

Prompted by concerns about Oracle's commitment to MySQL following its acquisition, one of MySQL's founders established a separate fork of MySQL, named MariaDB, which is developed and maintained by the community. MariaDB (available from `mariadb.org`) is intended as a drop-in replacement for MySQL, such that MySQL users are not adversely affected if Oracle chooses to withdraw MySQL in the future. All of the concepts and command line examples in this chapter should therefore be fully compatible with MariaDB.

2.4.4   Microsoft Access

Microsoft's Access program is present on many computers, but sadly under-utilized in favour of the more approachable Excel spreadsheet package. As a store of data, Excel is extremely limited and encourages a variety of complex worksheets and formulae to be created to achieve what would be very simple tasks within a database framework.

If you have Access installed on your computer, you may prefer to use this rather than install one of the other packages suggested here. Furthermore, if you work for a company, MySQL and PostgreSQL may not be on the list of approved software for your work machine. In such cases, Access may be your only choice for a locally installed database server with which to experiment.

Access is a fully featured RDBMS and has the advantage of looking and feeling similar to all other MS Office packages. It also has some advantages over a basic install of MySQL in that it natively features a graphical representation of tables, and hence implementing database schemas is straightforward as they may literally be 'drawn' into existence (although the MySQL Workbench software provides this functionality for MySQL, as we will see later). Using Excel to access data stored in the database in order to take advantage of the spreadsheet package's features is also very easy. This can also be achieved without too much difficulty using other RDBMSs through a system called ODBC—consult your RDBMS user guide for more information on this topic.

On the negative side, Access is a fairly slow system when compared to standard installations of MySQL, and there are likely to be unacceptable delays when running queries on moderate to large databases or when querying data from a number of linked tables. Furthermore, as a commercial piece of software, Access goes against the open source ethos of the bioinformatics community and ubiquitous availability cannot be assumed because it costs money, plus it is only available for Windows.

2.4.5 Big Data and NoSQL databases

You will no doubt already have seen the phrase *Big Data* used in the popular press, or heard it uttered by a technically minded friend or colleague. Slightly less used, but with as much hype in certain circles, is the term *NoSQL*. Bioinformaticians have, of course, been working with large data sets for some time, but Big Data has become a synonym for analytics performed on data sets so large that it is generally unfeasible to tackle them using the traditional methods that are covered in this book. There is no prerequisite for NoSQL databases to be a part of a Big Data solution, however they are often used for storage of data in this context.

Although the definition of Big Data above is rather self-fulfilling, it is worth covering a little about the concept here, to allow you to at least consider if the problem you would like to solve could be better addressed through a Big Data and/or NoSQL type approach than using a traditional RDBMS and scripting method. The most commonly used Big Data system is Hadoop, an Apache open-source project originally funded by Yahoo!. Hadoop is itself an implementation of the MapReduce framework developed by Google to allow them to perform faster searches against their indexes of the web. Although the names of the systems do not matter too much, the concepts they embody are fairly fundamental. At their core they allow large data sets to be split up into chunks that can comfortably fit onto commodity, that is cheap, servers with their own local storage. Data analysis is performed by each server on the data it holds—this part of the processing is the Map stage in MapReduce. The results are then gathered together and Reduced to a single result set. For analysis of large data sets, this type of distributed, parallel processing can be far quicker than trying to process a similar total volume of data through a single processing pipeline.

Although many RDBMSs can manage large volumes of data over multiple servers, through a functionality known as clustering, data access tends to slow down with increased volume due to the overhead of managing the data and the relationships between tables across these servers. By removing the need to manage relationships between individual pieces of data, MapReduce solves this issue as the data can effectively sit anywhere with relatively little oversight or management overhead, but it does shift the problem of ensuring that the results set is representative of the complete data set onto the Reduce phase of the approach. Since the data itself is now inherently non-relational, as there is no system to manage these relationships, there is little to be gained by imposing a rigid structure on it beyond some logical key-value pairing that allows the MapReduce type program to pick up and analyse the variables we are interested in. The use of SQL therefore becomes fairly redundant, and unstructured 'NoSQL' data storage methods are generally used in this context instead.

This very loose, unstructured approach may appear messy and difficult to understand, but many systems have been developed to make both handling the data and programmatic access to the data easier. Some even go so far as to add SQL-like querying back onto the data sets (though at a significant performance cost) and others provide APIs that make it easier for general programmers to access, manipulate, and perform operations on the data. Often, however, the Big Data approach requires a different way of thinking about the problem at hand, such that it is more amenable to this style of processing. As a practical example of this type of thinking in the bioinformatics space, Lewis *et al.* have used Hadoop to identify proteins from peptide keys (Lewis *et al.*, 2012).

We expect Big Data and NoSQL to become increasingly important as biological data sets continue to grow. If you are interested in finding out more, we recommend the books Big Data Now (O'Reilly, 2012), Big Data (Marz, 2013) and NoSQL Databases (Strauch, 2011). However, for the foreseeable future, most bioinformatics data management problems should be well suited to the more traditional SQL RDBMS solutions that form the remainder of this chapter.

2.5   Database access using SQL

Structured Query Language (SQL) is, as the name suggests, the language used primarily for querying your database. As this section will show, for almost everything you could want to do with your database the command should be straightforward to generate using just a few simple keywords. As with the structure of the database itself, it is not the implementation that's important, it is the design of your queries that matters. It is also at this stage that the quality of thought behind the design of your database is revealed. A poorly designed creation will result in poor performance when queried, or worse, incomplete or incorrect data being returned by functionally correct SQL commands.

By convention, all SQL commands are written in capital letters. This helps to distinguish them from non-SQL words, such as your table and field names, and as such the convention has been used in this book. However, there is no need to type them in capitals when querying your database if you don't want to—the end results will be just the same.

One common error to watch out for is the use of SQL reserved words (any word in the SQL vocabulary, such as TABLE or VALUES) as the names of your objects, such as databases, tables, and fields. Such usage will confuse the RDBMS and most likely result in it reporting a syntax error that may be hard to find as the query will not look obviously wrong. In such cases look carefully for conflicts between your naming and SQL words.

2.5.1 Compatibility between RDBMSs

In principle, it should be quite easy to move between RDBMSs that support SQL, as the database commands and syntax will be the same. So, examples provided in this chapter for MySQL should also work in, for example, Oracle. However, as we have already mentioned, names of data types can differ between RDBMSs. Also, each RDBMS can have additional commands that go beyond standard SQL, so care needs to be taken. In particular, we should warn you that the MySQL commands `SHOW`, `DESCRIBE`, and `LIMIT` introduced later in this section are not standard SQL commands so may not work in another RDBMS.

2.5.2 Error messages

It is very likely that you will make numerous typing errors when entering commands to the RDBMS. Often the error message that is returned can appear exceptionally unhelpful, but there are guiding clues that can help you identify where things went wrong. A very common error message is something like:

```
mysql> SELECT * FROM Supplier WHRE supplier_name LIKE 'Eps%';

ERROR 1064 (42000): You have an error in your SQL syntax;
check the manual that corresponds to your MySQL server
version for the right syntax to use near 'supplier_name like
'Eps%' at line 1
```

Here we are informed that there is an error in the SQL syntax. This means that MySQL cannot interpret something that we have typed. The problem might be that we have entered some keywords in an incorrect order, or that we have typed something incorrectly. The error message offers some guidance as to where the problem might lie, in this case just before where we wrote `supplier_name`. If we look in this region, we can quickly spot the typo in the SQL keyword, `WHERE`, here typed `WHRE`. If there is a keyword miss-entry, the location may be harder to spot as the error may well be earlier in the query than indicated.

Every RDBMS has a complete listing of error codes and help in interpreting them as part of their documentation. For MySQL, these are in the appendices to the reference manual (`dev.mysql.com/doc/#manual`).

2.5.3 Creating a database

In Section 2.3.2 we covered the command used to create an empty database with no tables from which to start. Let's create a database called `PCR_experiment`, in preparation for building a database with the design described in our earlier example. To do this it is necessary to log on to MySQL as root and issue the command below. You can then either continue to work in MySQL as root (not recommended) or grant another user access to the database as described above, then quit MySQL and log back in using that username.

```
CREATE DATABASE PCR_experiment;
```

Within the RDBMS you may have a number of databases. The first action to perform when preparing to interact with one of these is to tell the RDBMS which one you want. This is accomplished by the USE command:

```
USE PCR_experiment;
```

This will change the focus of your commands from the database currently being accessed, if any, to one called PCR_experiment.

A newly created database will be empty, with no tables defined and no data present. Before getting data into the database, it is necessary to define tables, for which the simplest, generic, command is:

```
CREATE TABLE tablename (
      field_1 type_1,
      field_2 type_2,
      ...
      field_n type_n
);
```

Returning to our example database schema, shown in Fig. 2.11, we have five tables to create. We could do this now for each of the tables, but we will also want to enforce the relationships between the tables at this point, for which a little more explanation is required.

2.5.4   Creating tables and enforcing referential integrity

The term *referential integrity* refers to the ability of the database to maintain relationships between the data held in different tables. This is primarily achieved at a design level through the use of the foreign keys discussed previously. Therefore, the responsibility for much of this area rests in your hands, as the database designer.

MySQL supports a number of different table types, of which MyISAM and InnoDB are the most often used, with MyISAM being its default. Different table types impact the way in which a RDBMS handles and stores data within a given database table and can have effects on performance and functions. We don't need to go into details about this here, suffice to say that newer versions of MySQL fully support referential integrity when the default MyISAM type is used, so it should not be necessary to specify the table type when defining tables.

There are a number of options that we can include within a CREATE TABLE statement. These include the table type, whether a field can hold a NULL value (whether it can be left empty), and also whether a field is part of a primary key or a foreign key to another table. Bearing in mind that foreign keys to other tables should logically be created after the table to which they refer, the first tables to be created should be those with no foreign keys. In our example, these are Scientist, Kit, and Supplier, created using:

```
CREATE TABLE Scientist (
      email VARCHAR(255) NOT NULL,
      given_name VARCHAR(255),
```

```
        family_name VARCHAR(255),
        title VARCHAR(255),
        department VARCHAR(255),
        tel_country INT,
        tel_area INT,
        tel_number INT,
        PRIMARY KEY (email)
);


CREATE TABLE Kit (
        manufacturer VARCHAR(255) NOT NULL,
        name VARCHAR(255) NOT NULL,
        kit_cost DECIMAL(6,2),
        buffer VARCHAR(255),
        buffer_conc FLOAT,
        enzyme VARCHAR(255),
        enzyme_conc FLOAT,
        nucl_mix VARCHAR(255),
        nucl_conc FLOAT,
        PRIMARY KEY (manufacturer, name)
);


CREATE TABLE Supplier (
        supplier_name VARCHAR(255) NOT NULL,
        supp_building VARCHAR(255),
        supp_street VARCHAR(255),
        supp_town VARCHAR(255),
        supp_city VARCHAR(255),
        supp_postcode VARCHAR(255),
        supp_country VARCHAR(255),
        PRIMARY KEY (supplier_name)
);
```

Note the simplicity of the method by which to create the compound primary key when creating Kit. There are other ways to signify which elements of a table are to be included in the primary key, but this method is consistent and clear.

If you wish to check that the tables have been created, you can issue the following MySQL command:

```
SHOW tables;
```

If you would like to see more detail about any of the tables, just enter the following command with the appropriate table name:

```
DESCRIBE tablename;
```

These two commands are further described in Section 2.5.8.

Now the remaining two tables may be created. There is an order that needs to be followed here too. Consider the relationship between Experiment and Kit_order. Because kit_order_number in Experiment is a foreign key on order_number in Kit_order, it follows that the table Kit_order must exist before Experiment can be created.

```
CREATE TABLE Kit_order (
     order_number CHAR(16) NOT NULL,
     manufacturer VARCHAR(255),
     kit_name VARCHAR(255),
     supplier VARCHAR(255),
     PRIMARY KEY (order_number),
     FOREIGN KEY (manufacturer, kit_name)
          REFERENCES Kit(manufacturer, name),
     FOREIGN KEY (supplier)
          REFERENCES Supplier(supplier_name)
);


CREATE TABLE Experiment (
     ID INT NOT NULL AUTO_INCREMENT,
     date DATE,
     time TIME,
     sequence VARCHAR(255),
     purpose TEXT,
     kit_order_number CHAR(16),
     denature_temp INT,
     denature_time INT,
     anneal_temp INT,
     anneal_time INT,
     elongate_temp INT,
     elongate_time INT,
     cycles INT,
     completion_temp INT,
     completion_time INT,
     primer_1 VARCHAR(255),
     primer_2 VARCHAR(255),
     primer_conc FLOAT,
     design_software VARCHAR(255),
     primer_supplier VARCHAR(255),
     primer_cost DECIMAL(6,2),
     gel_image BLOB,
     assessment TEXT,
     scientist_email VARCHAR(255),
     PRIMARY KEY (ID),
     FOREIGN KEY (kit_order_number)
```

```
                REFERENCES Kit_order(order_number),
        FOREIGN KEY (scientist_email)
                REFERENCES Scientist(email)
);
```

Once again, note the simplicity of creating the foreign keys whereby the field in the table being created is linked, *referenced*, to a field in a previously created table. It is just as simple to link multiple fields, where a compound key is required, as in the creation of `Kit_order`.

AUTO_INCREMENT

You will have no doubt noticed the new command `AUTO_INCREMENT` in the text used to create the `Experiment` table. Whenever new data is entered into the table, this command will populate the field `ID` with a value that is one greater than the previous highest ID in that field. In this way, the `ID` field, which is the primary key of this table, is created automatically, without you needing to know how many entries have been placed in the table before the current one.

2.5.5 Populating the database

All of the tables for this example have now been created and their primary and foreign keys have been defined. The tables are therefore now ready to take in data. This can be done in a number of ways. The simplest, but most cumbersome, is for us to enter the data using SQL from the command line, in a similar manner to creating the tables. This is done using the `INSERT INTO` command—once again bearing in mind that the order in which the tables can be populated is dependent on the structure of the foreign keying. To reiterate, `Experiment` cannot be populated with results from a newly ordered kit (i.e. a kit that is not already in the database), or performed by a new scientist unless their details have already been entered into the relevant tables (`Kit` and `Scientist` respectively). If, however, this information is already in the database, because the scientist has done other experiments and the same kit from a previous order is being used, then data can be entered into `Experiment` as its foreign key conditions will be met.

The generic form of the `INSERT INTO` statement is:

```
INSERT INTO tablename (field_1, field_2,…)
VALUES(value_1, value_2,…);
```

So, to enter a new scientist's details, the following statement could be used:

```
INSERT INTO Scientist (email, given_name, family_name, title,
department, tel_country, tel_area, tel_number)
VALUES('a.scientist@example.com', 'Andrew', 'Scientist', 'Dr.',
'Toxicology', 44, 0117, 4960808);
```

Note the single quotation marks surrounding all text entries, but not the numeric entries—this is an SQL convention that should be followed. If you forget, an error will be shown and the command will not complete.

It may seem ridiculous to have to enter all the fields of a table into a command when surely the database system knows that these fields are there. In this case you would be correct to think this, as we have entered values for all fields in the order in which they appear in the table. Because this is the case, we could shorten the above command to:

```
INSERT INTO Scientist
VALUES('a.scientist@example.com', 'Andrew', 'Scientist',
'Dr.', 'Toxicology', 44, 0117, 4960808);
```

This would achieve identical results to the above command, although, if you just tried it, you would notice that your database complained. This was because you already had a scientist in the table with an email address of `a.scientist@example.com`. This is the primary key for the table and therefore all values in this column must be unique—remember the definition introduced in Section 2.2.4.

You could try the command using a different e-mail address. Please note though that if you are not the only administrator/designer of this database, and you plan to use this shortened form of insert, you will need to check that the fields in your database have not changed prior to running your inserts this way, as extra or removed fields would cause your inserts to fail (as indeed would reordered database fields if someone had rebuilt your table in a different order).

If we wanted to only enter data into some fields of a table, or to enter data in an order that is not identical to that of the table, the field name specification in the `INSERT INTO` command may be used to guide the database—showing it where we want the data entered. For example, if we were to only want to enter a new scientist's e-mail address (required in all circumstances as this is the primary key) and their department, perhaps because they did not yet have a telephone number, we could use the following command:

```
INSERT INTO Scientist (department, email)
VALUES('Systems Lab','a.techie@example.org');
```

In this statement, we have only populated two fields and did this in a different order to that in which they appear in the table definition. However, because we explicitly told the database system what we wanted, it will have entered these values into the correct fields. If we wanted to make sure this was so, we would use the `DESCRIBE` command, explained later in Section 2.5.8.

Although we can considerably shorten the `INSERT INTO` command by omitting the field names, there will necessarily be a lot of typing to enter the details of just one experiment into the database. In reality this burden is likely to fall to the people using the system day to day and so will be split. Furthermore, the use of forms or well-presented programs used to provide access to the database will eliminate the need for the user to know any of the syntax needed to enter data into the database from the command line. The following chapter on programming in Perl and Chapter 5, on integrating these concepts with web-based systems, will allow you to create your own routes by which you and your users can enter data to the database without repeated and excessive typing. Having said that, in many bioinformatics applications, databases are not *populated* (filled) by people at all, but by programs (often written in Perl or other scripting languages). A typical example of this is where a database is used to store the results from an automated data analysis pipeline.

For test purposes, it is often convenient to use a *source file* to rapidly populate a database with reasonable quantities of representative data—this is discussed in Section 2.5.7.

2.5.6 Removing data and tables from the database

As well as adding data to a database, we might wish to remove data from the database. This might be because there was a user error when populating the system, or, quite likely when designing and testing a system, you wish to reset the database to a blank state. We may wish to remove an entire table from the system, or only parts of it. At the most extreme we may want to delete the entire database and start again. Each of these is possible and remarkably simple to achieve; and therein lies the rub—once you have done this, there is no going back[3] and the system will not ask if you are sure that you want to do this.

Deleting data from a table

The command below will delete everything from the table *tablename*. The structure of the table will remain intact so that we can enter new data there immediately, but all of the old data will have been removed.

```
DELETE FROM tablename;
```

More selectively, we can use:

```
DELETE FROM tablename WHERE field LIKE 'xyz';
```

This command will delete only those rows of the table for which the condition following the WHERE statement is met. For more discussion of the types of condition that can follow WHERE, see Section 2.5.8.

When deleting rows from tables, it should be remembered that there may be foreign keys to those rows from other tables. In these cases, multiple delete statements may be required in order to maintain the referential integrity of the database. If one or a number of rows in one table are foreign keyed to another table, for example all of one scientist's Experiment data will be linked to their entry in the Scientist table, then before deleting the parent row, all records in the child table(s) that refer to this key must also be deleted. For example, first delete from Experiment all entries attributed to that scientist before removing the scientist's information from Scientist.

---

3 This is generally the case, although if you use transaction handling (see Section 2.5.9) you would be able to recover to the last point at which you confirmed all database changes.

Deleting complete tables

If we want to remove an entire table, that is all of its data and its structure and name, from the database, this may be done using the `DROP TABLE` command:

```
DROP TABLE tablename;
```

Deleting a database

Similarly, an entire database may be removed from the RDBMS using the `DROP DATABASE` command:

```
DROP DATABASE databasename;
```

2.5.7   Creating and using source files

As we have seen, creating a database can involve a lot of typing at the command prompt. For this reason, it would be useful if, when we do need to replace a database entirely, or we want to easily reset all the tables to empty, there were a quick way of doing this that meant we only had to do the majority of typing once. Fortunately there is, through the use of a *source file*.

Source files are text files that contain SQL commands and/or data that we can call upon from within the database system. The text files should be plain text, which means that to create them we need to use a very basic text editor, such as Windows Notepad, to avoid storing all of the extra formatting information that word processors would place in the document. Section 3.1.5 in the next chapter briefly reviews text editors, should you want to find an alternative to the one that came with your operating system.

To create a simple source file that would create the example database discussed in this chapter, we could just copy the five SQL statements appearing in Section 2.5.4 and put them into a `.txt` file. This file, which we call `PCR_database_create.txt`, can be downloaded from this book's companion website, `www.bixsolutions.net`.

We can use the SQL contained in the source file by using the `SOURCE` command:

```
SOURCE filename;
```

For example, if we had downloaded the file `PCR_database_create.txt` into the directory `BBSfiles` on our E: drive, the following commands would create all the tables for our PCR database example within the, previously created, database `PCR_experiment`:

```
USE PCR_experiment;
SOURCE e:/BBSfiles/PCR_database_create.txt;
```

A useful file to have handy when building a database, which is going to act as the data store for a program that you are designing, is a source file to delete all the data from all tables while keeping the database structure intact. Once again this is a simple file to create. For this example it would look like the following:

```
DELETE FROM Experiment;
DELETE FROM Kit_order;
DELETE FROM Supplier;
DELETE FROM Kit;
DELETE FROM Scientist;
```

This file may be found at `www.bixsolutions.net`, entitled PCR_database_clean.txt.

If we wish to enter a set of data into the database using a large number of `INSERT` statements, we can do this with a source file too. As the source file just consists of SQL statements, these can be written in exactly the same way as if we were typing them, but with the added advantage that if a mistake is made part way through, the whole series of commands may be run again with little extra effort. A source file to populate the PCR database is provided at `www.bixsolutions.net`, under the name `PCR_database_populate.txt`. The file populates the database with three entries for scientist, three for supplier, three for kits, four for kit orders, and five for experiments. We recommend you populate the database using this script before proceeding to the examples of querying the database in the next section.

2.5.8 Querying the database

Finally we come on to accessing the data that is in a database. There are many ways in which to do this and the complexity of the queries may seem daunting at first. However, they are all built up from a few very simple concepts, explained below. The most important thing to keep in mind is to first think carefully about exactly what it is that you want to achieve and only then to try and write the query for this. As with the design of the database itself, it is the preparation that is the key to getting this right—with solid thinking, the implementation is often straightforward.

SHOW

Once you have focused on the database of your choice (USE *databasename*), you may well want to list the tables present. Here you can use the SHOW command:

```
SHOW tables;
```

This will produce a list of all the tables present in the database. SHOW may also be used to show all the databases to which you have access in your installation of MySQL, as in:

```
SHOW databases;
```

DESCRIBE

For any table within a database, you may want to see the field names and their associated data types. The DESCRIBE command will return this information.

```
DESCRIBE Scientist;
```

will therefore produce a table of results similar to the following:

```
mysql> DESCRIBE Scientist;
+-------------+--------------+------+-----+---------+-------+
| Field       | Type         | Null | Key | Default | Extra |
+-------------+--------------+------+-----+---------+-------+
| email       | varchar(255) | NO   | PRI |         |       |
| given_name  | varchar(255) | YES  |     | NULL    |       |
| family_name | varchar(255) | YES  |     | NULL    |       |
| title       | varchar(255) | YES  |     | NULL    |       |
| department  | varchar(255) | YES  |     | NULL    |       |
| tel_country | int(11)      | YES  |     | NULL    |       |
| tel_area    | int(11)      | YES  |     | NULL    |       |
| tel_number  | int(11)      | YES  |     | NULL    |       |
+-------------+--------------+------+-----+---------+-------+
8 rows in set (0.00 sec)
```

SELECT

The SELECT command is arguably the most useful, as it allows us to access the data held within the database. Used in conjunction with the other commands described below, it should allow you to access any set of information you require in any order, grouped in any way from any table or combination of tables.

The format of any SELECT query follows this basic convention:

```
SELECT field FROM Table;
```

So, if we wanted to extract a list of all the scientists' surnames that have performed PCR experiments in the laboratory, we could use the following command:

```
SELECT family_name FROM Scientist;
```

We may select multiple fields in an order that we specify using a similarly structured command, for example:

```
SELECT family_name, given_name FROM Scientist;
```

This returns a table of results that has all the surnames in the first column and all the first names in the second.

Sometimes we may want to return the entire contents of a table, in which case an asterisk may be placed in the *field* position, as in:

```
SELECT * from Scientist;
```

COUNT

Often we will want to know just how many records are going to be returned by a query. To do this we will use the COUNT command. For example, the command below would return the number of scientists that are present in the `Scientist` table.

```
SELECT COUNT(*) FROM Scientist;
```

DISTINCT

`DISTINCT` allows us to specify that we do not want repeated pieces of information to be returned by our queries. This command is often used along with `COUNT` to give the number of different elements within a table.

```
SELECT COUNT(DISTINCT field) FROM Table;
```

It may also be used to return a list without duplicate entries within fields, such as:

```
SELECT DISTINCT sequence FROM Experiment;
```

This query would return a list of sequences that had been investigated, but each sequence would appear only once, even though it may have been processed under many different conditions or using different PCR kits. The command below would therefore return the number of unique sequences that had been investigated.

```
SELECT COUNT(DISTINCT sequence) FROM Experiment;
```

ORDER BY . . . ASC / DESC

We cannot assume that results will be returned from a query in the order in which they were placed into the database, or in an order defined by an index. The `ORDER BY` command may be used to specify the order in which results are presented to suit the application. For example:

```
SELECT family_name FROM Scientist ORDER BY family_name ASC;
```

This command returns the list of family names of scientists alphabetically ordered, as this is a character-based field. The use of `ASC` ensures that they are returned in ascending order, that is from A to Z. To return them in the reverse order (Z to A), we would use `DESC`. The same syntax applies when sorting results from numeric fields.

LIMIT n

When creating queries, sometimes we may just want to see a sample of the results that would be produced. This can be quite useful when checking to make sure that the results are of the form that we think they should be. This is especially useful when creating complex queries that we are not quite sure are correct! In such cases, the `LIMIT` command may be used to return only the number of results that are required. For example, the command below will return the first three results retrieved from the table `Experiment`.

```
SELECT * FROM Experiment LIMIT 3;
```

Remember that these may not be the experiments numbered 1 to 3. If we wanted these fields in particular, an ORDER BY modifier would be required, as in:

```
SELECT * FROM Experiment ORDER BY ID ASC LIMIT 3;
```

WHERE

WHERE is an extremely useful command modifier as it allows for highly specific queries to be written, often just the sort of queries that the database was created to allow us to answer in the first place. It is often used to return a set of results for which a certain condition is achieved. For instance, to return from the table Experiment only those results that were generated after a certain date, the following statement could be used:

```
SELECT * FROM Experiment WHERE date > '2008-01-01';
```

This returns all the information about experiments that took place after 1 January 2008—note the use of quotation marks around the date, as the date (and time) types are not truly numeric. Numeric comparison symbols other than > may be used. The full list is given in Table 2.5. In the example below, all the experiment information for experiments performed with an annealing temperature set to equal or below 70 degrees would be displayed.

```
SELECT * FROM Experiment WHERE anneal_temp <= 70;
```

For textual and character comparison, we have the LIKE command, an example of which is:

```
SELECT tel_country, tel_area
FROM Scientist
WHERE department LIKE 'Toxicology';
```

This would return the telephone country and area codes for the toxicology department(s). If unsure of the exact string that we are looking for in a textual comparison, a wildcard character, %, can be used. For example:

Table 2.5 Numeric comparison operators in SQL

Symbol	Description
=	equal to
<>	not equal to
>	greater than
>=	greater than or equal to
<	less than
<=	less than or equal to

```
SELECT department, tel_country, tel_area
FROM Scientist
WHERE department LIKE '%tox%';
```

This returns the full name, country and area codes for all departments that contain the string 'tox' somewhere in their name. Note that these comparisons are not case sensitive.

AND and OR

Multiple constraints using the WHERE function may be considered together with Boolean logic statements, such as AND and OR.

For instance, to return all the experimental results performed by a certain scientist (the ubiquitous Darren Oakley) during January 2008, the following statement may be used:

```
SELECT *FROM Experiment
WHERE scientist_email LIKE 'd.oakley%'
AND date <= '2008-01-31'
AND date >= '2008-01-01';
```

To select all experiments using one of two specific order numbers, we could use:

```
SELECT *
FROM Experiment
WHERE kit_order_number = 115
OR kit_order_number = 121;
```

GROUP BY

Often we may want to view the results grouped by one particular feature of the data. The most common of these are for statistical measures, such as finding the average number of amino acids in all human proteins. This could be done manually by retrieving the sequences of all proteins, summing their lengths, and then dividing by the total number of proteins. Usefully, there are a number of GROUP BY functions that allow us to access this sort of information automatically. These include COUNT, MAX, MIN, AVG, and SUM.

To count the number of experiments that a scientist has performed, we can use:

```
SELECT scientist_email, COUNT(scientist_email)
FROM Experiment
GROUP BY scientist_email;
```

This will return a two-column table containing both the scientists' e-mail addresses and the number of times that each of these has appeared in the Experiment table.

If we wanted to know the maximum kit cost from each manufacturer, this method could require a lot of manual inspection of the table. Instead, we would use:

```
SELECT manufacturer, MAX(kit_cost)
FROM Kit
GROUP BY manufacturer;
```

This will return a two-column table; the first column containing the name of each manufacturer and the second the cost of their most expensive kit.

JOIN

Up to this point we have been considering queries against single tables only. However, as has been mentioned previously, one of the most powerful features of a relational database is that all tables are related to one another. If we are to only query single tables, why are these relationships useful? With experience it becomes apparent that the most useful queries involve more than one table, and therefore the queries that are used to access their data need some way to recognize this. The method by which this is done is called a *join*.

Often a join may be implicit in a SELECT query. Consider the following query.

```
SELECT manufacturer, Kit_order.supplier
FROM Kit_order, Supplier
WHERE Kit_order.supplier = Supplier.supplier_name
AND supp_city like 'Oxford';
```

This would return all the kit manufacturers that suppliers in the city of Oxford could supply. To break the query down, we have asked for two pieces of information, manufacturer and supplier, to be returned. Up until now, we have only been querying one table and so only that table has featured after the FROM keyword. In this case, we are looking to link this to information from another table, from Supplier. Therefore, Supplier now features after Kit_order in a comma-separated list. We should also be explicit in stating where each field we would like returned is coming from. In the above example, it is obvious that the field manufacturer comes from the table Kit_order because manufacturer does not feature in Supplier. However, very similarly named fields (supplier_name and supplier) appear in both tables, so it is good practice to dictate which table we expect to retrieve this field from. This is done using the dot (.) convention, whereby the format *tablename.field* uniquely determines the field we are interested in. Indeed, if we had chosen to give the two supplier fields the same name in both tables, the command simply would not work without specifying both the table and field names in this way.

The WHERE statement allows us to join these two tables using their shared field, in this case the field supplier in Kit_order which is equivalent to supplier_name in Supplier. This tells the system that every time these two fields are equal, those are the data we are interested in. It is sometimes easy to forget this WHERE statement as it seems obvious to us which two fields in two tables are the same. If you do forget to define this, then every row (*n*) in the first table will be joined with every row (*m*) in the second table. This results in a set of information that if displayed would have *n* times *m* rows. This can be a very large number and is likely to take a long time to run, only to get results that are unlikely to be what you were looking for. We then constrain the query results further to only those suppliers that are based in the city of Oxford.

This type of query features an implicit JOIN statement—we have joined the two tables together using their shared field and are expecting only results for which the condition is met. This statement may have been written with an explicit JOIN statement as:

```
SELECT manufacturer, Kit_order.supplier
FROM Kit_order JOIN Supplier
ON Kit_order.supplier = Supplier.supplier_name
WHERE supp_city like 'Oxford';
```

This form of JOIN is known as an INNER JOIN. In set theory terms, it will select the intersection (overlap) of the two tables, with any further constraints applied. Therefore, if there are manufacturers with no suppliers in Oxford, these will not appear. Similarly, if there are suppliers in Oxford who have not sold any PCR kits to this lab (say if the Supplier table contained all suppliers for all lab equipment), they will not appear.

We might, however, want some of this information. For instance we might want to know details for all of the suppliers of PCR kits in Oxford that we have used plus any other suppliers based in Oxford. For this query we would use a RIGHT JOIN.

```
SELECT *
FROM Kit_order RIGHT JOIN Supplier
ON Kit_order.supplier = Supplier.supplier_name
WHERE supp_city like 'Oxford';
```

In this case, whenever there is a supplier in Oxford that has not yet supplied a kit to the lab, columns from Kit_order will return NULL, otherwise all details will be returned.

It is also possible to use a LEFT JOIN, which in this case will give the same results as a JOIN as there are no kit orders that don't have a supplier. If, however, we swapped the order of each table name in the query, as shown below, we would get identical results to the previous RIGHT JOIN example.

```
SELECT *
FROM Supplier LEFT JOIN Kit_order
ON Supplier.supplier_name = Kit_order.supplier
WHERE supp_city like 'Oxford';
```

Additionally, it is also possible in some RDBMSs to combine the two, forming a full OUTER JOIN, but not (simply, at any rate) in MySQL. It is sometimes easier to picture these operations graphically, as in Fig. 2.12.

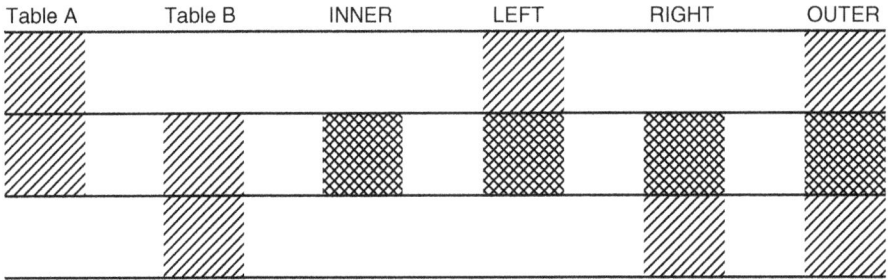

Fig. 2.12 Graphical depiction of the actions of JOIN statements. Table A and Table B are joined using the different types of join. The results of these combinations are shown. The INNER join results in the union of the two tables where there are equivalent fields in each. The LEFT join returns the INNER product and the remaining fields from Table A. The RIGHT join returns the INNER product and the remaining fields from Table B. Finally, the full OUTER join returns the contents of both tables, linked where appropriate.

2.5.9   Transaction handling

As mentioned above, you have complete control over your database when interacting with it either via the command line, or through the use of a computer program, as shall be described later. This can often place you within a few misplaced keystrokes, or bad lines of code, of disaster. You may mitigate this through a comprehensive backup program, but this would always run the risk of losing the results of valid activity that has occurred between backups.

Fortunately, the designers of database management systems also recognize this danger, and so have introduced the concept of transaction handling. Simply put, a transaction is any block of database activity, including `INSERT`, `DELETE`, `SELECT`, and any other SQL, or other, actions. The start of the block is defined using a specific command. At this point, all database activity resulting from the commands following the transaction start is not permanently written to the database. Upon reaching the end of the block, a test may be performed, in which a person, or an automated program, checks for errors. If this test is successful then the results from the block are written permanently to the database. If, however, the test reveals some unintended behaviour, the actions of the block are ignored and the database is rolled back to the point at which the transaction block began. In MySQL, this sequence of events is controlled using the commands described below.

AUTOCOMMIT, COMMIT, ROLLBACK

By default, MySQL runs with the `AUTOCOMMIT` mode turned on. This means that whenever you type, or otherwise run, a command, the results of this are immediately written to the database. This can be turned off using:

```
SET AUTOCOMMIT=0;
```

Once `AUTOCOMMIT` has been turned off, each command needs to be followed by a `COMMIT` statement in order for its effect to be written to the database. For instance, if you wanted to remove some of the rows from a table, you may have entered the following command:

```
DELETE FROM Experiment;
```

You then realize that what you really meant to do was only delete certain rows from the table that were entered by one particular scientist. If AUTOCOMMIT was on, as it would be as standard, then it is now too late; all the rows are gone. However, if you had turned AUTOCOMMIT off and had not issued a COMMIT statement, the choice is yours. In this case, you could ROLLBACK the statement:

```
ROLLBACK;
```

The database is now exactly as it was before you issued the DELETE statement. You may now enter the correct statement:

```
DELETE FROM Experiment
WHERE scientist_email LIKE 'i.shadforth@bixsolutions.net';
```

```
COMMIT;
```

The first statement is executed and then written to the database. You may enter any number of commands before issuing a COMMIT statement. All will then be permanently written to the database, or rolled back if you choose.

AUTOCOMMIT may be turned back on again using the command:

```
SET AUTOCOMMIT=1;
```

START TRANSACTION . . . COMMIT

You may wish to enter a block of commands that you wish to try out without setting AUTOCOMMIT to off. To do this, start the block with START TRANSACTION. This will temporarily suspend AUTOCOMMIT, which will be reinstated at the end of the block, as indicated by your use of either a COMMIT or a ROLLBACK.

```
START TRANSACTION;
DELETE FROM Experiment;
DELETE FROM Scientist;
COMMIT;
```

Note that when a START TRANSACTION command is issued, this implicitly issues a COMMIT command finishing any previous transaction that had not been committed, or rolled back.

2.5.10 Copying, moving, and backing up a database

We will often want to backup or move a database from one machine or location to another. To achieve this, MySQL has a simple set of commands.

The first of these, to backup your database, should be invoked from outside the MySQL environment, so type exit to leave MySQL and return to the normal terminal prompt. In Windows, to backup your database to a text file, which will contain all the information needed to recreate the database including table structures and data, type the following and enter the password when prompted.

```
mysqldump -u username -p databasename > c:\directory\backup.txt
```

If you have opted not to use a password for your chosen user account, then don't type the –p switch.

You will find your output file in the specified directory. If you are using Linux or Mac OS, the format of this command is similar, but navigating from a suitable point, for example:

```
mysqldump -u username -p databasename > /usr/mydirectory/
backup.txt
```

If you use a text editor to look at the file produced by `mysqldump` you will see that it is essentially a series of MySQL statements that define the database and, if it contained data, populate it. The database can therefore be reinstated on any MySQL server by using this file with the SOURCE command. To do this, you would need to start MySQL then first CREATE (or DROP and then CREATE) your blank database, change focus to that database with USE, and then use SOURCE to execute the SQL commands in the backup file:

```
SOURCE c:\directory\backup.txt;
```

This will drop all existing tables in your database, recreate them, and then populate them with any stored data.

2.6   MySQL Workbench: an alternative to the command line

We have focused in this chapter on creating and manipulating MySQL databases by typing SQL commands. This is because, as explained in Chapter 1, anything that can be typed at the command line can be automated by writing software — essential for bioinformatics applications where databases are large. However, while tasks like adding and retrieving data from a database need to be done frequently and are therefore routinely automated, some of the other tasks are less common. In particular, we will typically only design each database once, and we may only occasionally want to add users or perform other maintenance tasks. MySQL Workbench provides a convenient graphical interface to accomplish these tasks.

MySQL workbench is a free tool available from the MySQL website (`www.mysql.com`). If you selected one of the more complete packages when installing MySQL, such as 'Developer Default', then you will already have MySQL Workbench installed on your computer. It is not our intention to provide usage instructions for MySQL Workbench here, partly because its MySQL-specific nature clashes with our aim of making this part of the book about SQL in general, but more importantly because it should be mostly self-explanatory now that we have covered the basic RDBMS principles. However, it is worth briefly mentioning some of MySQL Workbench's capabilities because it is a great companion to MySQL, and its functionality is representative of that provided by similar tools for other RDBMSs.

MySQL Workbench is organized around three main themes: SQL development, data modelling, and server administration. Data modelling is arguably the most impressive and useful. This allows a database schema to be created graphically in the form of an enhanced entity-relationship (EER) diagram, by drawing tables and connecting them together. Within this graphical interface, fields can be added to tables, their datatypes can be defined, and other properties such as AUTO_ INCREMENT and primary key are set simply by ticking boxes. It is even possible to create an editable EER diagram from an existing database (so-called reverse engineering). As an example, Fig. 2.13 shows an EER created from the PCR_experiment database in just a couple of mouse clicks.

MySQL Workbench's server administration functionality covers things like user account management and backing up and restoring of databases. MySQL

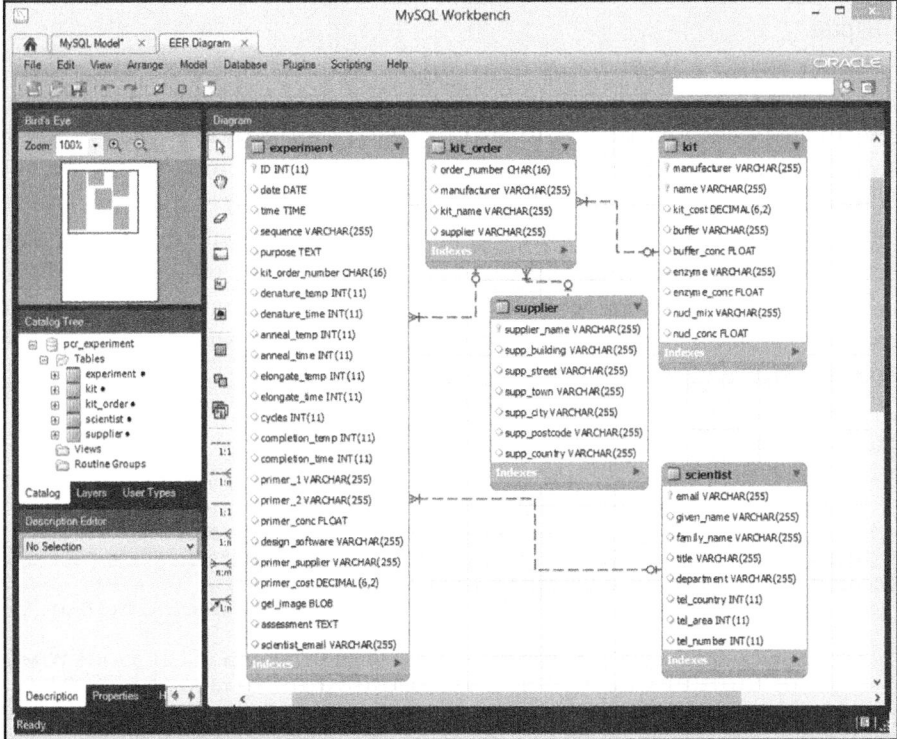

Fig. 2.13 An enhanced entity relationship (EER) diagram created from the PCR_ experiment database using MySQL Workbench. Within MySQL Workbench, the database is fully editable via this diagram - tables and fields can be added, removed and their properties modified. Note that the links between tables do not line up with shared fields as they did in our manually created diagrams (e.g. Figure 2.11) but the relevant fields are highlighted as the mouse cursor is moved over links in MySQL Workbench.

Workbench also allows editing of data within a database, but it is unlikely that you will make extensive use of this because databases are typically populated and queried through automated software or bespoke front ends.

2.7   Summary

This chapter has dealt with the basics of building databases, starting with design—which is the most important factor—through to getting your hands on a suitable system, creating, populating, and finally querying your data. This is a large topic and there is only space in this book to cover the basics of database design, creation, and use. There are many excellent books available that will guide you through more advanced topics, as referenced through the chapter. However, one that we would like to recommend in particular at this stage, for those using MySQL, is the *MySQL Cookbook* (DuBois, 2006). Having used this chapter to grasp the basics of MySQL and database design, you will find *MySQL Cookbook* is an excellent reference for ways in which to perform almost anything you might want to do with MySQL. There are similar works available for other RDBMSs and also for SQL in general, all of which should be accessible after working through this chapter.

It is likely that you will not want to interact with your database system directly once you have designed and built it, and you will almost certainly not want other users of your database to have to log in to the command line directly. There are also lots of tasks for which manual entry of data at the command line would be a bore. This is why all the popular bioinformatics databases mentioned in the previous chapter are accessed via a more intuitive web-based interface—with no knowledge of SQL required. This ease of use is achieved by producing a database front end, essentially a program that sits between the database server and the user. In bioinformatics, such interfaces are often written using Perl, which is the subject of the next chapter.

References

Codd, E. (1970). A relational model of data for large shared data banks. *Communications of the ACM*, 13 (6): 377–87.

Connolly, T. & Begg, C. (2003) *Database Solutions: A Step-by-Step Approach to Building Databases*. Addison-Wesley: Boston, USA.

DuBois, P. (2006). *MySQL Cookbook*. O'Reilly: Sebastapol, California, USA.

Hernandez, M. (2003) *Database Design for Mere Mortals: A Hands-On Guide to Relational Database Design*. Addison Wesley: Boston, USA.

Lewis, S., Csordas, A., Killcoyne, S., Hermjakob, H., Hoopmann, M.R., Moritz, R.L., Deutsch, E.W. & Boyle, J. (2012). Hydra: a scalable proteomic search engine which utilizes the Hadoop distributed computing framework. *BMC Bioinformatics*, **13**: 324.

Marz, N. (2013). *Big Data: Principles and Best Practices of Scalable Realtime Data Systems*. Manning Publications: Shelter Island, New York, USA.

O'Reilly Media Team (2012) *Big Data Now: 2012 Edition*. O'Reilly: Sebastapol, California, USA.

Simecek, N. (2007). *Development of a database with web-based user interface for taqman assay design*. MSc Thesis. Cranfield University.

Stephens, J. & Russell, C. (2004) *Beginning MySQL Database Design and Optimization: From Novice to Professional*. Apress: Berkeley, California, USA.

Strauch, C. (2011) *NoSQL Databases*. Review Paper. Stuttgart Media University.

CHAPTER 3

Beginning programming in Perl

Being able to program was once a prerequisite for doing bioinformatics, simply because there was very little bioinformatics software available. Although the situation is now different, with freely available bioinformatics tools released all the time, being able to write your own software is still invaluable for all but the simplest bioinformatics tasks. Not only does programming give you the flexibility to produce novel tools, it crucially allows you to automate processes, which overcomes the bottleneck of manual analysis that typically occurs when data throughput increases. Programming can also be used to link existing tools together to create powerful data analysis pipelines.

There is an array of programming languages available which can fulfil the requirements above, but the programming language of choice when it comes to bioinformatics tasks is often Perl (Practical Extraction and Retrieval Language). This is due to Perl's specific strengths—a very quick learning curve, relatively easy-to-read syntax, ability to add extra functionality by installing third-party modules, and very accomplished text manipulation abilities. Perl is now engrained in the bioinformatics community, with some major organizations using it as their standard development tool and vast amounts of Perl code out there for you to incorporate into your own programs. If you have any doubts as to Perl's importance, we recommend the paper *How Perl Saved the Human Genome Project* by Lincoln Stein, (an archived copy can be found via `goo.gl/qErlr`).

In this chapter we cover the basics of Perl programming, particularly those of specific relevance to bioinformatics. The level of detail in each section is limited to allow the maximum number of individual bioinformatics-related topics to be covered. If we were to cover everything in depth, we would need a separate book for Perl, and there are already plenty of good Perl programming books out there (Wall et al., 2000; Tisdall, 2001; Tisdall, 2003). There are also many excellent Perl resources on the web. As such, we cover the basics and give you enough information to enable you to get on with the technologies that we cover.

Building Bioinformatics Solutions. Second Edition. Conrad Bessant, Darren Oakley and Ian Shadforth.
© Conrad Bessant, Darren Oakley, and Ian Shadforth 2014. Published 2014 by Oxford University Press.

3.1   Downloading and installing Perl

Before getting started programming in Perl, you will need to have a Perl interpreter installed on your computer. The interpreter is what actually runs Perl programs, so although you could technically write a Perl program without it, you wouldn't be able to see what that program does. If you're using Linux or Mac OS, you already have it installed as it comes as part of the operating system. In order to confirm this, and to find out which version of Perl you have, go to your operating system's command line (see Appendix A) and type the following (followed by the Enter key):

```
perl -v
```

You should then be greeted with output similar to the following:

```
This is perl 5, version 12, subversion 3 (v5.12.3) built for
darwin-thread-multi-2level (with 2 registered patches, see
perl -V for more detail)

Copyright 1987-2010, Larry Wall

Perl may be copied only under the terms of either the
Artistic License or the GNU General Public License, which may
be found in the Perl 5 source kit.

Complete documentation for Perl, including FAQ lists, should
be found on this system using "man perl" or "perldoc perl".
If you have access to the Internet, point your browser at
http://www.perl.org/, the Perl Home Page.
```

This indicates that you have Perl version 5.12.3 installed on your system.

3.1.1   Older versions of Perl on Mac OS

For Mac OS users, if you do not have at least version 5.10.1 of Perl installed on your system, you will need to update your distribution of Perl otherwise you will not be able to run all of the examples in this book. The easiest way to achieve this is using a tool called Perlbrew which you can setup easily in a small number of steps:

1 Install Xcode and its 'Command Line Tools'—if you installed MySQL in Chapter 2 as instructed via homebrew you will already have this setup.

2 Visit the Perlbrew homepage (`perlbrew.pl`) and follow the installation instructions that use the command-line tool `curl`. Once the install is completed you will be told to add some commands into your `~/.bashrc` file—do this as instructed. Now close and re-open your terminal. You should now be able to type `perlbrew`, hit return and get some usage instructions.

3 View a list of available Perl versions with the command `perlbrew available`.

4 Install the latest stable release of Perl (denoted by an even second number), this was 5.16.3 at the time of writing—that is: `perlbrew install perl-5.16.3`.

5 Install `cpanm` (a utility for installing third-party libraries) via `perlbrew install-cpanm`.

6 Learn how to use Perlbrew by reading the documentation on the Perlbrew webpage (`perlbrew.pl`)—it is short and very concise.

We have also written up a slightly more detailed set of instructions on the `www.bixsolutions.net` forum (`www.bixsolutions.net/forum/thread-100.html`) please head over there if you need a touch more guidance or help.

3.1.2 Older versions of Perl on Linux

Linux users, if you do not have at least version 5.10.1 of Perl installed on your system, you should really upgrade to the latest release of whichever distribution you are running. This version of Perl is really quite old, and any recent Linux build will have a newer release of Perl.

If you really cannot update your Linux distribution, but do have an older build of Perl, you could try installing a more recent build of Perl via Perlbrew as directed for Mac users above, but note that supplied Perl module installation instructions later on in the book may not work for you should you go down this route; you will need to use a combination of your package manager to install any non-Perl dependencies, and `cpanm` to install Perl modules—instructions for this approach will not be documented through the rest of the book.

3.1.3 Installing Perl on Windows

The above commands (Section 3.1) will also work on a Windows based computer, but only if you have a Perl interpreter installed. Perl is not automatically installed as part of the operating system, so unless somebody has already done it for you, you will need to download and install it. The easiest way to do this is to use ActivePerl.

Head to `www.activestate.com/activeperl` and follow the very obvious download links to get the free 'Community Edition' of ActivePerl for your version of Windows (32-bit or 64-bit).[1] This edition of ActivePerl is free to download and use. You may be given the opportunity to enter your contact details while downloading, but this is optional. As long as you have administrator permissions on your system, all you have to do is open the downloaded package and go through the usual Windows install procedure. We recommend accepting the default installation options unless you have a very good reason not to.

3.1.4 Compilers and other developer tools

Later on in the chapter, we explain how you can use third-party code and libraries within your own programs (Section 3.10 Harnessing Existing Tools). Some of these

---

1 We recommend Perl 5.14, as some of the Perl modules/packages we use later on in the book have not been made ready for ActivePerl 5.16 at the time of writing.

tools need compilers and certain C libraries in order to install correctly, so it is best to get them installed now before we get started.

Windows users can breathe easy here—you have nothing to do. ActivePerl comes with everything you need already.

Mac users, if you chose not to use Perlbrew to install/manage your versions of Perl (Section 3.1.1), we have another small set of instructions on the `www.bixso-lutions.net` forum for you—head over to `www.bixsolutions.net/forum/thread-68.html` and follow the instructions there. This will set up a tool called `cpanm` (for installing Perl modules) and your environment, so all third-party Perl code is installed in your home directory and does not impact the base operating system.

Linux users, the vast majority of third-party Perl modules you are likely to want to try will already be available via your distribution's package manager, but just in case we ever do need to build/compile some modules ourselves, look for a package called 'build-essential' or something similar in your package manager and ensure that you have that installed.

3.1.5   Before getting started

Finally, before embarking on understanding and programming in Perl, there are a couple more things to get ready beforehand.

Picking a text editor

A Perl program is just a list of commands in a plain text file, so the first thing we need is a text editor in which to write programs. There are many different text editors available for the various operating systems. There is no 'best' editor and in the end it comes down to what you feel comfortable working with. Below is a list of some of the text editors that we recommend for Perl programming, organized by the operating system(s) they support:

Multi-platform

◆ **Sublime Text** (`www.sublimetext.com`). A very customisable and expandable (via plug-ins) programming editor useful for most programming languages and compatible with most operating systems (including Windows, Mac OS, and Linux). Sublime Text is free to evaluate, but a licence must be purchased for continued use.

◆ **Komodo** (`www.activestate.com/komodo-edit`). A programming editor developed specifically for dynamic languages such as Perl. A slimmed down, but still very good, edition of this editor (Komodo Edit) is available for free. The full Komodo IDE with a built-in debugger and other useful tools is a commercial package, for which payment is required.

◆ **Vim** and **Emacs**. As classic Unix/Linux tools, these editors (and variants such as gVim and XEmacs) can be installed from the Linux package manager. Ports of these editors are also available for Windows (via `www.vim.org`, `www.gnu.org/software/emacs`, or `www.xemacs.org`) and Mac (`code.google.com/p/macvim` or `aquamacs.org`).

Windows only

* **Programmer's Notepad** (`www.pnotepad.org`). A free, open-source, text editor. Suited to Perl and many other programming languages.

Mac OS only

* **TextMate** (`macromates.com`). A programming editor useful for most programming languages. This editor is also highly expandable and customisable through the use of plug-ins. Like Sublime Text, TextMate it is commercial product so a license should be purchased.

* **TextWrangler** (`www.barebones.com`). A free text editor suited for programming in many different languages. If you grow to like this editor, there is also a commercial product available with extra features called BBEdit.

Linux only

* **gedit** (`www.gnome.org/projects/gedit`). The default text editor for the Gnome desktop environment. This is a basic, general-purpose, and easy-to-use text editor.

* **Kate** (`kate-editor.org`). The default text editor in the K desktop environment (KDE). Like gedit, this is also a basic, general-purpose, and easy-to-use text editor.

If forced to recommend just one of the above editors for this book, we would opt for Komodo Edit, as it is free, available for all the major operating systems and, as well as being suitable for Perl, it can also be used with R and HTML which we cover in later chapters.

The command line

Although we will be writing our programs in a text editor, we will be executing these programs and interacting with Perl via the command line. If you're not familiar with the command line, we would refer you to Appendix A. Particularly important is which directory (or folder) you're positioned in (i.e. in a Windows 8 command line you will automatically be positioned somewhere like `C:\Users\Conrad`; on a Mac `/Users/UserName/`; on Linux `/home/UserName`). If you save your Perl programs anywhere other than your home directory, you must change your directory in the command line (using the `cd` command) to wherever you have saved your programs in order to run and use them.

3.2 Basic Perl syntax and logic

To get started, let's write a simple Perl program and see what happens. Open up your selected text editor and type the following code and save it as a file called `hello_world.pl`:

```
#! /usr/bin/env perl
# A Perl 'Hello, world!' program
print "Hello, world!\n";
print "=============\n";
```

Now, open up your command line terminal (ensuring that you have moved into the directory where you are saving your Perl scripts) and type the following:

```
perl hello_world.pl
```

If all went well, you should now have the famous words 'Hello, world!' (minus the quotes) printed to the screen in your terminal window. If not, and you get an error message instead, check the code above to see if you have mistyped something. In particular, make sure that you are using the right types of quotes in the `print` statement, and note that Perl is case sensitive, so `print` is not the same as `PRINT`, or even `Print`.

So, what went into that first program?

The first line in the program is known as the *shebang*[2]—this is a special command required to tell the operating system what is going to run the script (in this case, Perl) and it always starts with `#!`. The shebang also tells the operating system where to find the Perl interpreter—in this example we actually run a system command called `env` within the shebang, asking it to find the `perl` executable for us. In Windows, `env` is not available, so the shebang line should be set to point directly to your Perl interpreter, which is likely to be somewhere like `C:\Perl64\bin\perl`. You will notice that the program runs in Windows regardless of what the shebang says. This is because the shebang is not required in Windows, at least not when we are executing the program using the `perl` command itself. However, the shebang line is essential when we run scripts as generic executable programs, for example within a web server process, so it is good practice to include it regardless.

The second line of the program is a comment. Comments are lines or statements in programming code that are not read by the computer, they are just there to help you, the programmer, or any programmer who carries on your work, to annotate and make notes within the program code. In Perl, comments are defined with the use of a # symbol, and all text to the right of the # symbol until the end of the line is ignored by the Perl interpreter. (NB: This should not be confused with the shebang—#!—they are different beasts.)

The third line of the program is a `print` statement—this is the first part of the program that actually does something. Print statements are a way in which a program can let its user know what is happening. It is also a handy way of giving output from your script, including diagnostic information during program development and debugging. In the most basic form that we have used here, anything contained within the quotes is printed to the command line. Well, not quite everything is printed—for example the \n construct is a special code telling Perl to start a new line of text. Many such codes are available, and we shall cover more of these and their uses later.

---

2 The term *shebang* derives from the names of the characters # (*hash*) and ! (sometimes called *bang*).

The last line of the program is just another `print` statement, included to add some rudimentary underlining to our message. The reason we included this was to demonstrate a very basic principle common to most programming languages—programs generally run from the top of the code to the bottom. Indeed, simple programs like this one are sometimes referred to as *scripts*, as they are little more than lists of actions to be performed one after the other. We will see later in the chapter that Perl supports control structures that facilitate a much more complex program flow.

Note that the end of each line of Perl must be denoted by a semi-colon (;) character. Failure to do this will result in errors when you try to run the program, which often catches out beginners (and even some more experienced coders when they are up against a deadline!). The semi-colon is needed because, in more complex programs, commands may be split over several lines of text, so we need a way of telling the Perl interpreter where the command really ends rather than assuming that the start of a new line of text is the start of a new command. Because comment lines beginning with # are ignored by Perl, they do not require a semi-colon.

To avoid repetition, from here on we shall not mention or print the shebang, or other starting code, for our smaller examples. So, if you create new files to test the many snippets of code presented throughout this chapter, don't forget to add in the shebang, a quick comment to say what the program does, and to save the code with an appropriate name ending in `.pl` so that it is recognized as a Perl program. In longer examples, where we provide full programs, we have included the standard Linux shebang. If you're using Windows, you should change it to point to wherever you installed Perl.

3.2.1   Scalar variables

Bioinformatics is all about data, so being able to manipulate data within our programs is essential. As we have seen in previous chapters, databases and files are used for long-term data storage. In programming, we have the additional concept of *variables*, which are objects within a program within which data can be stored and manipulated whilst the program runs.

A scalar variable is the simplest type of data that Perl handles. This is typically a single number (e.g. 465 or 1.25) or strings of characters (e.g. 'carbon dioxide' or 'ATGGGCCGAT'). In most other programming languages, numbers and strings are handled separately (even different types of numbers are handled differently), in Perl, however, they are all treated nearly identically. Scalar data in Perl is indicated through the use of the dollar symbol in front of the variable name—for example $sequence is a scalar variable called 'sequence'.

Assigning values to scalar variables

One of the most basic operations for scalar variables is *assignment*—giving the variable a value. To do this we use the equals sign (=).

```
$num = 24;        # give variable $num, the number value 24

$dna = 'ACTG';    # give $dna the string value ACTG i.e. the
                  # string between the inverted commas

$bar = $num * 2;  # makes $bar equal to the value of $num (24)
                  # multiplied by 2 (48)

$bar = $bar + 12; # make $bar equal to $bar (48) plus 12 (60)
```

From the examples above, you can see that we assign the variable named on
the left, with the value defined on the right. Also, the last two statements show
that it is possible to overwrite variables with new values, even using the same
variable that we are about to replace (as in the last line where we use $bar twice).
Operations like this are quite common, so common in fact that there are useful
shortcut operators that help us do this. A few examples of these are shown below.

```
$foo = 5;         # make $foo equal to 5
$foo = $foo + 5;  # this is one way of increasing $foo by 5
$foo += 5;        # this is another way of doing it!
$bar = 5;         # make $bar equal to 5
$bar = $bar * 2;  # this multiplies $bar by 2
$bar *= 2;        # so does this!
```

This is not just applicable to numbers; similar operations can be carried out
on strings. One such operation is using the concatenate operator, a dot symbol
(.)—this gives us the ability to append one string onto the end of another string.

```
$dna = 'ACTGATCG';      # define a DNA sequence
$dna = $dna . 'AAAA';   # add a poly-A tail to our sequence
$dna .= 'AAAA';         # another way of adding the poly-A tail –
                        # the string is now ACTGATCGAAAAAAAA
```

Special attention for strings

Assigning numbers to scalars is quite straightforward—you just use an equals
sign, followed by the number. Strings, on the other hand, need a little extra care,
in that they need to be surrounded by quotes to let Perl know the start and the
end of the string. Furthermore, different types of quotes are used depending on
how you want Perl to treat the string.

Single-quoted text is the simplest way of defining a string in Perl—the string
is generally read exactly as typed into the variable. However, we need to make
special consideration for the quote (') and backslash (\) characters. The backslash
character is used to cancel out special characters in strings, for example to get a
quote character (') in a string we would have to use \' within the string assignment, otherwise the string assignment will simply stop at the quote. Similarly,
to get a backslash character at the end of a string, this must also be preceded by
another backslash (otherwise it would cancel out the closing quote). Examples
of single quoted string variables, and how they appear when printed, are shown
Table 3.1.

Table 3.1 Perl strings and their appearance when printed with single quotes

Code	Output
`'ACTG'`	ACTG
`'Homo Sapiens'`	Homo Sapiens
`'The Human\nGenome Project'`	The Human\nGenome Project
`'The Human Genome Project\'s Website'`	The Human Genome Project's Website
`'\\\\servername\\path'`	\\servername\path
`'Human\tand\tMouse'`	Human \tand\tMouse

Double-quoted text is another method for assigning text to variables in Perl. This is slightly more advanced than the single-quote method in that we can utilize special text characters and even other variables within our strings. Examples of this can be found in Table 3.2. As you can see, the backslash character has more power within double-quoted strings. There are many of these special characters that are quite useful—some of the more relevant are listed in Table 3.3.

Multiline strings

It is quite common to need to have strings that are made up of large blocks of text within your programs. There are a number of ways you can do this with Perl, using standard quotes (and double-quotes) as above and with structures known as 'here documents'. These are demonstrated and explained in the following code:

```
my $big_string = 'This
is a
perfectly
valid string';

my $big_string2 = "As
is
this";

my $big_string3 = <<'TXT';
This is a 'here document' - another way of creating strings in Perl.
This is equivalent to the single quoted method of string creation
(i.e. no variable interpolation). The TXT marker above can be any
uppercase string - the important thing is to match it to the one
below as this denotes the end of the string.
TXT

my $big_string4 = <<"FOO";
This is another 'here document', but this is equivalent to double
quoting your string as variable interpolation is possible. e.g.
here's \$big_string: $big_string
FOO
```

Table 3.2 Perl strings and their appearance when printed with double quotes

Code	Output
`"ACTG"`	ACTG
`"ACTG\nACTG"`	ACTG
	ACTG
`"Human\tand\tMouse"`	Human and Mouse

Table 3.3 Some of the most important special characters in Perl

Character	Meaning
`\n`	Newline
`\r`	Return
`\t`	Tab
`\\`	Backslash
`\'`	Single quote
`\"`	Double quote
`\l`	Lowercase next letter
`\L`	Lowercase all following letters until `\E`
`\u`	Uppercase next letter
`\U`	Uppercase all following letters until `\E`

We encourage you to take a moment to work out in your head what each of these assignments is doing, and then test your understanding by printing each variable to the screen.

Some useful scalar operations

As with other programming languages, Perl has many built-in functions and operators for manipulating variables of different types. Before we move on to explaining some of these, let's pause for a moment to explain what functions and operators are and how they relate to each other. Functions (often also known as subroutines, methods, procedures, or subprograms) are small portions of code that you can call upon for performing a specific task. An example of a function that you have used already would be `print`—this is a function that is used for generating output from a program. Operators, on the other hand, are a specific subset of functions that are typically used in direct manipulation of variables—examples of which would be arithmetic operators (+, -, etc.) and string operators such as the assignment operator (=) and substitution operator (s) which we will come to later.

Two useful functions that are worth committing to memory are `chomp()` and `chop()`. These two functions perform similar operations, namely removing characters from the end of strings. However, there is one very important difference:

chomp() will remove the end character from a string if it is—and only if it is—a newline (\n) character; chop(), on the other hand, will remove the end character from a string no matter what it is. Although these might not seem immediately useful, you will find chomp() invaluable later when we start receiving input from the command line and reading in files, as these always have newline characters at the end of them. Unhandled, these could cause unexpected results from your programs.

More information about chop(), chomp(), and indeed any other aspect of Perl can be found by searching the official Perl documentation (perldoc.perl.org). This excellent resource details every standard Perl function and operator, complete with example code.

String substitution

Another essential string operator is the substitution operator, s. This can be used to swap a specific segment (that you define) of your string to something else. A simple example is given below.

```perl
$string = 'I like Perl';    # create a new string
$string =~ s/like/love/;    # substitute the word 'like' for
                            # 'love'
print $string;              # would give: I love Perl
```

In the above snippet of code, we created a new string and then substituted one of the words in the string. The generic syntax for using the substitute operator is shown below—as in the previous chapter, italics are used to indicate placeholders for parameters that you need to specify:

```perl
$string =~ s/string_to_replace/replacement_string/modifiers;
```

Simply, we have our string variable, followed by =~ (this indicates that we are doing a pattern match operation), then the substitution operator that contains the part of our string that we are trying to match, followed by what we are going to replace it with (surrounded and separated by forward-slashes). At the end of this we also have a position for optional modifiers to affect our substitution. Examples of two such modifiers are the letters i and g, which, respectively, imply that our match is to be case insensitive and global. A global substitution means that the substitution would happen on every occurrence of the match— not just the first match that is found.

Here is another example of the substitution operator in action. In the following example we try to determine the reverse complement of a DNA sequence:

```perl
$dna = 'ACTGACC';       # assign DNA sequence to a string
$dna =~ s/A/T/ig;       # swap all the As for Ts
$dna =~ s/T/A/ig;       # swap all the Ts for As
$dna =~ s/C/G/ig;       # swap all the Cs for Gs
$dna =~ s/G/C/ig;       # swap all the Gs for Cs
$dna = reverse($dna);   # reverse the sequence
print $dna;             # output result
```

If you run this code, you will get the result CCACACA, which is not actually the reverse complement of the sequence we started with. This demonstrates an important limitation with the substitution operator—if you want to perform more than one substitution at a time (in this example we want to do four), it might not work in the way you expect as the substitutions happen in the sequence they are written, resulting in a chain where each substitution operates on the result of the previous one, giving an undesired result.

The good news is that this type of operation is possible in Perl, we just have to use another operator. This is known as the *transliteration* operator, or `tr` for short. The `tr` operator has a syntax structure almost identical to `s`:

```
$string =~ tr/string_to_replace/replacement_string/;
```

The main difference between transliteration and substitution is that the transliteration operator acts on individual characters at the same time—whereas the substitution operator acts on the contents of the first set of slashes as a whole.

Therefore, in the case of our reverse complement problem, our code would look like this:

```
$dna = 'ACTGACC';                 # create our DNA string again
$dna =~ tr/ACTGactg/TGACtgac/;    # change all the bases at
                                  # once
$dna = reverse($dna);             # reverse the sequence
print $dna;                       # output result
```

This time we get the intended result: GGTCAGT. There are many more string manipulation operations in Perl, which we shall come back to later in the chapter.

Printing strings

One other common activity that we often want to perform with string-based variables (or indeed any of the variable types that we cover in this chapter) is that of printing them out to the screen for users to see. This is achieved with the `print` function that we met in the first example. Examples of this are shown below.

```
$dna = 'ACTGACC';

print "$dna \n";    # This prints ACTGACC followed by a new-
                    # line
print $dna . "\n";  # This prints the same as above
print '$dna \n';    # This prints $dna \n
```

This shows once more that the choice of quotes used to surround a string can have an impact on the resulting output. If we use double quotes in our `print` statements, variables can be used intermixed with other text and special (e.g. newline) characters and, upon printing, the value of the string variable will be printed, not the name of the string. If, however, we were to use single quotes, the text inside the quotes will be printed—no variables or special characters will be interpreted. The last thing that we would like to note here is the use of the concatenation operator (.) in the second print statement. If you prefer to clearly detach your variables from text in your code (this can aid in the readability and in cleanliness of your code), you can use the concatenation operator to join variables with text strings in a single print statement. Here is a typical example of this in use:

```
# Demonstrates concatenation operator in print statements
print "This is our DNA string: " . $dna . "\n";
```

This would then print the text 'This is our DNA string: ACTGACC' followed by a newline character to the console.

3.2.2 Arrays

An array may simply be considered as a collection of scalars—we can loosely think of them as a list of scalar variables that can contain any type of scalar variable described previously. An important difference between Perl and other languages is that different data types (e.g. string and numbers) can be mixed together in a single array. Figure 3.1 shows a pictorial representation of a Perl array.

Creating and assigning values to arrays

As with many things in Perl, there are several ways to create and populate arrays. Here are some examples:

```
@dna_seqs = ("ACTG", "CCGGC", "CGCGC");   # a 3-element array
@more_dna = qw(ACTG CCGGC CGCGC ATGAAA);  # a 4-element array

@other_array = ();              # create an empty array
$other_array[0] = 'ACTG';       # add 'ACTG' as the 1st element in
                                # array
$other_array[2] = 'ENSMUSG01465';  # 'ENSMUSG01465' comes next
$other_array[3] = '56.7';          # then '56.7'
```

As you can see from the above examples, assigning values to arrays uses the same assignment operator (the equals sign) as used with scalar data. The most obvious difference is the use of the @ symbol to represent an array instead of the $ scalar indicator.

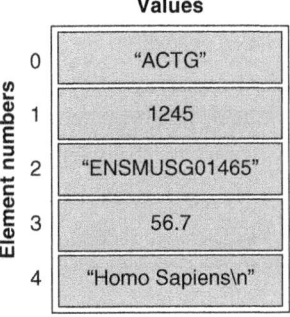

Values

Element numbers	
0	"ACTG"
1	1245
2	"ENSMUSG01465"
3	56.7
4	"Homo Sapiens\n"

Fig. 3.1 A Perl array containing five elements.

Let's look at these examples more closely, step by step:

```
@dna_seqs = ("ACTG", "CCGGC", "CGCGC");
```

This line of code creates a three-element array called `@dna_seqs`. The assignment of values to the array is done with the use of the brackets surrounding the new contents of the array, and each individual element is surrounded by quotes and separated by a comma. The next example uses a different approach to populate another array, `@more_dna`.

```
@more_dna = qw(ACTG CCGGC CGCGC ATGAAA);
```

The assignment for this is slightly different as there are no commas or quotes used within the braces—the elements are separated by white space—this is the syntax defined by the `qw()` function preceding the braces; specifically, the `qw()` function takes a list of space-separated values and returns the same list, comma separated, with each element surrounded in quotes.

```
@other_array = ();                         # create an empty array
$other_array[0] = 'ACTG';                  # add 'ACTG' as the 1st
                                           # element in array
$other_array[2] = 'ENSMUSG01465';  # 'ENSMUSG01465' comes next
$other_array[3] = '56.7';                  # then '56.7'
```

The above code demonstrates yet another way to create and populate an array. It seems long-winded, but introduces several concepts that are useful when writing programs that use arrays. In the first line we name and create an empty array, then on the following lines we add scalars to individual elements of the array using list interpolation—using the element numbers to enter scalars into specific elements of our array one element at a time. Note one of the important character changes in the above lines—the `$` sign is used instead of the `@` sign as we are accessing individual elements of the array. The individual elements are scalars—hence the use of the `$` prefix. You will also notice that we neglected to put anything in the second position of the array, element number 1. This is not a problem because arrays can have empty elements. Accessing this array element will return a null value, which will behave differently depending on context. Finally, note that the numbering of elements in an array starts at zero.

Printing and retrieving data from arrays

Having created and populated an array, it's quite natural that we will want to get some information back out of it. If you wish to copy a specific element of an array and put it into its own scalar variable, this is as simple as the following line of code:

```
@dna = ("ACTG", "CCGGC", "CGCGC");
$sequence = $dna[2];
```

This assigns the contents of the third element of the array to the scalar variable called `$sequence`. However, we need to note here that we have not done anything to change the third element of the array—it's still there, we have merely copied it. We shall discuss *removing* elements from arrays shortly.

As well as getting data out of arrays and into scalar variables, you might just wish to print either an element of an array or the array as a whole. This is really quite simple, but there is one small warning that needs to be given. Consider the snippet of code below.

```
@dna = qw(ACTG GGCG AAAA TTTG);
print "Here's a single element:\n";
print $dna[2] . "\n\n";
print "Here's the whole array:\n";
print "@dna\n";
```

This will give the following output:

```
Here's a single element:
AAAA


Here's the whole array:
ACTG GGCG AAAA TTTG
```

The above code demonstrates two different ways in which we can print out data from within arrays. The first section prints a single element on its own, concatenated with two line breaks—you will also notice that we left the array element to print outside the quotation marks, as mentioned when we covered printing string variables earlier; this is not entirely necessary and we could have used the following to get the same effect:

```
print "$dna[2]\n\n";
```

The second print statement in the example shows how to print out a whole array. As you can see, each element of the array is separated by a space, which makes the output very readable.

Special array operators

Arrays are slightly more advanced data types than scalars and there are a number of special operations that can be performed on them. Functions to perform these operations include `pop()`, `push()`, `shift()`, and `unshift()`. These allow us to add and remove data from the beginning and end of an array.

```
@array = qw(0 1 2 3 4 5 6 7 8 9 10);
$scalar = pop(@array);       # remove the last element from @array (10),
                             # and put it in $scalar
$scalar = shift(@array);     # remove the first element from @array (0),
                             # and put it in $scalar
push(@array,'ACG');     # 'ACG' is added to the end of @array
unshift(@array,'CGC'); # 'CGC' is inserted at the beginning of @array

# @array now contains ('CGC' 1 2 3 4 5 6 7 8 9 'ACG')
```

Using combinations of `pop()`, `push()`, `shift()`, and `unshift()` it is possible to quickly and easily add or remove data from either end of arrays.

There are two more array functions that are especially useful for the conversion of data: `split()` and `join()`. The first of these, `split()`, allows us to create an array from delimited data. Common examples of delimited data are comma-separated (known as `.csv` files) and tab-separated data.

An example of using the `split()` function follows:

```
$genestring = "ENSG00000058668,ENSG00000047457,
              ENSG00000067715";
@array = split(",",$genestring);
```

Although this is not something you would ever want to do in practice (why would we create a string on one line, only to split it on the next?), it does show how it works. The `@array` variable returned by this code contains three elements taken from the above string. A more common use of this would be to read data from a file. Getting data in and out of files is covered later in the chapter.

The generic syntax for the `split()` function is:

```
array_variable = split(delimiter,scalar_variable_to_split)
```

What if we have an array that we wish to print out to the screen, or a file as comma-separated or tab/space-separated data? This is essentially the opposite of the `split()` function—implemented in Perl as the `join()` function. Here is the syntax for the `join()` function followed by a quick example:

```
scalar_variable = join("delimiter",array_to_convert)
```

```
@genearray = qw(ENSG00000058668 ENSG00000047457
              ENSG0000006771);
my $output = join(",",@genearray);
print $output . "\n";
```

As you can see from the above code, the syntax for `join()` is essentially the same as `split()`.

One final useful array operator we would like to cover is `scalar()`, which is used to find out the size of an array, and can be used as follows:

```
@codon = qw(ACT CCG GGC AAA);
print "Our codon array has " . scalar(@dna) . " elements\n";
```

If you put this code into a file and run it, you will get the output 'Our codon array has 4 elements' which is correct and as we expected. It's worth noting that the function `scalar()` counts the elements of our array starting from 1, which is different from the actual number assignment of the elements (array numbering starts at zero, remember). This is important as we often use the number of elements in an array to control loops and other control structures, as discussed later in the chapter.

3.2.3 Hashes

The final Perl data structure that we want to highlight is the hash. A Perl hash is similar to an array in that it is a list of scalar variables, the difference is that a hash is not a number indexed list—the values within a hash are indexed using *names*. These names are more correctly referred to as *keys*. They are unique string variables that are completely arbitrary in form, that is they can be anything you like.

Figure 3.2 shows a hash containing five scalar variables that can be accessed using the given example keys. To create this hash in Perl:

```
# Please note - the extra spacing is optional...
%genbank_record = (
        'Official Symbol'      => 'BRCA1',
        'Official Full Name'   => 'breast cancer 1, early onset',
        'Primary Source'       => 'HGNC:1100',
        'Gene Type'            => 'protein coding'
        'Organism'             => 'Homo sapiens'
);
```

The declaration of a hash and its structure is clearly different to those of scalar or array variables. The first, and possibly most important, difference is the change in symbol to represent this type of variable; a percentage sign (%) is used to represent hashes. The other main difference is the way in which we put data into our hash. As we are no longer dealing with a single scalar, or an ordered list of scalars, we need a more structured approach: assigning the key/value pairs is done using the => construct, with the key on the left and the value on the right.

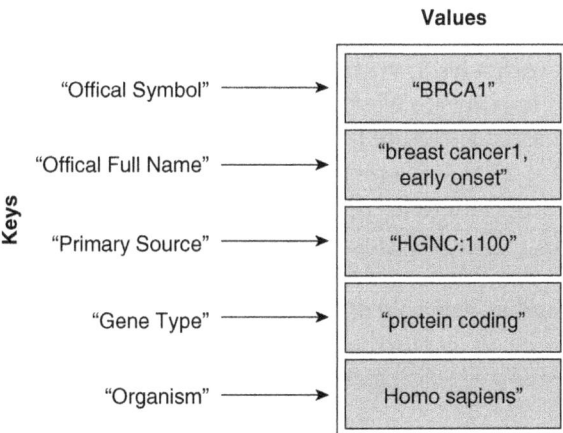

Fig. 3.2 A Perl hash containing five key/value pairs.

The above example is not the only way to create a hash, (as with most things in Perl—there is more than one way to do it) but it is the simplest to understand. We could have done the following:

```
%genbank_record = ( 'Official Symbol', 'BRCA1', 'Official
Full Name', 'breast cancer 1, early onset', 'Primary Source',
'HGNC:1100', 'Gene Type', 'protein coding', 'Organism', 'Homo
sapiens' );
```

In the above we have the key, followed by the value on the same line, then followed by the next key/value pair. The above code will give you exactly the same hash in the end—but it's by no means as easy to read, especially for larger hashes.

Like arrays, hashes do not need to be completely filled from the beginning—we can declare an empty hash, and then add/remove values as and when we wish. An example of this is shown below.

```
%rna_triplets;                    # declare empty hash

$rna_triplets{'UAG'} = 'stop'; # This adds the key/value of
                               # UAG/stop
$rna_triplets{'GCC'} = 'Ala';

delete $rna_triplets{'UAG'};    # deletes the UAG/stop entry
```

As you can see, adding and removing elements to and from a hash is quite similar to an array, we just use the scalar key to assign the value instead of the number index used in arrays, and curly braces ({}) are used instead of square brackets. Also like arrays, the individual elements of the hash are scalars (denoted by the $ symbol). This highlights an important consideration—hashes are unordered lists. The elements of an array have an order to them ranking each element from zero to the size of the array. In hashes there is no order to the keys as they are completely arbitrary scalar values. This will be important later when we cover looping through the values in data structures. In arrays we may loop through the array in a set order to access each element, whereas to achieve a similar result with a hash we might loop through an array of hash keys.

One last thing to be aware of when considering the similarities between hashes and arrays, is that arrays can only have one value for each element within them. So, if you declare a value for position [2] of an array, then re-declare that again later as something different, the first value will be overwritten. Hashes are exactly the same—your keys must be unique within the hash (as the numbers within the array are unique). If we re-declare a value within a hash, we overwrite the original value. An example of this is:

```
@ala_triplets = qw(GCU GCC GCA);
$ala_triplets[1] = 'GCG'; # This replaces 'GCC' with 'GCG'

%genetic_code = (
      'GCU' => 'Ala',
      'AAA' => 'Ala',
      'GCA' => 'Ala'
);
```

```
$genetic_code{'AAA'} = 'Lys'; # We replace 'Ala' with 'Lys'
                              # as this is the
                              # correct translation.
```

As you can see, replacing elements of a hash or array is easily accomplished, but this can also lead to accidental replacements that can be hard to spot.

Getting data out of hashes

As we have shown, getting data into a hash is analogous to getting data into an array. Getting data back out of a hash is equally straightforward.

```
%genetic_code = (
      'GCU' => 'Ala',
      'AAA' => 'Lys',
      'GCA' => 'Ala'
);
print $genetic_code{'AAA'}; # This prints the value 'Lys'
$alanine = $genetic_code{'GCU'}; # Makes $alanine equal 'Ala'
```

The above code shows just how easy it is to get data out of a hash if you know what the keys are. However, what happens if you do not know what the keys for your hash are? This can happen when you load data into a hash dynamically within your program. Thankfully, Perl's `keys` function is there to get this information for you.

```
@keys_from_my_hash = (keys %genetic_code);
```

The keys function takes all of the keys from the specified hash, and returns an array of the values that it has found. This then allows you to gain access to all elements of a hash in an ordered manner.

Finally, `keys` also has a complementary function, `values`, which returns all of the values in a hash. This may be useful in certain circumstances, such as when you would just like to know the values stored within a hash and are not particularly worried about the keys associated with them.

```
@values_from_my_hash = (values %genetic_code);
```

The scalar, array, and hash are all you will ever be likely to need in a Perl program to store and handle data. We will come back to data structures later in the chapter when we discuss the use of references, but for now we will move on to controlling the flow of our programs.

3.2.4 Control structures and logic operators

So far our Perl programs have been limited to simple *scripts*—lists of commands that execute in the sequence that they appear in the file. Scripts like this are good for automating otherwise tedious sequences of command-line input, but to produce real programs with more complex behaviour and the ability to respond to user interaction, we need to be able to implement loops and conditional statements to control the flow of our programs.

IF, ELSIF, and ELSE conditionals

The first control structures that we are going to consider are the `if`, `elsif`, and `else` control structures. These are used where you want a program to do something specific based on a test; if that test is not met, you might want the program to do something different. This is why we refer to them as conditionals; they allow your program to do different things based on a set of conditions that you define. That may sound complicated if you haven't done programming before, but be assured that they are not complicated and will become second nature to you after your first couple of programs as they will most likely be in all of your Perl programs. The reason for this is that conditionals allow our programs to respond differently according to the data they receive, or interactions with the user. Here is an example of how we might use an `if` conditional:

```
$species = 'Human';

if ($species eq 'Human') {
     print "Your species is Human";
}
```

As you will see in the first line of the above code, we create a new string variable that has the value 'Human'. The next line begins with the function `if` followed by a test—in this case `$species eq 'Human'`—this is a string comparison testing whether our variable is the same as 'Human', this is then followed by an opening brace. If the condition is met (i.e. if `$species` is exactly the same as 'Human'), the code in between the braces following `if` will be executed. If, however, the test does not return true (i.e. if `$species` is not equal to 'Human'), nothing will happen, the block of code defined by the braces (controlled by the `if`) will be ignored. Let's expand on this example:

```
$species = 'Mouse';

if ($species eq 'Human') {
     print "Your species is Human";
} elsif ($species eq 'Mouse') {
     print "Your species is Mouse";
}
```

In this example we introduce an extension to the `if` conditional—an `elsif` conditional. This is used in combination with `if` conditionals to give you more options within your programs. The conditionals are evaluated sequentially as we move down the program (as with everything else in Perl), therefore the `if` conditional is evaluated first. If this condition is met, the code block associated with it will be executed. However, if it is not met, the conditional will continue onto the `elsif` block and this will be evaluated as if it was another `if` conditional. What about when none of our conditionals are evaluated as true? In the above example, if `$species` was neither 'Human' or 'Mouse', the program would simply skip over the conditional blocks and move on; sometimes we need a backup plan. Let's expand on this idea once again:

```
$species = 'Zebrafish';
if ($species eq 'Human') {
     print "Your species is Human";
} elsif ($species eq 'Mouse') {
     print "Your species is Mouse";
} else {
     print "Your species is not Human or Mouse!";
}
```

In the above code we introduced the `else` conditional. This conditional is used with other conditionals to offer alternatives for execution should the previous conditionals fail (the `if` and `elsif`). In the above section of code, we once again create our variable `$species` and have the same `if` and `elsif` conditionals and braced code blocks. However, this time the value of `$species` is not equal to 'Human' or 'Mouse', (it's set as 'Zebrafish'), therefore the tests following the `if` and `elsif` statements will not return true, and the code blocks within the braces will not be executed. As the `if` and `elsif` statements are followed by an `else` conditional it's this code block that is executed in this case, so the output of this piece of code would be the line 'Your variable is not Human or Mouse!' printed to the command line. Table 3.4 shows other comparison operators that can be used in conditional tests.

Table 3.4 Common comparison operators for numbers and strings. When using comparison operators with strings the terms "less than" and "greater than" to refer to alphabetical order in the sense that 'a' is less than 'b' in value as it comes earlier in the alphabet. Note that upper case letters are considered to be 'less than' their lower-case counterparts, i.e. 'A' is considered less in value than 'a'

| Test Operator | | Description |
Numbers	Strings	
==	eq	Equal
!=	ne	NOT equal
<	lt	Less than
>	gt	Greater than
<=	le	Less than or equal
>=	ge	Greater than or equal

More advanced conditionals

The if, elsif, and else conditionals are fairly simple, but very useful. However, what if we want to test for more than one condition simultaneously? You could simply string together lots of if and elsif conditionals, but sometimes this can seem like a lot of work for simply testing a few variables, and the code can become difficult to read. On occasions such as this there is a potentially more efficient option, which is to combine a series of tests using logical operators such as and (which can also be shown as &&) or or (which can also be shown as | |). Here is an example of such use:

```
my $test_var = 18;

if (($test_var > 12) && ($test_var < 24)) {
      print "Your variable is between 12 and 24!";
}
```

The above test would look to see if our variable ($test_var) had a value between 12 and 24, as we test that the variable is above 12, and below 24. The basic rule of thumb when stringing together multiple tests is that all of the tests must be surrounded by a pair of brackets, and each individual test must be surrounded by their own set of brackets. There is no limit on the number of tests that you can string together, you can use as many as you need.

Creating loops with FOR, FOREACH, and WHILE

The next set of control structures that we are going to look at is loops. These are used in programs when you want to repeat an action multiple times, and are features of most programming languages. There are two different flavours of loop that can be used for different tasks; these are loops that are carried out for a defined number of iterations, and loops that repeat until a certain condition is met.

The first class of loops (those that repeat for a defined number of iterations) includes for and foreach loops, which are particularly useful if you have a list of variables, (either in an array or hash) and you would like to perform an operation on each element of the list. Let's start by looking at an example using a for loop:

```
@array = qw(one two three four five); # create a five element
                                       # array

for ($i=0; $i<scalar(@array); $i++) {
      print "We are looking at: $array[$i] \n";
}
```

In the above snippet of code we establish our for loop with the for statement, and iterate through each element of our array and print the contents of each element of the array. To explain this more clearly, consider the generic syntax of a for loop in slightly more detail:

```
for (iterator_variable; final_test; change_to_iterator) {
      operations_to_perform_upon_each_iteration;
}
```

So, the definition of a `for` loop is split into three parts. We first define an iterator variable (in the example above this was `$i`), in this case this variable contains a count of how many times we have been through the loop. The second parameter is the test to determine when we stop going through the loop. In the above example we first calculate the size of our array (using the `scalar` function), and state that we should keep going through the loop while `$i` is less than the size of the array. When `$i` is equal to the size of the array, or greater, the `for` loop will no longer be run and the program will continue onto the next section of code. It might seem more logical to use <= instead of < in our example to be sure not to miss the final array element, but the `scalar()` function gives the length of an array as if we had started counting from one, while we know arrays begin from zero (as does our iterator). So the use of < in the example stops us trying to use a non-existent array element.

The final parameter that we set in the `for` loop is the increment for our iterator variable with each iteration of the loop. In this example we increment our variable by one as the command `$i++` is equivalent to `$i = $i + 1`, and we want to access each element of our array. However, we are not restricted to incrementing iterators by one—sometimes it might be desirable to only look at every other element of the array, so you could simply put `$i + 2` at the end of the `for` loop definition.

The next type of loop that we are going to look at is the `foreach` loop. This is a more specialized version of the `for` loop, designed specifically for the common task of working with arrays. Bascially, the `foreach` loop is used to iterate over each element of an array—hence the name `foreach`—you can think of it as meaning: '*for each* element of this array, do this'. Here is an example of how we could use a `foreach` loop to do the same as the previous example:

```
@array = qw(one two three four five);

foreach $element (@array) {
        print "We are looking at: $element \n";
}
```

The above does exactly the same as our previous `for` loop, but the code is more concise and easier to read. Let's just look at the syntax of the `foreach` loop and explain exactly what is going on:

```
foreach element (array) {
        operations_to_perform_upon_each_iteration;
}
```

The major difference here between `for` and `foreach` loops is that `foreach` loops do not have an iterator variable, instead they have a variable that gets assigned the element value of the array that the loop is currently working on. In the above example, we call this variable `$element` and treat it as a normal scalar variable (as that it exactly what it is). Using the `foreach` loop makes moving through each element of any given array a very simple process.

However, there is one last thing to note about `foreach` loops—the element variable that we defined before as `$element` is an entirely optional argument in `foreach` loops. If you choose not to set a name for the element variable, Perl will use its default variable: `$_`. Here is an example of the use of `$_`:

```
@array = qw(one two three four five);

foreach (@array) {
     print "We are looking at: $_ \n";
}
```

As you can see, this example is identical to the previous one, we just use `$_` to access the element of the array that we are looking at instead of declaring another variable.

You may now be wondering why you would declare a named variable when you don't have to? There are instances when it is of benefit to name our variables, such as nested loops (one loop within another), as it makes it clear to us which array element we are looking at. Here is an example of how we would use nested loops (loops within loops) to print out a list of letter/number coordinates (possibly useful for indexing items in a grid):

```
@columns = qw(A B C D E F G);
@rows = qw(1 2 3 4 5);

foreach $row (@rows) {
     foreach $column (@columns) {
          print $column . $row . ' ';
     }
     print "\n";
}


# This produces the following output:
#
# A1 B1 C1 D1 E1 F1 G1
# A2 B2 C2 D2 E2 F2 G2
# A3 B3 C3 D3 E3 F3 G3 etc...
```

As you can see above, if we are using more than one loop structure, it is beneficial to name the variables for each loop as it makes it clear which variable is which. In fact, the arrangement we have demonstrated here is impossible to do without naming the loop variables, as the `$_` from the outer loop is not accessible using that name within the inner loop.

The next type of loop that we'll consider is the `while` loop. The main difference between this type of loop and the previous kind is that the previous loops (`for` and `foreach`) iterate a given number of times—defined either by your constraints in a `for` loop, or the size of your array in a `foreach` loop—whereas `while` loops do not have a set number of iterations. They are based on a logic test that you define (using the same syntax as the `if`/`else` conditionals earlier), and they keep going until the condition that we have defined is met. The generic syntax of a `while` loop is:

```
while (test_case) {
        operations_to_perform_upon_each_iteration;
}
```

Let's now look at a simple example of a `while` loop in use:

```
$test_var = 12;

while ($test_var < 45) {
        print "Variable is $test_var \n";
        $test_var = $test_var + 10;
}


# This will produce the following output:
#
# Variable is 12
# Variable is 22
# Variable is 32
# Variable is 42
```

As you can see from the example above, the `while` loop will keep repeating until the condition that we define at the beginning of it is met—this can be useful if you have no way of knowing how many times you need a loop to repeat, but you can set a goal, or condition, that you need to meet. The only thing that you need to watch out for is initiating an infinite loop. This occurs when your loop conditional is never met (i.e. if `$test_var` in the example above never becomes greater than 45), the loop will never exit and your program will never complete. This is why programs sometimes appear to 'hang' or 'freeze'—they are trapped in an infinite loop. Provided you look out for instances such as this when you use a `while` loop, they can be an indispensable tool.

3.2.5 Writing interactive programs—I/O basics

In this section, we introduce ways in which you can make your Perl programs interact with users—this is a particular subset of I/O (input/output) functionality. I/O also includes dealing with files, which is covered later in Section 3.6.

So, say you have created a program to reverse complement a DNA string, as we did earlier in the chapter. Here is a program to do this:

```
#! /usr/bin/env perl
# Program to convert a DNA string to its reverse complement

# Our DNA string
my $dna = 'ACTTTTGGGGCCCCAATGCATTTTAAAAA';
```

```
# First we reverse the DNA
my $revcom = reverse $dna;

# Now translate the DNA bases
$revcom =~ tr/ACGTacgt/TGCAtgca/;

# Then print out the results
print "Original DNA string: " . $dna . "\n";
print "Reverse complement: " . $revcom . "\n";
```

In this example, the DNA string to reverse complement is defined at the top of the program. However, what if we want to work out the reverse complement of another DNA sequence? Simple, we just change the DNA sequence within the program. What happens, though, if we would like to do this quite a number of times, or even as part of another program and call this program automatically from the command line? Having to go into the program and edit the sequence is fiddly, and we run the risk of accidentally modifying another part of the program in the process and introducing an error. This is where we can increase the functionality of our program by taking input from the command line.

There are two ways in which we can take command line input from users; real-time interaction and at run-time. By real-time interaction we mean that our program will ask the user for input as it runs, so when we need to get our DNA string, we get our program to pause and ask the user for a DNA string. This approach is useful for scripts that you would like other people to use, as you can prompt for input as it is needed. When we say run-time input, we mean that we declare all of our input for the program up front, as we run the program. This can be useful for users of your scripts as it means that they no longer have to edit program code to change variables, but the real benefit of run-time input is that no interaction from the user is needed from there—so if the program takes a long time to run, the user can just start the program off and walk away. If the input method is interactive, the user will need to sit there and watch the program run as they wait for the next input stage. Here is how we would use the interactive input method:

```
#! /usr/bin/env perl

print "What is your name?\n";  # Ask the user their name
$name = <>;                    # Get the user to type an answer
chomp($name);                  # Remove the newline
print "Hello $name!\n";        # Say hello
```

In the above example we show you how easy it is to ask the user for input—you simply declare a new scalar variable and mark the contents as <>, this makes Perl wait for the user input. You can do this for as many variables as you like in a given program—there are no limits to its use. You will also notice the use of the chomp() function that was discussed earlier in the chapter—chomp() is used to remove a newline character from the end of a string. The reason for this is that the user must hit the Enter key to signify that they have finished typing, a side effect of this is that a newline character is appended to the end of the input. Because this could potentially cause problems later on, if we were using the input for something more significant than printing back to screen, we use chomp() as a matter of course.

Accepting command-line parameters

Now let's have a look at using run-time inputs that are passed to Perl as command-line parameters. Command-line parameters are pieces of data that you type as part of the Perl command to run your program. Consider the program below.

```perl
#! /usr/bin/env perl

# Make sure that we have some command-line arguments
unless (@ARGV) {
        die "No input given!";
}

# Do something with our input...
print "Your first input was $ARGV[0]\n";
print "Your second input was $ARGV[1]\n";
```

If you save this program as input_example2.pl and run it with the following command:

```
perl input_example2.pl ACTGGG DNA
```

You will then get the following output to your screen:

```
Your first input was ACTGGG
Your second input was DNA
```

This shows how you can accept command-line parameters in your Perl scripts. The first thing that we must understand is the way in which Perl handles command-line input. Simply, your scripts can accept parameters separated by spaces, which are then passed into your Perl program as elements of the special array @ARGV. In the first section, we use an unless conditional to determine if we have any command-line parameters being passed to the program—if no parameters have been passed, then @ARGV would not exist and our program would then exit. This premature exiting of the program is achieved using the die function, which is a handy way of making the program stop wherever it is, write the specified message to the screen, and return control to the command line. If one or more parameters have been passed, we then simply print out the first two elements of @ARGV to screen.

This is a very basic example of what we can do with command-line input, serving only to demonstrate the concept and syntax. In reality there would be little point in writing the parameters back to the screen—more typically the input would be tested in a conditional statement to determine the behaviour of the program, or used to pass a file name or analysis parameter to the program.

As an example, let's return to the reverse complement program shown at the start of this section. We said that it would be beneficial for our users to not have to edit the Perl program each time they wish to run another DNA sequence through it. Well, now that we know how to get Perl to accept input from the command line (i.e. the DNA sequence to process), we can modify the program accordingly:

```
#! /usr/bin/env perl
# Program to convert a DNA string to its reverse complement

# Get the input string
unless (@ARGV) {
        die "You need to pass me a DNA sequence!";
}

# Our DNA string
my $dna = $ARGV[0];

# First we reverse the DNA
my $revcom = reverse $dna;

# Now translate the DNA bases
$revcom =~ tr/ACGTacgt/TGCAtgca/;

# Then print out the results
print "Original DNA string: " . $dna . "\n";
print "Reverse complement:  " . $revcom . "\n";
```

Basically, the only difference is that we first make sure that there are command-line arguments being passed to the program, then we make whatever the first variable the user gives us into our variable $dna, therefore completing our program. If you want to try this program out without typing it in, you can download it from www.bixsolutions.net (it's called revcomp.pl).

This is all we want to cover about getting information from your users for now, but later (in Chapter 6) we will detail how you can go about improving the process of getting command line input from your users and also displaying instructions of how your program works.

If a program needs large amounts of input, such as whole protein sequences, or a long list of gene IDs, expecting the user to enter the data by hand is not really feasible, and we would look to load the data from a file or database instead. This is discussed later (in Sections 3.6 and 3.9 respectively). Where user input remains essential, we may spare our users the discomfort of the command line by building a web-based front end through which users can interact with our programs—this is explained in some detail in Chapter 5. First, we shall look at some good bits of advice to consider as our programs begin to get slightly bigger and more complex.

3.2.6 Some good coding practice

We have shown thus far the basics of Perl programming, and we hope you will agree that it is really not too difficult to get along with. However, as Perl is a straightforward language that allows you many different ways to do things, mistakes can easily be made. We would therefore like to take a little time to introduce a few recommendations, based on our experience, which will help you avoid common pitfalls as you develop increasingly complex programs.

Use strict

There are several built-in modes (more correctly called *pragmas*) within Perl that assist you in making sure that your code is as *clean* as it can be. By clean, we mean that it is likely to be free of simple errors. Switching on these pragmas can result in Perl throwing up more error messages which, although somewhat off-putting, are of great value as they alert you to potential problems with your code. The first pragma we recommend is called `strict`. In order to use `strict`, we have to add the following to the top of our Perl programs (just under the shebang line):

```
use strict;
```

So what does `strict` do? Basically, `strict` forces you to code your Perl programs better as it checks for unsafe constructs and enforces variable namespacing (also known as the *scope* of a variable). So what is unsafe construct and namespacing?

Up until now, whenever we declared a variable we have been declaring *global* variables. This means that when we have subroutines (reusable pieces of code, sometimes called functions—we cover these in Section 3.4) in our scripts; if they contain a string, array, or hash that matches the name of one of your variables, they will overwrite your original variables without warning. This is why we use `strict`; with the `strict` pragma in use in our program the global variable would not be replaced by another variable of the same name, instead it would be temporarily put to one side whilst a *private* variable would be used within the function. Once the function has completed, your global variable would once again be available, unchanged from its previous state. We shall go into more detail about this when we look at subroutines later.

In order to be able to get away from using global variables (and for our programs to work whilst the `strict` pragma is in use), we need to use the `my` operator when we define our variables. The use of `my` declares a variable as a *local* variable (private) and is enough to satisfy the needs of the `strict` pragma. Some examples of this are shown below.

```
my $dna_string = 'ACTG';
my @asn_codons = qw(AAU AAC);
my %genetic_code = (
      'AAU' => 'Asn',
      'GCU' => 'Ala',
      'GAU' => 'Asp'
);
```

We shall be using this form of variable declaration as well as the `strict` pragma throughout the rest of this chapter. We would recommend that you never write another Perl program without using `strict`.

Warnings and diagnostics

Two other pragmas that are commonly used to assist you in writing your code are `warnings` and `diagnostics`. They can be included in your program in the same way as `strict`, by adding the following lines to the top of your code:

```
use warnings;
use diagnostics;
```

These two pragmas aid in the writing of problem free code and debugging code that is not behaving as you intended. The `warnings` pragma specifically gives you a text warning while your code is running at the command line whenever it finds something that is potentially wrong with your code. On the other hand, `diagnostics` only does something when your program goes wrong—it adds additional explanation to Perl's standard error messages and even suggests possible fixes to your code when an error occurs. Most of the time, people only tend to use the `strict` and `warnings` pragmas, but if you are having problems debugging a particularly difficult bit of code, `diagnostics` can be invaluable in pointing you in the right direction.

Variable naming and commenting

Our last tip is really just a piece of advice for the way in which to program. Basically, as with naming elements of a database, when naming your variables and programs, use common sense and give them names that relate to their function. This will aid you in the long run by allowing you to walk away from a program for quite some time, and the next time you come to it you will not have to figure out what `test_script1.pl` does or what information is contained in `$string1`. Here is a quick list of recommendations.

- When you name your program, name it according to its function. For example, if it's a program for parsing BLAST results, call it `parse_blast_results.pl` or something similar.

- When you name variables, name them sensibly—`$a` or `$string1` means nothing when you have forgotten exactly how your program worked. For example, if you have a string that is to contain a DNA sequence, call it `$dna`, or something more specific if possible.

- This will make more sense after reading the next section, but is more appropriately mentioned here. When naming and using references, the following tips can make your code a lot more readable.

  - Always name your reference `$something_ref`—the important part here is the `_ref` at the end, this immediately lets you know that this is a referenced variable and not just another scalar.

  - If you have a reference to an array or hash, precede the variable name with either an `a_` for an array reference or a `h_` for a hash. If you have more complex data structures, such as a reference to an array of hashes, use a combination of the two— `ah_` would indicate an array of hashes, whereas `hh_` would indicate a hash of hashes. Using this naming scheme can help you keep track of what is going on in the complicated soup of variables that you may find in a large and complex Perl program.

Following these guidelines—using the `strict`, `warnings`, and `diagnostics` pragmas and applying some common sense when it comes to naming things— can greatly help you in writing your code, and make it understandable to other people, or even yourself when you have been away from your program for a while. Finally, it is always good practice to insert copious comments in your programs as you write them—it may slow you down slightly but could save hours later when you need to return to the code to unravel how it works.

3.2.7 Summary

That brings us to the end of our coverage of the basics of syntax and logic for Perl. Since this has been a pretty big section, let's quickly recap the main points:

- The basics of putting together a Perl program. Remember to use the appropriate shebang so the operating system can find the Perl interpreter, and save your scripts with a `.pl` file extension.
- There are three different types of data structures/variables available in Perl:
  - The scalar (a number or string).
  - The array (a number indexed list of scalars).
  - The hash (an unordered list of scalars arranged in key/value pairs).
- Control structures, like loops including `for`, `foreach`, and `while`, and conditional tests using `if`, `elsif`, and `else`, combined with logic operators are the building blocks for programs that can automate tasks, respond to user input, and perform many other useful functions.
- The user can provide input to Perl programs interactively, or via command-line parameters.
- Finally we provided some basic advice to keep in mind when writing your Perl programs to try and keep things as understandable as possible.

If you understand these concepts, you can start writing original programs in Perl right away. Indeed, most of these concepts are common across all major programming languages, so if you decide to move on to another scripting language, like Ruby, Python, or even Java or C++, you will find you have a head start. In the remainder of this chapter, we focus on some more advanced topics, most of which are specific to Perl, and all of which are useful in bioinformatics.

3.3 References

A reference is a pointer to a variable, it is not actually a variable itself— it just points to a variable that already exists. If you're from a Linux background, think of references as Perl's equivalent of symbolic links. If you're from the Windows world, think of them as a shortcut—both of these appear to the file system as a normal file, but instead they just point to another file—Perl references act in the same way. They therefore appear to your code as scalar variables, although they are merely pointing to another variable.

As we have seen, scalars, arrays, and hashes are great for storing the data within simple Perl programs, so why do we need references? Essentially, references are used to extend the existing data structures—to make more complex arrangements possible that are not permitted with the standard variables. Such situations occur frequently in bioinformatics, hence the importance of learning this skill. The best way to understand this is with a few good examples.

3.3.1 Multidimensional arrays

For our first example, what if we wanted to store information that would be best arranged in a two-dimensional array (e.g. a set of experimental results, like you would normally record in a tabular form)? Maybe we could try this:

```
my @array1 = qw(24 48 56 12);
my @array2 = qw(25 48 55 12);
my @array3 = qw(23 49 54 11);
my @array4 = qw(24 48 55 12);

my @twodarray = (@array1, @array2, @array3, @array4);
```

If you put the above code into a program and try running it, it appears to work; no errors are given and our array is populated. However, all is not quite as well as it seems. If you check this array using the debugging technique in Section 3.3.3 (Data::Dumper), you would find that we have not actually produced a two-dimensional array, just one long list (single-dimension) array made up of the elements of each of the arrays we placed into it. This is not what we intended and will not be useful to us. To get round this problem we need to use references:

```
my @array1 = qw(24 48 56 12);
my @array2 = qw(25 48 55 12);
my @array3 = qw(23 49 54 11);
my @array4 = qw(24 48 55 12);

# This is how we create a reference to a variable - by using
# the '\' character. You must remember that
# references appear as scalar variables though.
my $a_array1_ref = \@array1;
my $a_array2_ref = \@array2;
my $a_array3_ref = \@array3;
my $a_array4_ref = \@array4;

my @twodarray = ($a_array1_ref,$a_array2_ref,$a_array3_
                 ref,$a_array4_ref);
```

The above code gives the desired result, thanks to the use of references to the arrays that contain our data. To the `@twodarray` variable these appear as scalars and are therefore permitted to create the array. This is quite a bit of extra code to write at the moment (as we have to first declare our array, then our reference), and might put some of the more lazy programmers off the idea of references. Thankfully, as with all things in Perl, there is more than one way to do it, so below are a couple of other ways we could have done the same. The way in which you use references is up to you—basically choose the method that makes the most sense to you.

```
my @array1 = qw(24 48 56 12);
my @array2 = qw(25 48 55 12);
my @array3 = qw(23 49 54 11);
my @array4 = qw(24 48 55 12);

my @twodarray = (\@array1, \@array2, \@array3, \@array4);
```

In the above example, we reference the arrays containing our data within the declaration of the `@twodarray`, this removes the extra effort of declaring our references explicitly and should make the code slightly easier to read. You can, however, take this even further.

```
my @twodarray = (
        ["24", "48", "56", "12"],
        ["25", "48", "55", "12"],
        ["23", "49", "54", "11"],
        ["24", "48", "55", "12"]
);
```

In the above we have taken our example to the most extreme and compact form possible, (without making the `@twodarray` a reference itself). In this instance we do not even create our arrays beforehand and reference them—they are created in situ as anonymous arrays. Because these arrays don't have names, it is impossible to access them elsewhere in the program, but this is fine if we have no intention of accessing them individually. The square brackets surrounding the values tell Perl that we are creating a reference to an array.

Once we have a two-dimensional array like this, we need to access the data contained within the structure. To do this we must remember that we are using referenced data (it's not the actual variable, but just a pointer to it), so to gain access to the data that we have referenced, we must first *de-reference* it. This can be achieved as follows:

```
# We have already created our @twodarray (using one of the
# methods above), so no need to show this again...

# Here we extract some single values
```

```
print "Patient1, var2: " . $twodarray [0]->[1]; # This will
                                                # print '48'
print "\n";
print "Patient4, var3: " . $twodarray [3]->[2]; # This will
                                                # print '55'
print "\n";

# Or we could just extract and dereference a whole array...

foreach (@twodarray) {
        # This is how we de-reference a whole array
        my @patient_data = @{$_};
        foreach my $var (@patient_data) {
                print $var . ' ';
        }
        print "\n";
}
```

In the above code we show how we can access the data within the multidimensional array that we created before (in any of the three examples—they all create the same final data structure), this is done by the process of de-referencing our data structure. The easiest way of explaining how this works is this:

• In a normal array, we access each (scalar) element of the array by using the index value of the element we want to retrieve—for example $array[1] gives us the second element in an array.

• In a referenced array we access data in the same way (using the index value), and then de-reference the scalar value by either using arrow symbols (->), or by using a double dollar symbol ($$). For example:

$array_ref->[1] returns the second element from an array reference.
$$array_ref[1] does the same.

The other method of de-referencing that we show above is where we de-reference a whole array at once, this is achieved by surrounding the array reference with @{ }. Finally, here is a brief example of how we would go about creating and accessing our data if we wanted to store all of the data within a reference (a reference to an array of arrays).

```
# Create our initial data structure as a reference...
my $aa_2d_array_ref = [
        ["24", "48", "56", "12"],
        ["25", "48", "55", "12"],
        ["23", "49", "54", "11"],
        ["24", "48", "55", "12"]
];
# And to access the data within...
```

```perl
print "Patient1, var2: " . $aa_2d_array_ref->[0]->[1] . "\n"; # '48'
print "Patient4, var3: " . $aa_2d_array_ref->[3]->[2] . "\n"; # '55'

# Or... (note the double '$$' to de-reference our data)
print "Patient1, var2: " . $$aa_2d_array_ref[0][1] . "\n"; # '48'
print "Patient4, var3: " . $$aa_2d_array_ref[3][2] . "\n"; # '55'

# Or access the whole structure...
foreach my $a_patient_ref (@{$aa_2d_array_ref}) {
        foreach my $measurement (@{$a_patient_ref}) {
                print $measurement . ' ';
        }
        print "\n";
}
```

3.3.2 Multidimensional hashes

We have looked at multidimensional arrays in our previous example, now, what if the data that you want to handle is slightly more complicated, such that it would be more ideally handled within a hash or even tree-type structure?

First, let's have a look at how you could create a hash containing arrays (as references) for the value part of each key/value pair. This uses the same techniques as in our previous example.

```perl
my %exp_results = (
        'patient1' => ["24", "48", "56", "12"],
        'patient2' => ["25", "48", "55", "12"],
        'patient3' => ["23", "49", "54", "11"],
        'patient4' => ["24", "48", "55", "12"]
);

# Now to retrieve some data...
print "Patient1, var1: " . $exp_results{patient1}->[0] . "\n";
print "Patient4, var3: " . $exp_results{patient4}->[2] . "\n";
```

The following code shows how to create the initial hash as a reference too:

```perl
my $ha_exp_results_ref = {
        'patient1' => ["24", "48", "56", "12"],
        'patient2' => ["25", "48", "55", "12"],
        'patient3' => ["23", "49", "54", "11"],
        'patient4' => ["24", "48", "55", "12"]
};

# Now to retrieve some data... (using '->' to de-reference)
print "Patient1, var1: " . $ha_exp_results_ref->{patient1}->[0] . "\n";
print "Patient4, var3: " . $ha_exp_results_ref->{patient4}->[2] . "\n";
# Or... (note the use of the '$$' again)
```

```
print "Patient1, var1: " . $$ha_exp_results_ref{patient1}[0] . "\n";
print "Patient4, var3: " . $$ha_exp_results_ref{patient4}[2] . "\n";

# Or access the whole hash...

# Just like 'normal' hashes, the keys function gives us the
# keys for a referenced hash - we just need to de-reference it first

foreach my $patient (keys %{$ha_exp_results_ref}) {
        print $patient . "\n" . "\t";
        # Now we can access all of the measurements in the inner array
        # reference by de-referencing it...

        foreach my $measurement (@{$ha_exp_results_ref->{$patient}}) {
                print $measurement . ' ';
        }

        print "\n";
}
```

As before, we created a reference to an anonymous array by creating our
initial array reference using square brackets, and we have also shown above
that it's possible to directly create a reference to an anonymous hash by
simply using curly braces. So, the rule for creating references to anonymous
data structures is: square braces for an array reference; curly braces for a hash
reference.

Having seen that we can put arrays within hashes through the use of references, let's have a quick look at an example of how we would embed numerous
hashes inside another hash (again using references). This is one of the most complicated examples in this chapter, so it's worth taking time to understand it.

```
my $hh_exp_results_ref = {
        'patient1' => {
                'var1' => 24,
                'var2' => 48,
                'var3' => 56,
                'var4' => 12
        },
        'patient2' => {
                'var1' => 25,
                'var2' => 48,
                'var3' => 55,
                'var4' => 12
        },
        'patient3' => {
                'var1' => 23,
```

```
                    'var2' => 49,
                    'var3' => 54,
                    'var4' => 11
            },
        'patient4' => {
                    'var1' => 24,
                    'var2' => 48,
                    'var3' => 55,
                    'var4' => 12

            }
};


# Now to retrieve some data...
print "Patient1, var1: " . $hh_exp_results_ref->{patient1}->{var1} . "\n";
print "Patient4, var3: " . $hh_exp_results_ref->{patient4}->{var3} . "\n";


# Or access the whole hash...


# Just like 'normal' hashes, the keys function gives us the
# keys for a referenced hash - we just need to de-reference it first

foreach my $patient (keys %{$hh_exp_results_ref}) {

    print $patient . ":\n";

    # Now we can access all of the keys to the measurements in
    # the inner hash reference by de-referencing it...

    foreach my $measurement (keys %{$hh_exp_results_ref->{$patient}}) {
        print "\t" . $measurement . ": ";
        print $hh_exp_results_ref->{$patient}->{$measurement} . "\n";
    }
    print "\n";
}
```

The above code looks pretty scary at first, but all we have done is to put to-gether each of the things that we have done previously. We first create a hash reference (using curly braces), and then the values in the key/value pairs for this reference are more hash references—this gives us a tree-like data structure that could be quite useful when building real bioinformatics solutions. Below this, we just de-reference the outer hash so that we can access the keys of the hash, and then do the same for each of the inner hash references.

The only data structure we have not shown an example of within this section is referenced hashes embedded within an array. Here is a quick example that should not really need any explanation, but shows a way we could handle related data from multiple experiments.

```
my $ah_exp_results_ref = [
    {
            'patient1' => ["24", "48", "56", "12"],
            'patient2' => ["25", "48", "55", "12"],
            'patient3' => ["23", "49", "54", "11"],
            'patient4' => ["24", "48", "55", "12"]
    },
    {
            'patient1' => ["22", "46", "54", "10"],
            'patient2' => ["24", "47", "54", "11"],
            'patient3' => ["25", "51", "56", "13"],
            'patient4' => ["24", "48", "55", "12"]
    }
];
```

This covers all of the data structures possible with Perl—obviously you are not limited to only one level of nesting within hashes and arrays, we just chose not to show more complex examples here. However, the syntax is exactly the same—you just have to remember that you have more than two levels when it comes to de-referencing.

3.3.3   Viewing data structures with Data::Dumper

One of the final things that we would like to touch on, with respect to references and complex data structures in Perl, is how to figure out what is in the data structures without going through complex de-referencing and lots of code. This can be a common task if you inherit poorly documented code from other people, or if you are just not sure how things are working. Thankfully Perl has a built-in module that can greatly aid in this event: Data::Dumper.

Data::Dumper, in its most straightforward use, simply allows you to print an entire data structure to the console including its structure and contents, thereby giving you vital clues as to how you can go about manipulating and using your data structure. Here is an example of Data::Dumper in use:

```
# First we create a data structure...
my $hh_exp_results_ref = {
    'patient1' => {
            'var1' => 24,
            'var2' => 48,
            'var3' => 56,
            'var4' => 12
    },
    'patient2' => {
            'var1' => 25,
            'var2' => 48,
            'var3' => 55,
            'var4' => 12
    },
```

```
        'patient3' => {
                'var1' => 23,
                'var2' => 49,
                'var3' => 54,
                'var4' => 11
        },
        'patient4' => {
                'var1' => 24,
                'var2' => 48,
                'var3' => 55,
                'var4' => 12
        }
};
# Now have a look at it...

use Data::Dumper; # Load the module
print Dumper($hh_exp_results_ref);# Print the structure using
                                  # Dumper()
```

This will produce the following output:

```
$VAR1 = {
        'patient1' => {
                        'var3' => 56,
                        'var1' => 24,
                        'var4' => 12,
                        'var2' => 48
                      },
        'patient4' => {
                        'var3' => 55,
                        'var1' => 24,
                        'var4' => 12,
                        'var2' => 48
                      },
        'patient3' => {
                        'var3' => 54,
                        'var1' => 23,
                        'var4' => 11,
                        'var2' => 49
                      },
        'patient2' => {
                        'var3' => 55,
                        'var1' => 25,
                        'var4' => 12,
                        'var2' => 48
                      }
        };
```

The above output clearly shows the structure of our data—we can see that it is a hash reference containing other hash references, as denoted by the curly braces. This technique can be applied to any data structure in Perl and is a very useful debugging tool to have around.

3.4   Subroutines and modules

Subroutines are pieces of re-usable code that are often called things like *methods* or *functions* in other programming languages. The typical use for a subroutine is when you have to repeat an operation several times in a program (e.g. to perform the same operation on a series of arrays or hashes). On such occasions you could copy and paste the same section of code several times, with the only difference being the variable names. However, not only does this lead to long programs, it also means that if you want to expand or modify the code you have to edit all of the copied sections. A much better solution is to write just one block of code—a *subroutine*—and call it each time you need to perform that operation. Let's consider the most basic syntax for defining a subroutine:

```
# The line below indicates the start of a subroutine...
sub subroutine_name {
        (input_variables) = @_; # Optional

        content_of_the_subroutine

        return subroutine_output # Optional
}
```

Basically, there are four parts that make up a subroutine:

- First is the name of the subroutine—this is the name you are going to use in the body of your program to call your subroutine.

- Second is the list of input variables that you have passed into your subroutine—this can be any Perl variable or data structure, and there may be any number of them. In fact, some subroutines do not need input variables at all—you could just write a subroutine to return a defined output (e.g. displaying instructions on how to use a program is a common application of this feature).

- Third is the content of the subroutine—this is the main body of code that performs the function of the subroutine.

- Finally we have the subroutine output—this is what gets passed back to the main body of your program on completion of the subroutine. Like the input variables, this part of a subroutine is optional, for example, the output from a subroutine could be information printed to the console, and therefore the returning of variables is not required.

As ever, we can better understand subroutines by studying an example:

```perl
#! /usr/bin/env perl

use strict;
use warnings;

my $dna = 'ACTGAAA';
print "My DNA string is " . $dna . "\n";

# Call on our subroutine...
print "The reverse complement is " . revcom($dna) . "\n";

# A subroutine to reverse complement DNA
sub revcom {
        # Get the DNA to be worked on...
        my ($dna) = @_;

        # First we reverse the DNA
        $dna = reverse $dna;

        # Now translate the DNA bases
        $dna =~ tr/ACGTacgt/TGCAtgca/;

        # Return the output
        return $dna;
}
```

The above code forms a complete program that will give us the reverse complement of a DNA string—something that we have done earlier (a couple of
times now)! The main difference in this instance is that producing the reverse
complement is done entirely within a subroutine, and all you need to do to use
it in your Perl program is to call it. This makes our code very portable—not only
can it be called multiple times from within our program but, if we ever need
the ability to calculate the reverse complement DNA of a sequence in another
program, we can simply copy the subroutine into the new program and use it
with no additional work. Better still, if it is something we do really regularly, we
could save it in a Perl *module* containing some of our most used subroutines so
that we can call them as they are needed without physically pasting them into
our programs. We will explain how to produce such modules shortly, but first
let's look in more detail at the above code as there are some things that need
explaining.

First, note the position of the subroutine within the program—we have placed
it at the very bottom of the code. This is the normal position for subroutines and
is where we would suggest that you place your subroutines. Although the subroutines are at the bottom of your code, they will not be executed at the end of your
program as the interpreter works its way through the lines of code—they are
separate 'subprograms' that are not run unless they are called within the body of

the main program, so it is safe to place them at the bottom of your code and call them as and when they are needed.

Second, note the variable declaration within the subroutine. Because we can send more than one input variable to a subroutine, the variables are passed in the form of an array (denoted by the @_ symbol). This is why the $dna variable is surrounded by round brackets, to acknowledge that it is an element of the array. If we had more than one input variable, we would simply have more than one variable declared within the brackets.

Third, as we are using the strict pragma, we are able to have two variables called $dna (we declare one in the main body of the program and the other within the subroutine). These two variables are completely separate entities as they were declared within different namespaces, as described when we first introduced strict—the first $dna variable is only usable within the main body of the program, and the second $dna variable is only available and usable within the body of the subroutine in which it was declared. If we were not using the strict pragma (and using my to declare our variables) we would now be getting into problems as our two separate variables would not be separate, they would be the same variable and would be overwriting each other without any warning. This concept is further explained in Fig. 3.3.

Before we move on to looking at Perl modules, let's consider another example of the use of a subroutine.

```
#! /usr/bin/env perl

use strict;
use warnings;

# Declare some variables
my @array1 = qw(AA BB CC DD EE);
my @array2 = qw(11 22 33 44 55);

# Now print the contents of the arrays
print_array(\@array1);
print_array(\@array2);

# The subroutine we use to print the arrays
sub print_array {
    my ($array_ref) = @_;

    foreach (@{$array_ref}) {
        print $_ . "\n";
    }
}
```

The above demonstrates a short example of the use of a simple subroutine to make the boring task of looping through and printing an array a simple call to

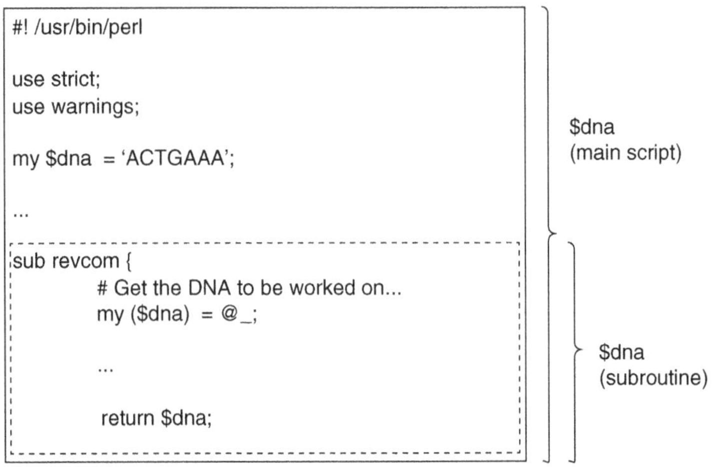

```
#! /usr/bin/perl

use strict;
use warnings;

my $dna = 'ACTGAAA';

...

sub revcom {
        # Get the DNA to be worked on...
        my ($dna) = @_;


        ...


        return $dna;
```

$dna
(main script)

$dna
(subroutine)

Fig. 3.3 A visual representation of a subroutine, and the concept of namespaces. When Perl's `strict` pragma is in use, any variable defined within a code block such as a subroutine will be a private variable – only available within that code block. Conversely, variables declared outside of any code block/subroutine will be available throughout the whole program, and indeed the subroutines if they are not overridden by private variables.

the subroutine. We also show the use of references in this example. We could have passed the entire array(s) to the subroutine and this would work, however, as we stated before (when first looking at references), if our arrays are very large, this method will perform much faster and use less memory on your computer as it does not have to copy a large array into the subroutine because it utilizes a reference to the original array. Such subtleties of coding may seem trivial, but ultimately these efficiencies can make the difference between a problem being solved or being totally intractable, particularly in bioinformatics where data sets can be massive.

3.4.1   Making a Perl module

Finally, we shall look at putting the above subroutines together into a Perl module. To put it simply, a Perl module is a text file that contains a collection of subroutines and has the file extension '.pm'—this can then be called from another Perl program allowing use of the contained subroutines as if they were in that program. Here is an example of the code for a Perl module that contains the two subroutines that we have just looked at.

```
# This file will be called 'MySubs.pm' and is
# saved in the same directory as our other Perl programs.

sub revcom {
     # Get the DNA to be worked on...
     my ($dna) = @_;
```

```perl
        # First we reverse the DNA
        my $revcom = reverse $dna;

        # Now translate the DNA bases
        $revcom =~ tr/ACGTacgt/TGCAtgca/;

        # Return the output
        return $revcom;
}

sub print_array {
        my ($array_ref) = @_;

        foreach (@{$array_ref}) {
                print $_ . "\n";
        }
}


# Note: this line below is needed for the
# Perl module to work, and you MUST ALWAYS end
# your Perl modules this way.

1;
```

That is our Perl module written, now here is an example of how we can use these subroutines in another Perl program (note, for simplicity, in this example the Perl program must be in the same directory as our Perl module for this to work).

```perl
#! /usr/bin/env perl

use strict;
use warnings;

# This is how we call our Perl module
use MySubs;

my $dna = 'ACTGAAA';
print "My DNA string is " . $dna . "\n";

# Call on our subroutine...
my $revcom_dna = revcom($dna);
print "The reverse complement is " . $revcom_dna . "\n";
```

All you have to add to your program is the line use *name_of_module* and Perl takes care of making the subroutines in that module available to your program.

From then on you just have to call your methods as if they were at the bottom of your program. This is a great way of re-using code for repetitive and common tasks. Best of all, there are thousands of Perl modules freely available via the web, many of which contain functions relevant to bioinformatics. Provided you can find a module that does what you want (we give you some pointers in Section 3.10), you can just call subroutines from it as if they were part of your own program.

This is all there is to cover about subroutines and Perl modules, the only other piece of advice that we can offer is that if you have to type the same piece of code more than once within the same program, you really should consider moving that bit of code into a subroutine. Like a lot of programming practice, it may take longer in the short term but could pay dividends later on.

3.5   Regular expressions

We are now going to look at one of the biggest selling points of Perl—the ease with which we can use regular expressions. A regular expression can be defined as a string that is used to describe or match a set of strings—it is a tool that can be used to look for specific pieces of text or patterns in strings. The use of regular expressions in Perl is very straightforward, and is one of the reasons that Perl has become the default language for bioinformaticians—as the majority of bioinformatics tasks involve some form of text manipulation.

So, how do we get started with regular expressions? Well, we already started back in Section 3.2.1 of this chapter, when creating the reverse complement of a DNA string using the substitution and transliteration operators. The examples so far have been very simple, just looking for and replacing single characters, but regular expressions can be far more powerful than this. Covering everything there is to know about regular expressions is just not possible here, the topic is almost like a programming language of its own and there are whole books dedicated to this subject (e.g. Stubblebine, 2007). What we aim to provide here is an introduction to the concepts of regular expressions and some of their more common uses in bioinformatics.

3.5.1   Defining regular expressions

The best way to get started with regular expressions is to use them in conditional statements—to be more precise, an `if` statement. This allows us to easily tell whether our regular expressions work or not, as when a regular expression gets a match, it returns true, if it does not get a match, it returns false—there is no middle ground here. Consider the example below.

```
#! /usr/bin/env perl

use strict;
use warnings;

my $dna = 'ACTGCCGTAAACCCTG';
```

```
if ($dna =~ /CCG/) {
      print "CCG present in sequence.\n";
} else {
      print "No match found.\n";
}
```

The above example uses a regular expression to test for a pattern of three letters within a string. First we declared our string $dna, from here we set up our conditional with the test case $dna =~ /CCG/—a basic use of a regular expression that needs a little more explanation:

• The first thing to note is the use of the binding operator =~ that makes our regular expression act on (bind to) whatever variable is to the left of it—in our case $dna. It is important to remember this binding operator, as without it the regular expressions will not work as intended.

• The second part is the test for the regular expression itself—this is the string housed between two forward-slash characters (/)—in our example we are looking for the three base sequence 'CCG' within our DNA string.

Metacharacters

In Section 3.2.1 we listed special codes for use within double-quoted strings, such as \n for newline characters and \t for tab characters. These can be used in exactly the same way within regular expressions so that we can look for these special characters if we need to. Indeed, there are further special codes, listed in Table 3.5, that can be included in regular expressions to match certain types of character.

Another character that is of use within regular expressions is the dot (.) character. This acts as a wildcard character within Perl's regular expressions, matching any single character except the newline character (\n). Therefore the pattern match /CC./ would return a positive match for CCG, CCA, CCT, CCC when used on a DNA string.

However, what if we would like to search for the dot character? How can we achieve this without it becoming a wildcard character? This is done by *escaping* the wildcard action by preceding the dot character with a backslash (i.e. \.), our regular expression will then look for the dot character. The same is true of the

Table 3.5 Codes for use in regular expressions

Character	Meaning
\w	Word characters. Matches any alphanumeric character and the underscore (_) character.
\W	Matches any non-word characters.
\s	Matches a whitespace character, i.e. space or tab.
\S	Matches a non-whitespace character.
\d	Matches a digit character.
\D	Matches a non-digit character.

forward-slash character—if we want to match on this we need to escape it with a backslash character first (\/), otherwise we'll end our regular expression early.

Repeating values

What happens if we want to match more than one wildcard character or any other character or letter for that matter? We could just repeat the given character (or wildcard) the required number of times, but that is not very flexible as it requires us to know the exact number of times we would expect our character to appear. In this instance, we should use yet another special operator, the asterisk (*). This is a *quantifier* that tells the regular expression to match a character *zero or more* times, and it is placed after the character upon which you wish it to act. For example, the regular expression /CCGA*/ will match CCG, CCGA, CCGAA, CCGAAA, and so on. You can also use the asterisk symbol in conjunction with the wildcard operator or any other special character, so . * is often used to ignore uninteresting pieces of string before or after the patterns that you are looking for.

A related quantifier is the plus symbol (+)—this tells the regular expression that you would like to match a given character *one* or more times and is used in exactly the same way as the asterisk character.

Grouping patterns

It is possible to group patterns together within Perl's regular expressions by enclosing them in brackets. This allows us to look for more intricate patterns, for example, /CCG+/ would look for CCGGGGG or any other number of trailing Gs, however this might not be the most useful pattern for us. On the other hand, /(CCG)+/ would look for CCGCCGCCG or any number of repetitions of the CCG triplet—most useful if our aim is to seek out CCG repeats.

What happens if we use the asterisk instead? This is something to watch out for, as it is a common mistake. The problem is that an expression such as /(CCG)*/ would match anything, not just repetitions of CCG, due to the use of the 'zero or more' asterisk character.

Using OR in regular expressions

The logical operator *or* in the form of the vertical bar (|) character can also be used within regular expressions when you want to look for alternatives within your pattern matching. Here are some examples of its use:

+ /CCG|GGC/ would match either CCG or GGC.

+ /CC(G|C)/ would match CCG or CCC.

+ /CC(G+|C+)/ would match CCGGGGGG or CCCCCCCCCC or any variation on the number of repeated Gs or Cs following the initial CC.

3.5.2 More advanced regular expressions

Using the concepts and characters introduced in the previous section, it is easy to produce a mind-bogglingly large number of fairly complex regular expressions, but Perl allows us to go further still.

Character classes

Character classes are a list of possible characters for use in our pattern match surrounded with square brackets ([|]) that would return true if any character within the brackets was found.

For example, a character class [ACTG] would match any of the four nucleic acids. You can also specify ranges within character classes, so [0-9] would match the digits zero to nine, [A-Z] would match any upper case letter of the alphabet. You could use [a-zA-Z] to match both upper and lower case.

Character classes would not typically be used as a regular expression on their own, but would be used as part of a larger regular expression. For example, the program below checks a string to see if it looks like an Ensembl gene ID (i.e. is it a number preceded by the letters ENSG?).

```
#! /usr/bin/env perl

use strict;
use warnings;

my $gene_id = 'ENSG000041';

if ($gene_id =~ /ENSG[0-9]+/) {
        print "Our string is an Ensembl Gene ID.\n";
}
```

More quantifiers

We have already seen that the quantifiers * and + match 'zero or more', or 'one or more' times respectively. It's possible to be more specific with the number of repeated characters, or sequences of characters if you are using grouping. This is achieved through the use of curly braces ({}) and a pair of numbers within these braces defining how many repetitions we will accept. Here are a few examples:

- /\w{1,}/ would match one or more 'word' characters.
- /\w{1,10}/ would match anything between one or ten 'word' characters.
- /(CCG){3,}/ would match three or more repetitions of the triplet CCG—for example CCGCCGCCG, CCGCCGCCGCCG, etc.

Anchors

Anchors are special characters that allow us to tie our pattern match to certain sections of the test string. The two anchors that we shall discuss here are ^ and $—when used in regular expressions these tie a pattern match to either the start (^) or the end ($) of the test string. For example:

```
#! /usr/bin/env perl

use strict;
use warnings;
```

```
my $id = ' ENSG000041';
if ($id =~ /^ENSG[0-9]+/) {
    print "Our string is an Ensembl Gene ID.\n";
}
```

The above test would return false as there is a whitespace character in the first position of our test string. Here is another example:

```
#! /usr/bin/env perl

use strict;
use warnings;

my $id = 'ENSG000041_A';

if ($id =~ /^ENSG[0-9]+$/) {
    print "Our string is an Ensembl Gene ID.\n";
}
```

This test would also return false as we have now tied the numbers at the end of our pattern match to the end of the string—at the end of our string we have a letter (A).

Although both of these examples result in a failure of our regular expression tests, they do show how the two anchors can increase the specificity of regular expressions.

3.5.3   Regular expressions in practice

Now that we have covered the basic syntax of regular expressions, in this final section we can cover some extra pieces of information that are useful in their practical use.

The pattern-match operator

So far we have been placing our regular expressions inside two forward-slash characters—this is actually a built-in shortcut for the m// pattern-match operator. The reason we mention this is not just for completeness, it is useful for cleaning up your regular expressions. Take, for example, a regular expression that tries to identify a URL:

```
/http:\/\/www.*/
```

This is a pretty straightforward example, but demonstrates how messy something like this can become when you need lots of backslash characters to stop the regular expression ending prematurely. One of the benefits of using the m operator is that you do not necessarily have to use the forward-slash characters to delimit your regular expressions—in fact, you can use any set of delimiters. For example, you could use `m()`, `m{ }`, `m[ ]`, `m<>`, `m! !`, `m%%`, and so on—thus making your original regular expression much easier to read:

```
m(http://www.*)
```

The basic rule is to choose a set of delimiters that make your regular expression as easy to understand as possible.

Modifiers

As described in Section 3.2.1 when looking at string substitution, regular expressions allow the use of various modifiers to change the way in which they operate (the substitution operator is one form of regular expression). These come after the closing / character (or whatever character you may be using to delimit your expression), and are single lower case letters. Two important modifiers are `i` and `s`.

- `i`, as described earlier with the substitution operator, makes the pattern-match case insensitive.

- `s` modifies the action of the wildcard dot (.) character. Normally the wildcard character matches anything but the newline (\n) character. With the `s` modifier, the dot character will also match the newline character.

As in the substitution example, all you need to do is place the modifier(s) required after the closing forward-slash of the regular expression and they will modify the regular expression accordingly.

Match variables

Quite often when using regular expressions, it is useful to find out the exact snippet of string that is causing the match to occur; this is useful if you are trying to retrieve something from a string—it can also aid in debugging a rather complicated regular expression.

By default there are three automatic match variables with Perl's regular expressions. When used these return the section of string before the match, the match itself, and then the section of string after the match—these can be accessed using the following special variables: `$``, `$&`, and `$'` (before, match, and after, respectively). Here is an example of their use:

```perl
#! /usr/bin/env perl

use strict;
use warnings;

my $string = 'The Human Gene ID is ENSG000041 revision 1';

if ($string =~ /ENSG[0-9]+/) {
    print "The gene ID is: '" . $& . "'\n";
    print "The text found before was: '" . $` . "'\n";
    print "The text found after was: '" . $' . "'\n";
}
```

In addition to these automatic match variables, it is also possible to define your own match variables to return specific portions of your pattern matches. To do this we make use of pattern grouping, using brackets, as described earlier, to encase the section of the regular expression that you would like to access. To access these values after the regular expression has been matched, we use the special Perl variables $1, $2, $3, and so on, where the number refers to the bracket grouping when counting from the left. The best way to understand this is by example:

```
#! /usr/bin/env perl

use strict;
use warnings;

my $string = 'The Human Gene ID is ENSG000041 revision 1';

if ($string =~ /^The (.*) Gene.*(ENSG[0-9]+).*$/) {
    print "Our gene ID is: " . $2 . "\n";
    print "Our species is: " . $1 . "\n";
}
```

It is also worth noting here that, even when you are using self-defined match variables (using $1 etc.), Perl's default match variables are still available to you.

Substitutions with s///

We used the substitution operator back in Section 3.2.1 to substitute individual characters, but we can also use this operator on more than one character at a time. In fact, it is possible to replace large chunks of text and use any of the regular expression syntax described above. For example:

```
#! /usr/bin/env perl

use strict;
use warnings;

my $string = 'ACTGCCGTGCCGCCGCCGTTGAC';

$string =~ s/CCG/---/g;

print $string . "\n";

# This would return...
# 'ACTG---TG---------TTGAC'
```

3.6 File handling and directory operations

As we have just discussed, one of the most common tasks performed with Perl is some form of text manipulation using regular expressions. However, what we have not discussed is where this text (e.g. DNA or protein sequence) has come from. So far we have been typing our text strings directly into our programs, often termed *hard-coding* data into programs. If we want to apply our program to other data, we would have to alter the Perl code itself. In real applications this is not normal practice as it is not practical to type, or cut and paste, large chunks of DNA sequence (or whatever data you are working with) into your programs or onto the command line. An alternative is reading in data from a local text file, while other options would be reading in files from the Internet, or taking data directly from a relational database. We shall look at interacting with databases later in the chapter—for now we shall consider using text files with Perl.

3.6.1　Reading text files

First, we will consider opening and reading local files that already exist. Here is the basic syntax for opening a file:

```
open(FILEHANDLE,filename);
```

The basic function we call for interacting with files is `open`, to which we pass two arguments; the *file handle* and the *file name*. The file name is simply the name of the file with which you wish to interact. The file handle is the name used to interact with the file in the rest of our program. Here is a snippet of code to read in a FASTA file containing the genomic sequence for the gene BRCA1 `BRCA1.fasta`—this can be downloaded from our companion website (`www.bixsolutions.net/BRCA1.fasta`) or you can use a FASTA file of your own if you have one to hand. Copy the file into your current working directory and try this snippet of code in a new program.

```
open(FILE,"BRCA1.fasta");    # Open the fasta file
my @file_text = <FILE>;      # Read the entire file into an array
close FILE;                  # Close the file
print "@file_text \n";       # Print the contents of the file
```

On the first line we open the file—giving it the file handle `FILE`. We then read the entire contents of the file into an array by referring to this file handle. Note the use of an array here—if we had used a scalar variable we would only get the first line of the file. On the next line of the code we close the file, again using the file handle to specify the file to close. On the final line, we print the contents of the array to the console. It is important to always use the `close` function as soon as you have finished with a file, as leaving it open longer than necessary can leave the file open to potential damage. In this example, the file handle may seem somewhat unnecessary, but these become very important in more complex programs where we have multiple files open simultaneously, as is the case when we are comparing files or moving data from one file to another.

This approach is all you need to read the entire contents of a file entirely into the computer's memory, ready for Perl to manipulate. If you are using small files, this works fine but with larger files, like those often found in bioinformatics, it may be impractical to load the whole file in memory at once (as some files can be larger than the memory capacity (RAM) of your computer). Instead, you would read them in one line at a time, as opposed to all at once, thus keeping the memory of your computer relatively empty. Consider the following:

```
open(FILE, "whole_human_genome.fasta");
while (<FILE>) {
    print $_;
}
close FILE;
```

In the above code, the opening and closing of the file are the same as shown previously, as is the end result of the program—it prints the contents of the file to the console. The difference here is that instead of reading the file completely into an array (and therefore into the computer's memory), we keep the file open and read it in one line at a time through the use of a `while` loop. The advantage of this approach is that it allows you to extract data from files of any size without ever exceeding the memory capacity of your computer. As we only read one line of the file into memory at a time, perform whatever operations we would like to perform, and then move onto the next line of the file, this method is suitable no matter what the file size is. The most important thing to consider when using this second approach, is that all operations that you would like to carry out on your file must be carried out within the `while` loop as this is the only section of your code where each line of your file will be available within your program (unless you save the whole file or portions of it into a variable).

3.6.2 Writing text files

Just as Perl can read from text files, so we can create and write to text files. This is invaluable for permanently recording the output of our Perl programs. The good news is that there is no need to learn a new function to do this; we can write files using the `open()` function that we used previously to read files. The following code shows us how to do just that:

```
open(FILE,">output.txt");              # Open the file for writing
print FILE "Some example text \n";     # Put something in the file
close FILE;                            # Close the file
```

The above code will create a text file in your current directory called `output.txt`, containing the text 'Some example text'. The differences in this snippet of code that allow us to write, rather than read, a file are found within the first two lines: the first change is that we added the > symbol in front of the file name when opening the file. The second difference is the use of the file handle—instead of using it to insert the contents of the file into an array, we use the file handle along with the now familiar `print` command to print data into the file, just as if we were printing to the command line. A word of warning when creating and writing to files in this manner: if there is already a file in your working directory with the same name as the file you wish to write to, Perl will overwrite it and you will lose the contents of the original file. There are no warnings or 'are you sure?' questions to make sure you want to overwrite a file, Perl will just go ahead and delete the file before creating its own.

This is not the only way to write to files in Perl. There is a slightly less destructive approach known as appending data to a file. This allows you to add text to the end of a pre-existing file. An example of the code used to achieve this is shown below.

```
open(FILE,">>output.txt");              # Open the file for
                                        # APPENDING text
print FILE "Yet more example text \n";  # Add something
close FILE;                             # Close the file again
```

If you create and run the above program, then look in the file `output.txt,` you will find the line from the previous program as well as the new line of text: 'Yet more example text'. This was achieved by using two > symbols in front of the file name—this instructs Perl to add the given text onto the end of an already existing file (although, if no file exists with the specified name, Perl will create a new one). Sometimes you might want to overwrite an existing file, if that is the case use the first approach we showed for writing files (using a single >), otherwise it might be good practice to use the append approach by default as you can always delete text from a file, but you cannot retrieve a file that you have inadvertently overwritten.

3.6.3 Directory operations

Reading and writing files in the current directory is a great way to learn how to handle files in our programs; however, this soon becomes limiting when we start tackling real bioinformatics problems. Thankfully, Perl has many built-in functions for navigating your computer's directory structure.

Creating directories

A typical scenario might be that you are performing operations on a large number of files and the end result of your program is even more files. Doing all this in one directory is feasible, but it could make your life difficult, as you have to manually sort out the files afterwards. Another option would be to have your Perl program create a directory in which to store its new results. This can be achieved using the `mkdir()` function:

```
mkdir('MyResults') # Creates a new directory called MyResults
```

The code above will create a new directory with the name `MyResults`.

Changing directories

Let's say that we want to interact with some files, but they are not in our current directory. You could move them into your current directory, or even move the Perl program into the same directory as your files and run it from there. Obviously this is a hassle to do manually, so instead we would use Perl's `chdir()` function to change the directory:

```
chdir('/home/user') # Moves us to /home/user
chdir('..')         # Moves up one directory
chdir($ENV{HOME})   # Moves into our 'home' directory (unix only)
```

As you can see from the code above, `chdir()` is quite simple to use, you just put the name of the directory that you want to change to (either the full path or a path relative to your current directory) in the brackets after the command. On Linux systems you can even use the built-in environment variables (such as $HOME) that would immediately get you the full directory path for your home directory.

Getting the contents of a directory (globbing)

Another common task is to retrieve a list of all of the files in a given directory. This can be achieved by using the `glob` function:

```
my @files = glob "*"      # Gets the entire contents of our
                          # current
                          # directory (directories included)
my @html = glob "*.html"  # Only gets the HTML files
my @root = glob "/*"      # Gets the contents of the root
                          # directory
my @html2 = <*.html>      # Another way of using glob!
```

The above code shows us the basic syntax for using `glob`, and the two ways we can go about using it to retrieve a list of files in a given directory. The one thing that you must remember when using `glob` is that it does not differentiate between directories and files, so you must be careful with what you try to do with the resulting list. For example, if you tried to open a directory for editing as if it were a text file, this would throw an error. When such errors occur, we can handle these as explained in the next section.

3.7   Error handling

If you are new to programming, you might think that a well-written program should never throw an error while it is running. Surely if the program code has been carefully written, tested, and debugged then nothing untoward should ever occur? However, most programs rely on interactions with external resources, such as user input, data files, computer hardware, and network resources. Because these outside factors are beyond the programmer's control, things can sometimes go wrong and Perl will throw an error. In this section we outline a couple of steps you can take within your programs to watch for, catch and deal with errors.

The 'or die' approach

One of the simplest ways of dealing with errors in your Perl scripts is to terminate your program in the event of an error—this is done by using the `die` command that we met earlier. Here are two examples of using this approach:

```
chdir('/home/public') || die "Can't move to /home/public: $! \n";
open(FILE, ">newfile.txt") || die "Can't create file: $! \n";
```

In the above two lines of code, we use the 'or die' approach to error handling twice.[3] In the first line, we try to change our current directory to /home/public. If the move to the directory works, the Perl program continues as normal; if, however, something goes wrong (e.g. the directory does not exist or we do not have the correct permissions to view the directory), the `chdir()` function will return false because as it didn't work and the program will exit via the `die` command. To make it apparent where and why the program died, it will also print out the message specified after the `die` statement—the message also includes a message from the Perl interpreter telling us why the error occurred, this is printed using the $! shortcut variable. In the second line of code, we try to open or create a file for writing. In the event of something going wrong (e.g. we do not have write permissions for the file or the directory), the program exits and tells us where it all went wrong and the specific Perl error message again.

The 'or die' approach to error handling is not just used in file operations; you can use it anywhere in your program when you use a function call. This approach to catching errors, whilst useful, is very basic, and crucially doesn't allow your program to keep on running. There are, however, more versatile ways of catching an error and letting your program continue to run—we shall look at these next, but if the sole purpose of your program is to create and write files to a certain directory on your system, and if your program cannot access the directory in the first place, the best thing it could do is exit very quickly and tell you why.

Catch and deal with errors using 'eval'

If we want our program to do more than just stop when an error occurs, we use `eval` (short for evaluate). Consider the example below.

```
# evaluate this section of code...
eval {
    open (FILE, ">myoutput.txt") || die;
    print FILE "Our output.";
}; # note the ; at the end of the eval block - this is required!

# if an error occurs, do the following...
if ($@) {
    print "Ooops, an error occurred: " . $! . "\n";
}
# the program carries on running from here
```

3 We use the syntax || for consistency with previous examples, but the word `or` can be used instead.

In the above example we call the function `eval` on a code block that has the potential to throw up errors (in this example, opening/creating and printing to a file). This section of code is run, but if an error occurs, the program does not exit as with `die` on its own, instead it just exits the `eval` block and carries on, but it then sets the value of a special variable `$@` to true to indicate that an error occurred. We then put some fallback code after the `eval` block to test if an error occurred (using `$@`), so if an error did occur we could do something about it—if not we ignore the fallback code.

Notice in the above code the combination of both `eval` and `die`. This is the most common approach to error trapping and handling in Perl. We still try to catch all of our errors on every line possible via the use of `die`, but we surround blocks of code that we would like to handle more gracefully with `eval`, so in the event of an error the program does not exit, it carries on using a pre-determined back-up plan.

We shall be using error catching for the rest of the chapter in our examples as appropriate, and we will touch more on the uses of `eval` shortly whilst talking about transaction handling with databases. We just have one final warning about using `eval`. Although it allows a program to continue running, the state of variables, file handles, and so on can be unpredictable after an error has been caught if you are not sure exactly where in the `eval` code block the error occurred. So, the program should continue with caution.

3.8 Retrieving files from the Internet

So, we can read data from local files and handle any errors that may occur. In bioinformatics the files we need are often stored remotely, somewhere on the Internet, and we need to write programs to go off and retrieve them. One way to do this is through the use of the built-in Perl module `LWP::Simple` and its function `get()`. This function performs a HTTP GET request (as a web browser would do) to a given URL and returns the contents of the file that it finds at the given URL as a string.

Before we can use `LWP::Simple` however, we must install it as it is not part of the standard Perl distribution—it is part of a package called `libwww-perl`.

Windows users

The Windows distribution of Perl (ActiveState Perl) comes with a package manager called the Perl Package Manager, or PPM for short. We shall discuss this in more detail later in this chapter, but for now we just need to use PPM to install `LWP::Simple`. PPM is called and used from the Windows command line. To launch the PPM interface, simply type `ppm` and press Enter—the PPM graphical user interface will then appear and you can search for and install `libwww-perl` (the interface is simple enough to need no explanation). Alternatively, you can use PPM entirely from the command line by issuing this command:

```
ppm install libwww-perl
```

Linux

`libwww-perl` will be found for installation in your distributions package manager. Simply search for 'libwww-perl' and install the relevant packages and their dependencies.

Mac OS

Mac OS users can use the `cpanm` tool from the command line: `cpanm LWP::Simple`. This shall be described in more detail later on in Section 3.10 (as well as some useful setup instructions), we would recommend quickly skipping forward in the chapter before installing this, and then come back.

Now that we are setup with `LWP::Simple`, below is an example program using it to copy a CSV file from `www.bixsolutions.net`. The file contains a table of data, which is described in more detail in the next chapter. After acquiring the file, the program performs various manipulations on the data, and prints the resulting data object to the screen. Much of this should be familiar—the only new technology is the use of `LWP::Simple` and `get()`. You can find the program itself at `www.bixsolutions.net`—it's called `get_example.pl`.

```perl
#! /usr/bin/env perl

use strict;
use warnings;
use LWP::Simple;
use Data::Dumper;

# This is the URL of the file we wish to fetch, this could
# even just be a webpage - we would then fetch the HTML code.
# (for more information on HTML, see Chapter 5)

my $file_data = get('http://bixsolutions.net/profiles.csv')
  or die "Unable to fetch file! \n";

# Now to play with the data, first let's split lines up...
# NOTE: we use \r\n here to split the file line by line as
# this is a file generated on a Windows machine. If it
# was a file generated on a Linux / Unix / Mac OS
# machine we would simply use \n

my @data = split("\r\n", $file_data);

# Now remove and process the header line
my $header_line = shift(@data);
my @headings = split(",", $header_line);
```

```
# Create an empty array to hold our sample information
my @sample_data;

# Then process the samples
foreach (@data) {
    my @sample = split(",", $_);

        # Now we convert this to a hash using the
        # column headings as the keys and put it into the
        # @sample_data array as a refernce...

    my %sample_hash;

    for (my $i=0; $i<scalar(@sample); $i++) {
        $sample_hash{$headings[$i]} = $sample[$i];
    }
    push(@sample_data, \%sample_hash);
}

# Now look at the resulting data structure...

print "Here is our data: \n";
print Dumper(@sample_data);

exit;
```

This demonstrates how easy it is to pull data off of the Internet. As stated in some of the comments in the program, this can be used for any textual file type, or even a web page itself. Other common uses for this technique are retrieving things such as GenBank files or FASTA files containing sequence data needed for analysis. The `get()` function can also be used for a technique called *screen scraping*, where a program is used to extract information directly from web pages by downloading the HTML code that makes up the pages (see Chapter 5) and searching through the code for the information of interest using regular expressions.

3.8.1 Utilizing NCBI's eUtilities

As another example of Perl's ability to retrieve and interpret files from the Internet, we would like to give you a brief introduction to automated querying of the databases available at the NCBI—this is achieved through the use of NCBI's Entrez Programming Utilities (also known as eUtils).

The eUtils are a web-based service that allows efficient searching of the databases through programmatic means, instead of via a web browser. You communicate with eUtils by making a HTTP connection (via `LWP::Simple`) to a given URL that defines our search. The information that is returned is in XML, and is therefore easily readable by machines. There is comprehensive information about eUtils at NCBI (`www.ncbi.nlm.nih.gov/books/NBK25500/`), so in this chapter we restrict ourselves to demonstrating the capability of eUtils by considering a single case study—retrieving bibliographic information from PubMed. A program to do this is shown below (it's also available on our website as `eutils_example.pl`).

```perl
#! /usr/bin/env perl

use strict;
use warnings;
use LWP::Simple;

# Set up the query URL
my $utils = 'http://www.ncbi.nlm.nih.gov/entrez/eutils';
my $db = 'Pubmed';
my $query = 'BRCA1';

# Set up a search out to the eSearch program:
# - we set the 'db' param to our database (pubmed)
# - and set the number of results we want as 1 (retmax)
# - leave the search term blank for now (term)
my $esearch = $utils . '/esearch.fcgi?db=' . $db .
              '&retmax=1&term=';

# Now submit the search and retrieve the XML based results
my $esearch_result = get( $esearch . $query );

print "----------------------\n";
print "--- eSearch Results ---\n";
print "----------------------\n\n";
print $esearch_result . "\n";

# Get the ID for the paper that we have found
$esearch_result =~ m|.*<Id>(.*)</Id>.*|s;
my $id = $1;

# Now set up a request to the eFetch program to retrieve our
# paper

my $report = 'abstract'; # we only want to fetch the abstract
my $mode = 'text';       # we want a text output, not XML

my $efetch =
    $utils . '/efetch.fcgi?db=' . $db.
    '&rettype=' . $report . '&retmode=' . $mode.
    '&id=' . $id;
```

```
# Get our paper
my $efetch_result = get($efetch);

print "----------------------\n";
print "--- eFetch Results ---\n";
print "----------------------\n\n";
print $efetch_result . "\n";
```

This program connects directly to NCBI to search PubMed for a paper about the breast cancer gene BRCA1. This is done using two of the NCBI eUtils programs: eSearch and eFetch.

- We first build up the URL for an eSearch query. This is basically how you use the NCBI's eUtils—you define the program that you wish to use and the parameters that you wish to send in a URL.

- Once we have built up our eSearch query URL, we send a HTTP request (using the get() method), which returns an XML response.

- We then use a regular expression to extract the returned id from eSearch, this is the PubMed ID of the paper that was returned from our search. This takes advantage of the fact that the ID is known to be contained between the <Id> and </Id> XML tags.

- Next we prepare our URL for eFetch—this is used to retrieve more information about the paper of interest. In this URL, we define the information that we want to retrieve (the paper's abstract in this case), and how we would like to get it—options are HTML (the default), XML, and plain text (in the example, we select the latter). We also add the id from the eSearch query to denote the paper that we would like to get.

- Finally, we send our HTTP request to the eFetch URL. This returns the paper's abstract to us, in plain text, as requested.

This is the generic approach that you can use to query the NCBI's databases via eUtils. In addition to eSearch and eFetch, there are several other programs available for tasks such as inter-database links. If you are ever likely to need to automatically interrogate the databases at the NCBI, we would most definitely recommend considering the use of the NCBI eUtils. For other database interactions, we often connect to the database directly using DBI, which is introduced below.

3.9 Accessing relational databases using Perl DBI

As discussed in the previous chapter, relational databases are the storage method of choice when you have large quantities of data that you need to arrange and query, as is often the case in bioinformatics. So, it is essential that Perl allow us to easily interact with such databases. This provides a platform for automated database manipulation, which forms the basis of many bioinformatics applications, from high throughput data analysis to web-based tools.

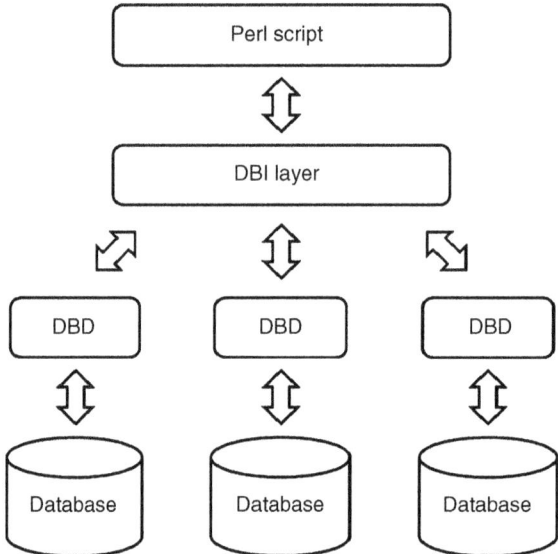

Fig. 3.4 An overview of how Perl DBI works. The DBI layer sits between your Perl code and the database, using a database-specific driver to talk to the database server.

The easiest way of dealing with databases in Perl is with the DBI (database interface) module, which comes as standard with Perl installations. Perl DBI works by adding a database interaction layer to Perl. This means that your Perl program interacts with the DBI layer and from there, the DBI layer talks to your database by using an appropriate database driver. A graphical overview of how Perl DBI works can be seen in Fig. 3.4. The benefit of this approach is that your Perl code is pretty much database independent and should work no matter what the database server is (MySQL, Oracle, PostgreSQL, etc.), the only thing that you would have to change would be the drivers that you use—these are known as DBD drivers (database dependant drivers).

3.9.1 Installing `DBD::MySQL`

Before we move on to look at an example of connecting Perl to a MySQL database, we must first install the DBD driver (`DBD::MySQL`) on our computer so we can use it to connect to our MySQL server.

Windows users

Like `LWP::Simple`, you will need to use PPM to install `DBD::MySQL`, either use the GUI supplied or install it from the command line by issuing this command:

```
ppm install DBD::MySQL
```

Linux

`DBD::MySQL` will be found in your distributions package manager. Simply search for 'dbd-mysql' or '`DBD::MySQL`' and install the relevant packages and their dependencies.

Mac OS

Unfortunately the install of `DBD::MySQL` is not quite as straightforward in Mac OS—it's still quite simple, but takes a little more time to explain. Rather than wasting precious space here explaining how to install it we will direct you to `www.bixsolutions.net` where you can find a short guide to getting `DBD::MySQL` installed in the forums— `www.bixsolutions.net/forum/forum-10.html`.

3.9.2 Connecting to a database

The first thing that we need to do when working with a database in Perl is to set up the connection details and establish a connection to our database so that the rest of our program knows where to send database requests. The program below uses Perl DBI to connect to the MySQL database created in the previous chapter.

```
#! /usr/bin/env perl

use strict;
use warnings;

use DBI;        # Load in the DBI module
use DBD::MySQL; # Load the MySQL driver

# First, the connection details of the MySQL server

my $ds = "DBI:mysql:PCR_experiment:localhost";
my $user = "user_name"; # Our MySQL username
my $passwd = "pass"; # Our MySQL password

# Now to connect to the database...

my $dbh = DBI->connect($ds,$user,$passwd) || die "Can't Connect!";
```

So, when using DBI in your Perl scripts, you need to use two specific modules. The first is DBI itself (the database independent layer that you and your program interact with) and a DBD driver (which is the database specific part that actually talks to our database server). In the example code above, we have loaded in `DBI` and the `DBD::MySQL` driver as we are using MySQL databases.

The next section of the above code lays out the connection details for the database that we are using. The first variable that we establish, `$ds` (short for datasource) is the most detailed, and therefore requires some explanation. The datasource is made up of four colon-separated arguments: the first argument is simple—we are stating that we are using DBI; the second argument is the DBD driver that we wish to use (in our case—MySQL); the third argument is the database name; and the final argument is the database server's host name. So the general syntax of the datasource is as follows:

```
DBI:DBD Driver:Database Name:Database Host
```

Following this, we then specified our username and password to be passed onto the database server. The final line of code actually connects to the database:

```
my $dbh = DBI->connect($ds,$user,$passwd) || die "Can't Connect!";
```

This tells DBI to create a connection to the specified datasource, and store this connection in the variable that we have called $dbh (dbh is short for *database handle*, and is analogous to the file handles introduced earlier). If there is any sort of error on connection to the server, we catch this using a die statement—there is little point in the program continuing if it cannot access the database. If the connection is successful, all subsequent interaction with the database is done through the database handle variable ($dbh).

3.9.3   Querying the database

Now that we have connected to our database, the next thing we might want to do is get some data out of it. We have seen how to do this with SQL queries in the last chapter. Similar functionality is available in Perl. Querying a database using DBI is a three-step process:

◆ Prepare an SQL statement.

◆ Execute the SQL statement.

◆ Retrieve the results.

The following code snippets demonstrate these three steps. Here is the preparation of the SQL:

```
my $sth = $dbh->prepare("SELECT id, sequence
                         FROM Experiment
                         WHERE design_software LIKE ?");
```

There are three things to note from the above code. First is that we are creating a new variable ($sth)—known as the *statement handle*, creating this variable enables us to refer to this specific statement when we run queries against the database. The second thing to observe is that we are using our database handle ($dbh) to prepare our SQL statement—this ties the SQL statement to our database handle (this is important to note in case you ever have more than one database connection open at once—you need to specify the database connection that your SQL statement is for). The final thing to note is the SQL statement itself—this is a standard SQL SELECT statement, with the addition of a question mark (?) at the end of the statement. This question mark is essentially a placeholder that allows us to use different arguments as part of our database query when we run the statement against the database, as we will now:

```
my $query_variable = 'Primer3';
$sth->execute($query_variable);
```

The above line of code runs the SQL query against the database. We use the $sth variable to specify the statement, which is run by the execute function.

We also pass on an argument to the `execute` method—a variable called `$query_variable`—this string variable replaces the `?` in the SQL statement. (Alternatively, we could pass a hard-coded string constant to `execute`.) Using this approach allows us to re-use our prepared SQL statements with different query variables. If we wish to use more than one query variable, we simply put more question marks within our SQL statements and then pass the corresponding number of arguments to the `execute` method, in the right order, separated by commas.

The third and final step in querying a database with DBI is retrieving the results of our database query. An example of this is shown below.

```
while (my @val = $sth->fetchrow_array()) {
    print "id: $val[0], sequence: $val[1]\n";
}
```

The above section of code uses the method `fetchrow_array()` to get results from the database. This returns us results from the database one row at a time, in this case with each field of the database table that we have selected as an element of an array. So, in the case of our SELECT query, we asked to get the `id` and `sequence` fields, so these are returned as the first and second element of the array.

At this point we should issue a warning about the use of SELECT * FROM when accessing a database via Perl—this is a very bad idea. When you use * you have no idea how many fields are going to be returned, and in what order, which makes programming for it impossible, so the advice here is to select each field that you want returned from your table explicitly—never use SELECT *.

There are several other methods to retrieve data returned from the database, as alternatives to `fetchrow_array()`. We will not be going through all of them as this is far beyond the scope of this introduction, but it is worth touching on two other popular methods: `fetchrow_arrayref()` and `fetchrow_hashref()`. Here is an example of how we would retrieve our statement using `fetchrow_arrayref()`:

```
while (my $ref = $sth->fetchrow_arrayref()) {
    print "id: $ref->[0], sequence: $ref->[1]\n";
}
```

The difference between `fetchrow_array` and `fetchrow_arrayref` is that `fetchrow_arrayref` returns each result from your query as a reference to an array, instead of a normal array. This method has the benefit of some small improvements in execution speed and memory use on larger queries due to the use of referenced arrays.

The next example uses `fetchrow_hashref` to retrieve the query results:

```
while (my $ref = $sth->fetchrow_hashref()) {
    print "id: $ref->{id}, sequence: $ref->{sequence}\n";
}
```

This returns each row of results as a referenced hash variable. The benefit of this approach is that you can now call each returned field by their actual field names (in this case `id` and `sequence`) instead of using the numbers that you have been using with arrays—thus making your code a little easier to understand. The unfortunate downside of this method is that it is both slower and uses more memory when compared to the previous two methods.

One thing to note if you plan on using either of these other two methods for returning your data, is that they return referenced variables; therefore, in order to get at the actual data, we must de-reference the variables. In the above example code this is done using the `->` operator, which was introduced back in Section 3.3.

Now that we have the results of our database query, and have looked at several ways to get this result set, we no longer need our connection to the database, so must finish our SQL session and disconnect cleanly from the database. This is done with the following two lines of code:

```
$sth->finish;
$dbh->disconnect;
```

Please note that you must never forget these lines when working with databases in Perl, as failure to cleanly disconnect and close your session with the database could cause problems with the database and possibly even lead to data loss.

3.9.4 Populating the database

We have looked at getting information out of a database, now let's look at the opposite action—inserting data into and updating a database. This procedure is much the same as selecting data from a database. An example program is shown below.

```perl
#! /usr/bin/env perl

use strict;
use warnings;

use DBI;
use DBD::MySQL;

my $ds = "DBI:mysql:PCR_experiment:localhost";
my $user = "user_name";
my $passwd = "pass";

my $dbh = DBI->connect($ds,$user,$passwd) || die "Can't
        Connect!";

# Prepare our insert statement...
my $sth = $dbh->prepare(
 "INSERT INTO Scientist (email, given_name, family_name)
VALUES (?,?,?)"
);
```

```
# Perform a couple of inserts...
$sth->execute('b.flemming@bixsolutions.net','Bob','Flemming');
$sth->execute('e.hunt@bixsolutions.net','Ethan','Hunt');

# Finish up
$sth->finish;
$dbh->disconnect;
```

The top portion of the program follows the same pattern as the previous examples in which we selected data. Even the preparation of the SQL is the same, albeit that we use an INSERT SQL statement here because we are inserting data rather than extracting it. As before, the ? characters get replaced by the specified parameters when we execute the SQL command. The main difference is that we are not retrieving data from the database, so the third step where we get results out is not necessary. To do an UPDATE action on a database table, the process and code is the same, except that we would use an UPDATE SQL statement in the prepare() function.

So that is how you interact with your databases using Perl and DBI. Basically, if you can interact with a database using SQL, you can interact with a database using Perl and DBI. However, now that we are looking at both putting data in as well as getting it out of our databases, we need to consider the inevitable—something going wrong! We therefore need to prepare for and cope with errors whilst carrying out operations on a database.

3.9.5 Database transactions and error handling

The normal way of handling errors that occur when working with databases is by using the concept of *transactions*, as discussed in the previous chapter. The same approaches can be used in Perl. However, by default, Perl DBI makes permanent changes to the database on the completion of each line of code, so we have to add a small bit of extra code and error handling to ensure that the intended database transactions are carried out correctly and safely. Below is a sample program that shows how to deal with correct transaction handling.

```
#! /usr/bin/env perl

use strict;
use warnings;

use DBI;
use DBD::MySQL;

my $ds = "DBI:mysql:PCR_experiment:localhost";
my $user = "user_name";
my $passwd = "pass";

my $dbh = DBI->connect($ds,$user,$passwd) || die "Can't Connect!";
$dbh->{'AutoCommit'} = 0; # Turn off AutoCommit
```

```perl
my $sth = $dbh->prepare(
 "INSERT INTO Scientist (email, given_name, family_name)
  VALUES (?,?,?)"
);

# Now let us check for errors as we insert data
eval {
     $sth->execute('b.flemming@bixsolutions.net','Bob','Flemming');
     $sth->execute('e.hunt@bixsolutions.net','Ethan','Hunt');
};

# Check for any errors in the above
if ($@) {
     $dbh->rollback;
} else {
     $dbh->commit;
}

# Finish up
$sth->finish;
$dbh->disconnect;
```

This performs the same actions as the previous INSERT example, but now has error and transaction handling added in to ensure that our database activities are carried out cleanly. There are three extra additions to the code, the first is the setting of the AutoCommit option for DBI—basically we are turning it off so that we explicitly have to tell the database server to commit our changes to the database (by default DBI would commit immediately). The second and third changes are where we use the eval method of error trapping to run our INSERT statements. If the statements are executed without any problems, we then tell the database to commit the changes to storage, but if there are errors we rollback the current session so that everything we have just tried to do does not impact the database, therefore keeping in line with standard transaction handling in databases.

That concludes the basics of using Perl DBI, but there is much more to learn. For more details we would recommend the DBI website (dbi.perl.org) and the Perl DBI CPAN page (search.cpan.org/~timb/DBI/DBI.pm).

3.10 Harnessing existing tools

When creating programs in Perl—or any other language for that matter—there will always be tasks that are common between your scripts. If these are things that are bespoke and only done within your group then you have got some subroutine programming to do. But often, if the activity is quite common to programming, or even common in a specific field like bioinformatics, there is a chance that someone will have done this already and possibly released some code on the Internet.

This is the great thing about programming within the bioinformatics community—many people like to share their ideas and techniques with each other when they can. If you ever need to do something, but are not quite sure how to get started, the answers are probably only a quick search away. There are many programming blogs and forums on the Internet where people dispense advice and give help, tips, and tricks to aid less experienced programmers. In addition to the blogs and forums, there are also code repositories and toolkits that can potentially supply you with ready-made tools to perform your task. In particular, we would draw your attention to two projects that will most definitely help you along your way: CPAN and BioPerl.

3.10.1 CPAN

The Comprehensive Perl Archive Network (CPAN) is a central worldwide repository for Perl code. Within CPAN you will find thousands of Perl modules that you can download and install on your system—it is possibly the richest library of extensions to a programming language available today. CPAN is one of the reasons that Perl is so popular—it can make developing software with Perl much more efficient.

The basic premise of CPAN is that people supply open source Perl modules to CPAN for anyone to use within their projects. You can simply download and install these on your system, and all of the functionality of the modules you have downloaded will be available to your install of Perl on your system, as discussed back in Section 3.4.1. The variety of modules available is far too large to even try to list here, so the best advice we can give is to head over to the CPAN website (www.cpan.org) and read the CPAN FAQ to try get yourself comfortable with how CPAN works and what is available.

When you come to install your first CPAN Perl modules, we first recommend reading the 'How to install CPAN modules' guide (www.cpan.org/modules/INSTALL.html), then, once you have understood that, follow these small bits of advice:

- If you are Windows user, your distribution of Perl (ActiveState Perl) comes with PPM (Perl Package Manager) as used previously in this chapter. To use this program, simply type ppm at the Windows command line and hit Enter—this will launch an easy to use graphical application via which you can install Perl modules on your system. If you cannot find the particular module you are looking for within the lists, look at the install documents for the module on CPAN—there might be a special repository that you need to point PPM to. If this is not the case (there are no special instructions for Windows), then use the cpanm command-line tool described in the 'Quick Start' section of the 'How to install CPAN modules' guide.

- If you are a Linux user, first look through the package management system that came with your distribution—you will more than likely find the vast majority of CPAN modules available and only a click or two away. If the module you are looking for cannot be found there, use the cpanm command-line tool described in the 'Quick Start' section of the 'How to install CPAN modules' guide.

- If you are a Mac user your best (and only, really) choice it to use the `cpanm` command-line tool described in the 'Quick Start' section of the 'How to install CPAN modules' guide. We have supplied a write up of how to get `cpanm` installed, and a well-configured environment for Perl on the `bixsolutions.net` forums— `www.bixsolutions.net/forum/thread-68.html`. However, if you are using Perlbrew, you do not need to do this as Perlbrew does this for you automatically.

3.10.2 BioPerl

BioPerl is a set of Perl libraries and subroutines similar to the ones that you would find in CPAN as described above. However, as the 'Bio' in BioPerl might suggest, these libraries are designed specifically for use in bioinformatics tasks. Learning about the functionality of BioPerl and understanding how to harness it will save you massive amounts of work over time. Just take a look at the example code below (adapted from the BioPerl wiki) where we create a DNA sequence object, print some details about it, and then save the DNA sequence in a FASTA formatted text file:

```
#! /usr/bin/env perl

use strict;
use warnings;

use Bio::Seq;
use Bio::SeqIO;

# create a sequence object of some DNA
my $seq = Bio::Seq->new(-id => 'testseq', -seq => 'CATGTAGATAG');

# print out some details about it
print "seq is ", $seq->length, " bases long\n";
print "revcom seq is ", $seq->revcom->seq, "\n";

# write it to a file in Fasta format
my $out = Bio::SeqIO->new(-file => '>testseq.fasta', -format =>
          'Fasta');
$out->write_seq($seq);
```

As you can see, the BioPerl functions `Seq` and `SeqIO` do a lot of the tedious work for you, making your programs easier to read and—most importantly— faster to write. With the basic understanding of Perl provided in this chapter, you should be ready to start exploiting BioPerl in your projects. To get started, visit the website (`www.bioperl.org`).

3.10.3   System commands

As well as using Perl programming libraries in your programs, it is also possible to integrate entire other command-line driven programs into your Perl programs. This is done through the use of the `system()` function. With this function we can call any other program or command that it's possible to call from the operating system command line. This function is of great use for pipelining several other programs together—a very common activity in bioinformatics—or for incorporating algorithms that have been tried, tested, and optimized for speed (e.g. BLAST).

3.11   Object-oriented programming

Object-oriented programming (OOP) is a design philosophy used in many programming languages. The examples of code and techniques we have shown you thus far are what you would call *procedural programming*—in that your program pretty much runs a list of tasks from the top to the bottom of the code. An object-oriented program can be viewed as a collection of interacting objects, where each object is capable of receiving messages, processing data, and sending messages to other objects. In many ways, OOP is much closer to the way we deal with the real world. It is particularly convenient when dealing with things that we know are objects in the real world, like proteins or patients, or where we have graphical objects such as windows, buttons, and scroll bars. However, in some applications it can be tricky to understand how to benefit from OOP, especially if you already have a lot of experience in procedural programming.

Classes

The first thing we need to understand when looking at OOP is classes; a class is a way of modelling some data and functions that can be applied to that data. At its simplest, a class can be thought of as a user defined data type, a bit like a Perl array, that can hold various pieces of data together in a single easily referenced entity. However, unlike an array, a class can also contain units of program code, called *methods*, that operate on the data inside.

One such example for a simple class that we could start with is that of a dog. A very basic dog class could have a name (a string variable), a breed (another string), and an age (an integer)—these would be the variables that make up the data within the class. The class could also have a number of methods, such as one to print out a description of the dog, and another to calculate the dog's equivalent age in human years.

The diagram in Fig. 3.5 could represent the design for our dog class. The style of this diagram will be familiar from Chapter 2, as it uses the same UML (Unified Modelling Language) approach.

Inheritance

Now, what if we wanted to make an object that represents a cat? We could model this in exactly the same way as we did the dog—create a cat object that has a

Dog
name: String
breed: String
age: Integer
info()
human_age()

Fig. 3.5 A UML representation of our Dog class. The top box contains the name of the object, the second box contains the variables that are repeated within the class, and the final box lists the methods (functions) that can be called on the object.

name, breed, and age, plus the same few methods, but then what happens when we need to model yet another animal—there is a lot of potential for repetition here. This is where another core concept of OOP comes in: *inheritance*.

Inheritance is used when you have several classes that have many (but not all) things in common, as it allows you to create a *super-class* that represents the common features that you are trying to model. Our other classes then inherit this behaviour from the super-class, meaning that you don't need to repeat yourself when it comes to describing the other objects. In our example, an appropriate super-class would be an `Animal` class that both the `Dog` and `Cat` class can inherit characteristics from (depicted in Fig. 3.6). Then, at any point the specifics of a particular animal differ from the base class (`Animal`), we simply *override* the given method or attribute in the child object.

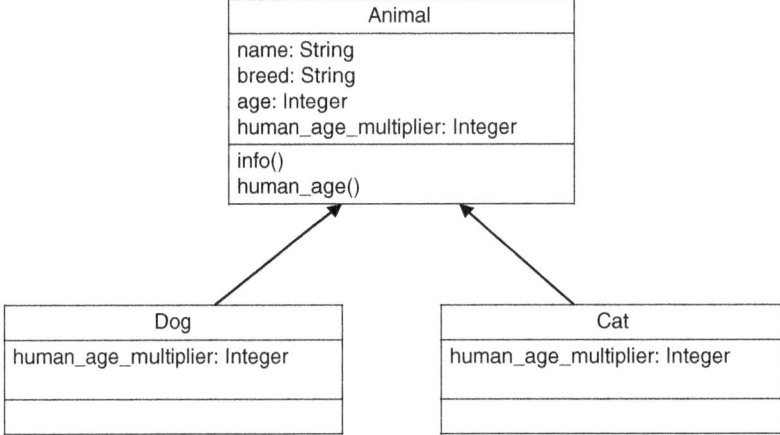

Fig. 3.6 A UML representation of how we can model our Dog and Cat classes by using a single Animal "super class" that both the Dog and Cat inherit their attributes and methods from. Note however that `human_age_multiplier` attributes are also found on both the Dog and Cat models as this is different for each type of animal, so these override the value found on the generic Animal object.

From classes to objects

So far we have only mentioned classes—what about the objects that give OOP its name? The term *object* is used to describe an instantiation or *instance* of a class. Consider our example. The `Dog` class can be considered as a blueprint for objects of a particular type—dogs. We can use this class to create any number of dog objects, each of which will have all the key characteristics of a dog, but also has its own individual properties (e.g. name and breed). To put it another way, a class could be considered as some kind of Perl type (just like a string). There is only one string type in Perl, but a single program can contain a great many strings. Similarly, we need only define one dog class to generate any number of different dog objects within a program.

Traits and shared behaviour

The final OOP concept we are going to introduce in this section is that of *traits*. These can also be known as *roles*, and sometimes people can refer to them as *mixins*, but they are something slightly different (we introduce and describe mixins later, in Section 6.5.2, when we introduce the Ruby programming language).

Traits are in fact a collection of methods that are *consumed* by a class to describe an interface or ability that the class will exhibit—a couple of examples of this type of behaviour would be:

- A 'Printable' trait on a class could enforce that the class must declare a method called `print` or `to_string`, so its content can be viewed at the command line.

- An 'Equal' or 'Eq' trait on a class could enforce that the class must create an `equal` method to compare itself to another instance of a class and return true/false depending on the outcome.

This is a fairly simple concept, but used properly it enables code- re-use throughout your applications as many different, completely un-related classes can share functionality and code through common traits.

3.11.1 Object-oriented programming in Perl using Moose

To see OOP in practice, we are going to look at some simple example programs in Perl. Perl is not an inherently object-oriented language, but OOP is possible in Perl (especially if we install a package for this) and it makes sense to stick with Perl for the moment so that the differences between OO and procedural programs are clear. We introduce some truly object-oriented languages in Chapter 6.

Although OOP is possible with Perl without the use of any external modules, it is not friendly or straightforward, so instead we will be using Moose (`moose.perl.org`), a popular module that makes Perl OOP much easier and cleaner.

Before getting started, you will need to install Moose. It can be installed through PPM (if you are on Windows), your distribution's package manager (on Linux), or via `cpanm` (on OS X). The package to look for is simply called `Moose`.

Creating our first objects

We are now going to use Moose to turn our dog and cat example into reality within Perl. So, our first class was the `Dog`. Here is a complete (and heavily

commented) example program that defines the Dog class and shows how you
would interact with it for both running methods and setting variables.

```perl
#! /usr/bin/env perl

# BEGIN: Dog object.
package Dog; # this says we are working in the namespace 'Dog'
            # (i.e. declaring a class called Dog)
use Moose;   # tell Perl to use Moose
            # - this also switches on 'strict' and 'warnings'
# declare variables
has 'name' => (
  is       => 'rw',  # is readable/writable
  isa      => 'Str', # is a string variable
  required => 1      # is required when we create an instance of
                     # a dog
);

# more variables (optional as we haven't said they're required)
has 'breed' => (is => 'rw', isa => 'Str');
has 'age'   => (is => 'rw', isa => 'Int');

# declare methods
sub info {
  my $self = shift;
  return
    $self->name." is a ".$self->breed." and is ".$self->age."
               years old.";
}

sub human_age {
  my $self = shift;
  return $self->age * 7;
}

no Moose;
# END: Dog object

# The following is only needed if we're declaring a class
# in the same file as our 'regular' code.
package main;

# now, create an instance of the class Dog...
my $sam = Dog->new(
  name  => 'Sam',
  breed => 'Yorkshire Terrier',
```

```
    age    => 5
);


# create another instance of Dog, note that you don't have to
# populate all
# the variables when you call new() - only the required ones
my $charlie = Dog->new(name => 'Charlie');
$charlie->breed('Cocker Spaniel');
$charlie->age(3);


# now call one of the methods on both instances
print $sam->info . " \n";
print $charlie->info . " \n";
```

This is a complete standalone program, in which there is quite a lot going on. Let's take a look at the first section of code (note, we've removed the comments/annotations to make the code slightly more clear and readable):

```
package Dog;
use Moose;

has 'name'  => (is => 'rw', isa => 'Str', required => 1);
has 'breed' => (is => 'rw', isa => 'Str');
has 'age'   => (is => 'rw', isa => 'Int');

sub info {
  my $self = shift;
  return
    $self->name." is a ".$self->breed." and is ".$self->age."
                 years old.";
}

sub human_age {
  my $self = shift;
  return $self->age * 7;
}

no Moose;
```

In this first section of code, we describe the `Dog` class. The beginning line uses the `package` declaration—this is a namespace command and means that all variables and methods declared after it will be tied to that namespace (and thus not clash with other variables or functions in Perl itself, or your code if used as part of a larger codebase). The second line says that we're using Moose in this namespace.

Following the opening two lines we then have the variable declarations, that is:

```
has 'name' => (is => 'rw', isa => 'Str', required => 1);
```

These lines use some Moose magic to save us a lot of boilerplate code for setting and getting (and checking) variable/attribute contents. The keyword in use on these lines is `has`—this Moose helper states that our class has an attribute (or *property*) called whatever is passed to it. The rest of the line is the options hash that is passed into the `has` declaration, these options are described in detail in Table 3.6.

Following the attribute declarations, we have the method declarations for our object. These should be familiar as they are standard Perl subroutines; the only slight difference you may notice is they all start with the following line:

```
my $self = shift;
```

This is standard practice when doing OOP in Perl, what this does is sets the current instance of the object to the variable `$self`—so that it can be used within the subroutine.

The final line of the code (`no Moose`) says that we're finished using Moose in this namespace and tells Perl to remove the Moose helpers from its memory. This is the final line in the declaration of our `Dog` class.

The next section of code is where we create actual instances (objects) of our `Dog` class. Here is that code again, with comments removed:

```
package main;

my $sam = Dog->new(
    name  => 'Sam',
    breed => 'Yorkshire Terrier',
    age   => 5
);

my $charlie = Dog->new(name => 'Charlie');
$charlie->breed('Cocker Spaniel');
$charlie->age(3);

print $sam->info . " \n";
print $charlie->info . " \n";
```

The first line of code is another `package` (namespace) declaration. As we created our `Dog` class in the same file we moved the namespace earlier in the code (to `Dog`), so we are simply moving us back to the default Perl namespace, `main`. This is really just a technicality that we shouldn't get distracted by too much—just know that it is good practice to do this if you declare classes in the same file as your procedural code (the code that makes use of your classes).

The following line creates the first instance of our `Dog` class—a virtual 5-year-old Yorkshire Terrier called Sam. This is done by calling the special method `new` on the `Dog` class—this is a standard Perl function (also used in other OO languages) for creating a new instance of a class, and you simply pass it the required (and optional) arguments it needs to create a new instance of the class in question. The

Table 3.6 Basic options for the Moose `has` command

Option	Description	Accepted Values
`is`	Specifies how the attributed can be accessed, i.e. is it a value that can be set or read by external calls?	`'rw'` – read/write `'ro'` – read only `'wo'` – write only
`isa`	Specifies what type of value to expect when setting the attribute. This type is also checked (by Moose), so if an attribute is supposed to be an integer, and it gets passed a hash, an error will be raised.	`'Int'` (integer) `'Num'` (floating-point number) `'Hash'`, `'HashRef'`, `'Array'`, `'ArrayRef'`, `'Bool'` (boolean) and many more...
`required`	Is this attribute required when the `new` method is called to create an instance of the class?	`1` – yes `0` – no
`default`	This is not shown in the code example, but if you want to specify a default value for the attribute, you can do it with this argument.	Any value that satisfies the `isa` declaration.

next line creates another instance of our `Dog` class, but with only the minimum amount of data needed to create a new `Dog`. We then pass in the rest of the values in the following two lines. These two examples show you the different ways you can create instances of classes and set their attributes (either all in the `new` declaration, or later).

The final two lines simply call the `info` method on both of the dog objects that we have created. This method (declared as part of the class) prints a string of information about the specified object to the terminal.

In summary, we have produced two objects, each of which is a particular instance of `Dog`, and we are able to manipulate these objects. However, this demonstrates only a small portion of the OOP concepts that we considered earlier.

Inheritance

Next up we are going to look at how to implement inheritance with Moose. The code below follows our example, creating a generic `Animal` class and then making `Cat` and `Dog` classes that inherit shared behaviour from it.

```perl
#! /usr/bin/env perl

# Animal class
package Animal;
use Moose;

has 'name'  => (is => 'rw', isa => 'Str', required => 1);
has 'breed' => (is => 'rw', isa => 'Str');
has 'age'   => (is => 'rw', isa => 'Int');
```

```perl
has 'human_years_multiplier' => (is => 'ro', isa => 'Int',
    default => 1);

sub info {
  my $self = shift;
  $self->name." is a ".$self->breed." and is ".$self->age."
            years old.";
}

sub human_age {
  my $self = shift;
  $self->age * $self->human_years_multiplier;
}

# Dog class
package Dog;
use Moose;
extends 'Animal';

has 'human_years_multiplier' => (is => 'ro', isa => 'Int',
    default => 7);

# Cat class
package Cat;
use Moose;
extends 'Animal';

has 'human_years_multiplier' => (is => 'ro', isa => 'Int',
    default => 5);

# End of class declarations

no Moose;
package main;

my $sam = Dog->new(
  name  => 'Sam',
  breed => 'Yorkshire Terrier',
  age   => 5
);

my $tom = Cat->new(
  name  => 'Gorbypuff',
  breed => 'Persian',
  age   => 5
);
```

```
print $sam->info . " (Which is " . $sam->human_age . " in
                    human years).\n";
print $tom->info . " (Which is " . $tom->human_age . " in
                    human years).\n";
```

As you can see, this code example is not much longer than the previous one where we only described a single class, whereas in this example we declare the generic `Animal` class, and then two child classes (`Cat` and `Dog`) that inherit their behaviour from `Animal`. This is an obvious example of the efficiencies made possible by inheritance.

The actual syntax and program code in use here is very similar to before but with two subtle additions (one syntactic and on conceptual); the first (syntactic) addition is the use of the Moose keyword `extends` (used in both the `Cat` and `Dog` classes). The `extends` function in Moose is used for inheritance; it takes a class name as an argument, and uses that class as a base for constructing the new class that your code is currently building. The other (conceptual) addition we show is the overriding of methods and attributes in child classes—in this example the `human_years_multiplier` attribute is overridden in both child classes as they have different values[4] in the different animals (from the parent class). It is also possible to add completely new methods to the child classes that are specific to them (and only them, so they do not belong in the super-class).

Using Moose roles (traits)

In order to show you the usefulness of roles in Moose, we are going to step away from our pet examples and show you an example of the utility of Moose roles using a class to represent a DNA sequence. This example builds two roles: `Eq` and `Printable`, and applies them to a `DNA` class.

```
#! /usr/bin/env perl

# Define an 'Eq' role.
package Eq;
use Moose::Role;
requires 'equal_to';

sub not_equal_to {
  my ( $self, $other ) = @_;
  not $self->equal_to($other);
}

# Define a 'Printable' role.
package Printable;
use Moose::Role;
```

---

4 We are aware that the relationship between an animal's age and the human equivalent is not really linear, but it serves our purpose for this simple example.

```perl
requires 'to_string';

# Define a DNA class that consumes these roles
package DNA;
use Moose;
with 'Eq','Printable';

has 'sequence' => ( is => 'rw', 'isa' => 'Str' );

sub equal_to {
  my ( $self, $other ) = @_;
  $self->sequence eq $other->sequence;
}

sub to_string {
  my $self = shift;
  $self->sequence;
}

no Moose;
package main;

my $a = DNA->new(sequence => 'ACTG');
my $b = DNA->new(sequence => 'AAAA');

if ($a->equal_to($b)) {
  print "They are equal!\n";
} else {
  print "They are NOT equal!\n";
}
```

As we said, this code declares two roles and a class. Let's take a look at the first role declaration in more detail.

```perl
package Eq;
use Moose::Role;
requires 'equal_to';

sub not_equal_to {
  my ( $self, $other ) = @_;
  not $self->equal_to($other);
}
```

This first section of code declares the role called `Eq`. We know it is a role as it uses `Moose::Role`, rather than just plain `Moose`, we also have a new Moose keyword `requires`. This says that any class that consumes this role must provide a method called `equal_to`. The `Eq` role also defines a `not_equal_to` method that uses the required `equal_to` method—meaning that the consuming code only has to implement the one method.

The final role declared in the example is the `Printable` role. This role is what is known as an *interface* role, in that it does not provide a consuming class with any additional functionality, it just enforces a specific set of methods or interfaces. This programming pattern is quite useful in larger codebases as it can enforce method naming/construction consistency between classes.

In the final section of the program, we declare a class called DNA, which consumes both the `Eq` and `Printable` roles (via the with keyword), and provides the necessary `equal_to` and `to_string` functions (as defined by the roles).

As you can see, there is a lot of re-usable code in this example. Many other classes over a large codebase could easily make use of these generic roles to define interfaces and enhance abilities. This is the power of roles (or traits).

Tidying up code with MooseX::Declare

Although powerful, Moose can be quite verbose when there is a lot going on in one program, and this can affect readability. To mitigate this there is a Moose extension called `MooseX::Declare` (available on CPAN, alongside many other Moose extensions—look for MooseX in the name) that gives Moose a much more direct, compact, and declarative syntax.

Let us re-visit some of our earlier code (the pet example, but with the additional use of a role) but use the `MooseX::Declare` style of syntax.

```perl
#! /usr/bin/env perl

use MooseX::Declare;

role Printable {
  requires 'to_string';
}

class Animal with Printable {
  has 'name'  => (is => 'rw', isa => 'Str', required => 1);
  has 'breed' => (is => 'rw', isa => 'Str');
  has 'age' => (is => 'rw', isa  => 'Int');
  has 'human_years_multiplier' => (is => 'ro', default => 1);
  method human_age {
    $self->age * $self->human_years_multiplier;
  }
  method to_string {
    $self->name." is a ".$self->breed." and is ".$self->age." years
              old.";
  }
}
```

```
class Dog extends Animal {
  has 'human_years_multiplier' => (is => 'ro', default => 7);
}

class Cat extends Animal {
  has 'human_years_multiplier' => (is => 'ro', default => 5);
}

my $sam = Dog->new(name => 'Sam', breed => 'Terrier', age => 5);
my $tom = Cat->new(name => 'Gorbypuff', breed => 'Persian', age => 5);

print $sam->to_string . "\n";
print $tom->to_string . "\n";
```

And now let's revisit the DNA example used to demonstrate roles.

```
#! /usr/bin/env perl

use MooseX::Declare;

role Eq {
  requires 'equal_to';

  method not_equal_to($other) {
    not $self->equal_to($other);
  }
}

role Printable {
  requires 'to_string';
}

class DNA with (Eq, Printable) {
  has 'sequence' => ( is => 'rw', isa => 'Str' );
  method equal_to($other) {
    $self->sequence eq $other->sequence;
  }

  method to_string {
    $self->sequence;}
}

my $a = DNA->new(sequence => 'ACTG');
my $b = DNA->new(sequence => 'AAAA');
```

```
if ($a->equal_to($b)) {
  print "They are equal!\n";
} else {
  print "They are NOT equal!\n";
}
```

These code examples need little explanation—all of the concepts and logic are the same as in the previous example code, but the syntax is clearer and easier to follow. We would just like to note a few key things that `MooseX::Declare` provides:

- `class` and `role` keywords, with automatic namespacing (no need for the `package` command anymore).

- Inheritance (`extends`) and role consumption (`with`) declarations are given on the same line as the `class`/`role` declaration.

- The `method` keyword is used for method/function declaration inside classes/ roles. This also automatically assigns the current class instance to `$self` in these methods.

This style of code is not to everyone's liking. Some people have complained that it is not very 'Perl-like' as it makes OOP in Perl look more like other programming languages (you can judge for yourself in Chapter 6). On the other hand, some people like this style of syntax exactly for those reasons. Ultimately it is your choice, but we like it and will be using `MooseX::Declare` in some code examples later on in the book.

Going further with Moose and OOP

The simple examples above have demonstrated how object-oriented programming can be done in Perl, and hopefully helped you see why such an approach to programming is useful in terms of code re-use and supporting good programming practice. Although the concepts of OOP are simple, beginners often struggle to understand how they might take advantage of it in a particular application. The best way to learn is by experimenting and studying examples. For further reading, we would recommend the Moose manual (`search.cpan.org/~doy/Moose/lib/Moose/Manual.pod`), **Moose Cookbook** (`search.cpan.org/~doy/Moose/lib/Moose/Cookbook.pod`), and *Modern Perl* (`modernperlbooks.com/books/modern_perl`), which includes a chapter about object-oriented programming using Moose.

3.12 Summary

That concludes our introduction to programming in Perl. Clearly, we have not been able to cover every facet of the language, but we have covered the fundamentals and highlighted those features of Perl that make it such a popular choice for writing bioinformatics software. Perl is not the only language used in bioinformatics—R, Python, Ruby, and Java are also very popular. We will take a look at these languages in subsequent chapters, where you will realize that, despite differences in emphasis and syntax, the fundamental programming principles are common among all these languages. If you master these principles in Perl you will be well placed to begin programming in most other languages. We shall return to Perl in Chapter 5 where we use it as a tool for programming for the web, and in Chapter 6 where it is used to demonstrate key software engineering principles.

References

Stubblebine, T. (2007). *Regular Expression Pocket Reference: Regular Expressions for Perl, Ruby, PHP, Python, C, Java and .NET*. O'Reilly: Sebastapol, California, USA.
Tisdall, J. (2001). *Beginning Perl for Bioinformatics*. O'Reilly: Sebastapol, California, USA.
Tisdall, J. (2003). *Mastering Perl for Bioinformatics*. O'Reilly: Sebastapol, California, USA.
Wall, L., Christiansen, T., & Orwant, J. (2000). *Programming Perl*. O'Reilly: Sebastapol, California, USA.

CHAPTER 4

Analysis and visualisation of data using R

Mathematics and statistical processing of data, and the visualization of the results of such processing, are a key part of many bioinformatics applications. The most obvious applications are in domains such as transcriptomics and metabolomics, where there is a wealth of quantitative data, but even when the acquired data is non-numeric, as in sequence analysis, there is a necessity to perform statistical analysis to determine, for example, the significance of the results acquired.

In recent years the statistical programming language, R, has become the tool of choice for such analysis. The aim of this chapter is to introduce R, to explain how it can be used to analyse and visualize biological data, and to begin to explain how R functionality can be integrated into bioinformatics solutions. Naturally, in a chapter of this length it is not possible to cover every facet of R. To get some idea of what that might involve, the PDF reference manual provided with R runs to more than 3,000 pages, and that doesn't even cover the many third-party add-on packages available. The intention of this chapter is therefore to introduce the key concepts of R, to provide a starting point, and to illustrate functionality that is most applicable to bioinformatics. We particularly emphasize the use of R as a software development tool as it can be integrated with Perl and MySQL, facilitating the creation of very sophisticated bioinformatics software. Directions to further information, such as R's useful help system, are provided to allow you to then develop your understanding in whatever way your particular applications require.

In terms of the data analysis techniques covered here, this also cannot be exhaustive because the bioinformatics community has spawned hundreds of algorithms for data analysis, and more are being published every month. For this chapter, we have chosen examples of analyses that can be applied to a wide range of data sets, and to which R is particularly well suited. We also emphasize the commonalities between different types of data set; although bioinformaticians often classify themselves according to the type of data they work with, experience shows that many analysis methods can each be applied to data from a range of different analytical platforms. If you do want to find out about a specific algorithm, it should be apparent where to find information about it once you've read this chapter, and quite possibly there is an R package out there to do exactly what you want.

Building Bioinformatics Solutions. Second Edition. Conrad Bessant, Darren Oakley and Ian Shadforth.
© Conrad Bessant, Darren Oakley, and Ian Shadforth 2014. Published 2014 by Oxford University Press.

One area we have purposefully avoided including in this chapter is univariate statistical methods, such as significance tests and ANOVA. The main reason for this is that such techniques are dealt with at length in most other introductions to R, so we would refer you to those. Furthermore, if you only wanted to do this type of analysis you might choose a less flexible but more user-friendly software package, such as Microsoft Excel, GenStat, or Statistica. Such packages are very capable tools for numerical data analysis and visualization, but they suffer from three key limitations. First, they are not particularly flexible—adapting or adding a new algorithm or graph type may be possible, but it is not always easy. Second, integration with other tools is not straightforward—if we want to incorporate some data analysis functionality into a web-based analysis tool, this is not the way to go. Finally, these packages are designed around the concept of a user sitting down and performing the analysis via a graphical user interface. This is fine for small analyses, but in bioinformatics, where data sets can be large and analysis repetitive, there is often a need for automation. Some packages, such as Microsoft Excel, have built-in programming languages for automation, and as we saw in Chapter 3, programming languages such as Perl are ideal for automating repetitive procedures on large amounts of data. Such programming languages are, however, designed to be general purpose, and lack the native data structures, built-in functions, and often the performance to tackle mathematical problems efficiently. In a general-purpose language like Perl, all but the simplest mathematical procedures have to be implemented from a fairly low level, which can lead to long development times and concerns over the veracity of the implementation, and ultimately programs that are painfully slow when applied to large data sets.

4.1   Introduction to R

Several software packages have been developed to fill the gap between programming languages and point-and-click analysis packages, and R is foremost among these in the bioinformatics community. R is an entirely free, open source package, which has at its core an implementation of the statistical programming language S, which is also the basis for the commercial S-Plus software package. R supports the installation of add-ons, called *packages*, to extend its basic functionality into specialist areas, and is available for a range of operating systems, including Windows, Linux, and Mac OS. R has become the data analysis tool of choice in bioinformatics partly due to the open-source and platform-independent ethos of bioinformatics, but also because many high quality add-on packages have been produced by the bioinformatics community (for example, the BioConductor packages described in Section 4.3.1). The popularity of R in the bioinformatics community is therefore perpetuated, as anyone planning to release data analysis tools to the community will generally want to do this as an R package because there is a large potential audience with R installed, so curious users will not be put off by having to change operating system or buy commercial software in order to try it out.

It is for these reasons that we have chosen R as the platform for the numerical data analysis part of this book.

4.1.1 Downloading and installing R

The process of getting hold of R may seem a little less slick than you might be used to for other software, but it is nevertheless a fairly painless process. R's home on the web is `www.r-project.org`. Here you can find some background information about R, links to manuals, and other documentation. The software itself is hosted on the Comprehensive R Archive Network (CRAN)—a worldwide network of servers from which R, R packages, and other related files can be downloaded. Clicking on the CRAN link on the R homepage brings up a list of CRAN servers, from which it makes sense to select the one closest to you. There is a range of different forms in which to download R, but for beginners the pre-compiled binary setup program for the base R system is the file to go for. This is available via the front page of any of the CRAN mirrors simply by clicking on your operating system in the 'Download and Install R' pane.

- For Windows, this file will be called something like `R-3.0.1-win.exe`, depending on the version number (in this case version 3.0.1). Executing this file launches a familiar Windows Setup Wizard that will install R on your computer. The default installation options should be fine for most people. Once installed, you can start R by clicking on the icon on the desktop, or in the Start menu or Start screen, just like any other Windows application.

- If you are running Mac OS, the file you need will be called something like `R-3.0.1.pkg` (again this depends on the current R version number). Download and run this installation package—just accept the default options and you will be ready to go. Note that the interface through which you can interact with R will be installed into your Applications folder.

- Installing R on Linux is ideally done using the package manager that came with your Linux distribution (look for the `r-base` package). Package managers are quite different between the various flavours of Linux, so we don't have the space here to go into how you would install R on each type of Linux. If you have problems using your package manager, or R is not available there, pre-built installation files can be downloaded from the R website for some of the more common Linux distributions. These can be found in the same section as the pre-built Windows and Mac OS files. You will also find some basic install instructions with the files to help you. Once installed, R can be started simply by typing `R` at the Linux command prompt. All subsequent commands entered will then be processed by R, until you execute the `quit()` function.

In use, R is very similar across all three platforms, so we won't hear much more about specific operating systems in this chapter. However, add-on packages that have been written using features specific to a particular operating system may not work properly on other systems.

4.1.2 Basic R concepts and syntax

The first thing people notice after installing and firing up R is the rather archaic looking user interface, the R *Console*. It is through this console that most interactions with R take place and, as we will see, while it may not look as inviting as commercial software, it is actually the lack of a graphical user interface that gives R a lot of its power. R's user interface is another example of a command-line interface—commands are entered into the console at the prompt (in this case a > symbol), and results are returned in the console window, or in a separate window in the case of graphical output. Depending on your operating system and particular installation of R, the console may exist within something called the *RGui*. The Gui part of the name stands for graphical user interface, which infers some level of sophistication, but at the time of writing RGui only provides very basic features, most of which have command line equivalents.

Because R is essentially a programming language, many of the concepts are the same as in other programming languages such as Perl. So, for example, there are variables and there are functions. One difference with R is that, as well as writing programs, we can work with these variables and execute these functions interactively in real time, right there in the console. This is useful for prototyping or doing a one-off analysis. As an example, let's consider a right-angled triangle, with sides of length x and y, and hypotenuse of length z. We can assign specific values to the variables of x and y by typing the following two commands at the R command line (hit Enter after each command):

```
x <- 3
y <- 4
```

The backwards arrow formed from the less than character (<) and the minus character (-) indicates the flow of data, that is values on the right are being assigned to variables on the left. (Values can also be assigned to variables using the = character, as in Perl, but <- is more commonly seen in R.) We can check the values assigned to these variables simply by typing the variable name at the command line and then pressing Enter. An example of this action is shown below.

```
> x
[1] 3
```

As expected, the variable x contains the value 3. The [1] part of R's output indicates that 3 is the first element in x. This is not very important in this case as x only has one element, but if the output was a list of values spanning more than one line then a number would appear in square brackets at the start of each line to indicate the element number of the first value on that line.

Having assigned values to the x and y variables we can manipulate them using operators and functions, just like in any other programming language. Also like other languages, we can append comments to our commands to make them more easily understandable. In R (like Perl) such comments must be preceded by the # character, which causes R to ignore everything that follows on that line. An example session with R, using the x and y variables defined above, is shown below.

```
> x+y                  # add the two variables together
[1] 7
> x^2                  # the ^ operator raises a number to a
                       # power (in
                       # this case the power of two)
[1] 9
> sqrt(y)              # the built-in sqrt function returns
                       # the square
                       # root of the number passed to it
[1] 2
> z <- sqrt(x^2+y^2)   # calculate the length, z, of the
                       # hypotenuse of a
                       # right angle triangle using
                       # Pythagoras' theorm
> z                    # see what value z has been set to
[1] 5
```

An important point to make at this stage is that R, like Perl and other programming languages, is case sensitive so a variable Y would be distinct from the variable y—each could have a different value. Similarly, typing SQRT instead of sqrt would cause R to reply with Error: could not find function "SQRT". To avoid this kind of confusion, function names are generally all lower case, as are most variable names, but this custom is not enforced and sometimes there are legitimate reasons for using upper case. To avoid confusion, refer to our general advice in Chapter 1 (Section 1.6.3).

All the variables (and other objects that we will learn about later) that you define are stored by R in its *workspace*. The content of this workspace can be seen using the objects command:

```
> objects()
[1] "x" "y" "z"
```

If you try to quit R (by typing quit() in the console, or closing the R window), you may (depending on your particular installation of R) be asked if you would like to save the workspace image. If you do, you will be able to continue where you left off with these variables next time you start R. Conversely, objects can be removed from the workspace using the remove() function:

```
remove(x)                  # remove variable x from the
                           # workspace
remove(list = objects())   # remove all objects from the
                           # workspace
```

Note that many R functions can be abbreviated, for example `rm` can be used in place of `remove`, or `q` in place of `quit`. A useful tool for finding out the abbreviated form of a particular function, or indeed for finding out what a function does, is the built in `help()` function. This is invoked simply by passing `help` the name of the function in question, for example `help(remove)`. This help system is an invaluable aid for learning and using the many functions available in R, and it is recommended that you consult the help system on all the functions and operators introduced in this chapter, as space does not permit us to explore each function in detail. Note that if you enter a function name at the command line without the brackets that should follow it, R will not execute the function but will instead show you the R source code for the function. This is pretty scary for beginners but can be useful in some circumstances, once you get more familiar with R.

4.1.3   Vectors and data frames

Of course, a major motivation for using R in bioinformatics is that we want to deal with large biological data sets, not simple variables like those in the example above. To get an idea of how R handles larger data sets, let's consider the data shown in Table 4.1. This data is typical of the type of results that would be acquired from measuring the concentration of various compounds in whole blood samples from a number of patients during a clinical study. In reality, we would doubtless have been monitoring more patients, and maybe more compounds, but for the purpose of illustrating concepts a small data set is more convenient. Real data sets are introduced later in the chapter.

In R, we can generate a variable for each patient, in which we can store the biochemical profile for that patient in terms of the concentrations of the compounds from Table 4.1. We do this using the `<-` assignment operator as before, except that this time we combine, or *concatenate*, a series of numbers into a list using the `c()` function before the assignment takes place.

```
profile <- c(3, 1150,750,310)
```

In mathematical parlance, a list of numbers like this is referred to as a *vector*. In this particular case, and indeed in many of the applications that we come across

Table 4.1 Concentration of key metabolites in five patients. The urea concentration has not been recorded for patient 3

Patient Number	Concentration of compound in whole blood (g/m³)			
	Bilirubin	**Cholesterol**	**Glucose**	**Urea**
1	3.0	1150	750	310
2	4.5	1650	2200	200
3	5.0	2150	260	–
4	14.0	1200	650	270
5	3.5	2000	700	320

in bioinformatics, this is referred to as a *measurement vector* or *sample vector* as it captures the list of measurements acquired from a sample. We can check that this new vector contains the correct information by typing its name.

```
> profile
[1] 3 1150 750 310
```

This is a bit like a Perl array, but storing data in this way in R is particularly convenient, as we can perform operations on the whole vector in a single command. For example, we can convert this data from the units of g/m^3 to the more commonly used g/cm^3 by multiplying by a scaling factor of 10^{-6}.

```
> profile * 1e-6
[1] 0.000003 0.001150 0.000750 0.000310
```

Essentially, you should be able to use any relevant R function or operator on a multi-element variable of this type. This is not typically the case in general purpose programming languages, such as Perl, and is one of the features that make R so well suited to numerical analysis.

What about getting the whole table of results into R? Well, R has a special type of object for representing tabular data, called a *data frame*. The example below shows one way of generating a data frame that captures the first two columns of Table 4.1.

```
> bilirubin <- c(3, 4.5, 5, 14, 3.5)      # bilirubin column
                                          # values

> cholesterol <- c(1150,1650,2150,
  1200,2000)                              # cholesterol column
                                          # values

> results <- data.frame(bilirubin,
  cholesterol)                            # combine in data
                                          # frame

> results                                 # check data frame
                                          # content

  bilirubin cholesterol
1       3.0        1150
2       4.5        1650
3       5.0        2150
4      14.0        1200
5       3.5        2000
```

So now we have half of our table stored in the R workspace, easily accessible under the name `results`. Essentially, a data frame is like a spreadsheet, and indeed we can view, edit, and add to the contents of this data frame in a familiar spreadsheet-like view using R's built-in `edit()` function. Figure 4.1 shows the data editor, launched by the command below, after the glucose and urea columns have been added.

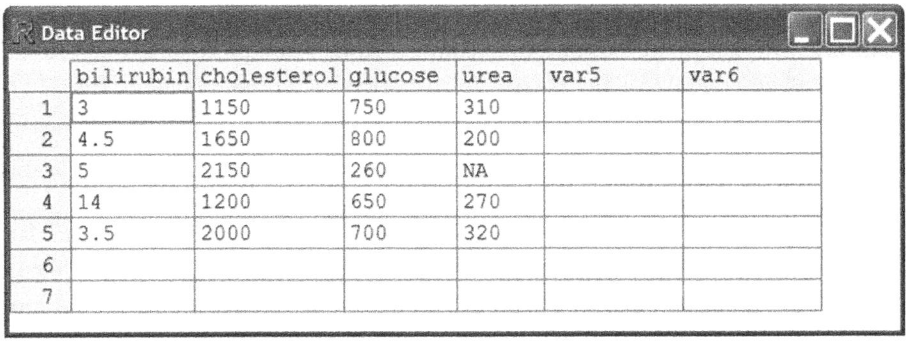

	bilirubin	cholesterol	glucose	urea	var5	var6
1	3	1150	750	310		
2	4.5	1650	800	200		
3	5	2150	260	NA		
4	14	1200	650	270		
5	3.5	2000	700	320		
6						
7						

Fig. 4.1 Editing the content of a data frame in Windows.

```
results <- edit(results)     # allow editing values in results
                               data frame
```

The `edit()` function does not change the original data frame, but returns a copy of the frame, including any changes made, when the editor window is closed. To update the data frame with the new edits, the output of the `edit()` function must be assigned back to the data frame. No urea value was entered for patient 3, and the editor has placed NA in the empty cell—this is the code that R uses to indicate missing values. Missing values are common in bioinformatics, and various methods for dealing with them are described in the literature. Typically, these methods rely on substituting missing values with statistically expected values. For now, it is just good to know that we can flag up such values instead of having to make up a placeholder value such as 0 or -999, which could easily be overlooked and accidentally treated like real data. Before moving on, fill out your `results` data frame with the remaining values.

Just as in the earlier vector example, we can apply most R operators and functions directly to a data frame so, for example, we could multiply the whole frame by a scale factor, just as we did in the previous section. One particularly convenient function is `summary()`, which provides a basic statistical overview of the data contained in a frame. In this case we can see the range of concentrations of each metabolite across the five patients, as well as mean and median averages.

```
> summary(results)
  bilirubin      cholesterol       glucose          urea
Min.   : 3.0    Min.   :1150    Min.   :260    Min.   :200.0
1st Qu.: 3.5    1st Qu.:1200    1st Qu.:650    1st Qu.:252.5
Median : 4.5    Median :1650    Median :700    Median :290.0
Mean   : 6.0    Mean   :1630    Mean   :632    Mean   :275.0
3rd Qu.: 5.0    3rd Qu.:2000    3rd Qu.:750    3rd Qu.:312.5
Max.   :14.0    Max.   :2150    Max.   :800    Max.   :320.0
                                               NA's   :  1.0
```

As well as operating on the whole table, it is possible to extract rows, columns, or individual elements by specifying specific parts of the table in square brackets ([]) immediately after the name of the data frame. Some examples are shown below. There are a couple of counterintuitive things to note in these examples. First, when a single column is extracted the result looks like a row. Second, when specifying both a row and column, the row is specified first, which seems odd if you are used to working with x, y coordinates, where x is the horizontal position and y the vertical.

```
> results[4,]              # return just the values in row 4
  bilirubin    cholesterol    glucose urea
4        14           1200        650  270
> results[4,1]             # return the value at row 4, column 1
[1] 14
> results[,2]             # return all values from column 2
                         # (cholesterol)
[1] 1150 1650 2150 1200 2000
> results[1:3,]           # return rows 1 to 3
  bilirubin  cholesterol   glucose   urea
1       3.0         1150       750    310
2       4.5         1650       800    200
3       5.0         2150       260     NA
> results[c(2,4),]        # return rows 2 and 4
  bilirubin  cholesterol   glucose   urea
2       4.5         1650       800    200
4      14.0         1200       650    270
```

Of course, entering data manually as we have done so far is tedious, time consuming, and prone to error. In most real applications data is imported directly from a file or database. This is described later in this chapter. However, the example that we have just worked through gives an indication of how we interact with the R command line. We only used a tiny fraction of R's functionality, but the principle of issuing commands using functions, operators, and multi-element variables is really what R is all about, it is just that the functions get more powerful and the variables get larger as you get deeper into R. In the remainder of the chapter, we will look at the general methodology used in biological data analysis, how R can assist us with this, and how we can build programs in R.

4.1.4 The nature of experimental data

A table of numerical data like the `results` data frame used in the previous example is referred to in mathematics as a *matrix*, and one of the big breakthroughs in becoming competent in data analysis is realizing that almost all experimental data can be considered in the form of a data matrix. This is because experiments typically entail analysis of more than one sample, and involve the determination of more than one parameter for each sample. We have already seen how a matrix can be used to store metabolic data in the blood analysis example in the previous section. Although that data set was small in terms of the number of metabolites and samples, it would clearly be trivial to extend the number of rows and columns to accommodate the larger data sets that typify metabolomics studies. The general approach is equally applicable to other areas of post genomics, such as transcriptomics and proteomics.

A matrix is a good way of representing experimental data regardless of the type of analytical platform, the number of samples, or the number of measured variables. This general way of representing data is shown in Fig. 4.2. Considering the case of a microarray experiment, each variable is a particular gene, and the gene expression levels for each sample can be represented by a vector of expression measurements, denoted mathematically as x_i, where i is the sample number. Within this vector, each element x_{ij} represents the expression level of gene j on the microarray. The expression level of each gene can be considered to be a variable, of which there are many, hence microarray data is referred to as being *multivariate*. Multivariate data analysis techniques, introduced later in Section 4.2, are therefore required to interpret such data.

The vectors representing the gene expression levels from individual samples can be amalgamated into an $I \times J$ data matrix, X, where I is the total number of samples considered and J is the number of genes per microarray. Each row of the X matrix therefore represents an individual microarray, while each column indicates the expression level of each specific gene over all samples. A single data matrix is therefore sufficient to describe all the samples analysed in a given experiment comprising any number of individual arrays.

In transcriptomics more generally, the variables would be the expression level of individual genes, regardless of the analytical platform used. The number of variables considered would be dependent on the platform, ranging from a handful if the data was from traditional PCR, through to many thousands from a microarray experiment. The number of samples would obviously depend on the size of the study being carried out, so may range from a handful through to many hundreds or thousands.

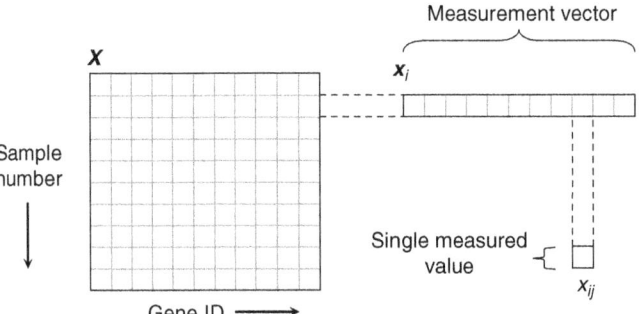

Fig. 4.2 Organisation of gene experimental data into a data matrix. For data from a gene expression microarray, the row vector x_i would be the gene expression profile over all genes for a specific sample. Each column of the matrix captures the variation of an individual gene over all samples.

Table 4.2 Web server statistics for a very basic web site

	Number of downloads		
	index.html	welcome.png	paper.pdf
Monday	15	13	2
Tuesday	27	26	1
Wednesday	34	30	3
Thursday	10	10	0
Friday	9	7	1

Proteomics data sets can also be considered as a matrix, although exactly what form the data takes depends on the particular proteomics protocol used. If quantitative proteomics has been carried out, then the data will be superficially similar to that seen in transcriptomics, except that the value in each element of the matrix will indicate the level of protein expression (either relative or absolute, depending on the protocol) rather than gene expression. However, at the time of writing, many proteomics protocols are only capable of indicating whether proteins are present or absent from a sample. For this type of data, we could use some kind of coding scheme, such as 1 for protein present, 0 for protein absent, or R's *factor* data type, which is described later.

There are a lot of different names that can be given to data which is in this form, such as matrix, array, table, or even spreadsheet. As we have seen, R even has its own way of capturing such data—the *data frame*. Exactly which of these classes a data set belongs to can be technically very important, especially in R as we will see later, but conceptually they are all the same—they are all matrices. The great benefit of this is that mathematicians have spent many years working with matrices and developing algorithms for manipulating, analysing, and extracting information from them. The algorithms are there for us to apply to our bioinformatics problems.

To see some matrix mathematics in operation, consider Table 4.2, which summarizes the number of times various files were downloaded from a web server. Each of the files is of a specific size: the front page index.html is just text, taking up 1624 bytes, welcome.png is a 23,172 byte image on that front page, and paper.pdf is a fairly substantial research paper taking up 1,234,065 bytes that can be downloaded via a link in index.html. We can input all this data into R using the commands below. Note that this time, the matrix() function is used to generate a matrix, instead of the data frame used in the previous example (the subtle differences between a matrix and a data frame will be explained in the next section).

```
> index = c(15, 27, 34, 10, 9)
> welcome = c(13, 26, 30, 10, 7)
> paper = c(2, 1, 3, 0, 1)
> days = c("mon", "tues", "wed", "thurs", "fri")
```

```
> filenames = c("index.html", "welcome.png", "paper.pdf")
> downloads = matrix(c(index, welcome, paper), nrow=5,
dimnames=list(days,filenames))
> downloads        # check what's in the data frame
      index.html welcome.png paper.pdf
mon           15          13         2
tues          27          26         1
wed           34          30         3
thurs         10          10         0
fri            9           7         1
> filesizes = c(1624, 23172, 1234065)
```

The useful part of this is that R has a built in operator, `%*%` for multiplying matrices. So, to get the total number of bytes downloaded per day, we can simply multiply the `downloads` matrix by the `filesizes` vector.

```
> downloads %*% filesizes
         [,1]
mon    2793726
tues   1880385
wed    4452571
thurs   247960
fri    1410885
```

This is exactly the type of calculation that we do frequently in bioinformatics. For example, we might want to multiply measured values in a matrix by a weighting vector in order to come up with some overall *score* for each sample. R allows us to do this calculation in just one line of code. This is just one example of the matrix functionality built into R—functionality that has great utility in bioinformatics.

Matrices as images—and vice versa

Bitmapped images, such as those coming from microscopes or from scanners of gels or microarrays, are essentially data matrices, in which the value of each element in the matrix represents the colour of the point at that position. Image analysis can therefore be carried out in R by loading images into a data matrix and working with that matrix. This is not as straightforward as it should be, because bitmapped images are not normally stored in a native R format, but in various dedicated image formats (TIF, JPG, PNG etc.). An appropriate R package, or some serious programming, is therefore required to convert these files into something that R can handle.

Conversely, experimental matrices can be displayed as images, and this is sometimes a useful way of getting a quick overview of a given data set. Such images are generically referred to as *heatmaps* (although in R a heatmap tends to have additional features). For example, we can generate an image representing the blood analysis results from Section 4.1.3 using the `image()` command below.

```
image(as.matrix(results))    # plot results matrix as bitmapped
                             # image
```

There is more about heat maps later in this chapter (Section 4.3.1).

4.1.5   R modes, objects, lists, classes, and methods

Before going much further, it is worth taking time to find out more about how data is stored in the R workspace, and how the various data objects (e.g. vectors, matrices, and data frames) are related.

As we saw in Chapter 2, it is common to consider a particular piece of data as one of several fundamental data types, typically numeric (of which there may be subtypes, such as integer and floating point), logical (of which the two-state Boolean type is most common), and character string. R is no exception, and every piece of data has to be of one particular basic type: *numeric*, *complex*, *logical*, *character*, or *raw*. In R parlance, these types are sometimes called *modes*. Because some functions can only work with certain data types, R allows conversion from one type to another if necessary. This is done using functions of the form as .type(), of which an example is shown below.

```
> x <- 2.17              # assign numeric value 2.17 to x
> y <- as.character(x)   # convert number in x to character
                         # string in y
> z <- as.numeric(y)     # convert text in y to number and
                         # assign to z
> y                      # show content of y
[1] "2.17"
> z                      # show content of z
[1] 2.17
```

Being limited to these basic data types would be a little restrictive in data analysis applications so R has *objects*; much more complex structures built from these basic data types. Objects are a crucial part of R, and it is impossible to feel comfortable in R without a reasonable understanding of the various classes of object, how to use them, and how they are related. To start with, let's consider the key built-in classes of object in R.

Vectors

We have already come across these, as well as single values (scalars) which are just a special single element type of vector. A key feature of a vector is that every element must be of the same type, for example all numbers or all text.

Matrices

Matrices are the next step up from vectors, the only difference being that they can have more than one dimension. As we have seen, two-dimensional matrices are common in bioinformatics, and can be considered to be a table of numbers, but a matrix can in fact have a third dimension, which transforms it into a cube of numbers. Such a matrix might be used to handle GC-MS data, with the three dimensions of the matrix being sample number, elution time, and *m/z* ratio. Indeed, a matrix may have any number of dimensions if an application requires it, although in most bioinformatics applications two dimensions is enough. As with vectors, it is not possible to mix data of different types (e.g. numbers and text) within a single matrix.

Factors

Factors are R's solution to handling categorical data. A good example of this is capturing qualitative proteomic data where, instead of having a value recorded for each protein in each sample, there is simply an indication of whether the protein is present or not. One solution is to use the Boolean type, with `TRUE` representing protein presence and `FALSE` indicating absence of a protein.

```
> proteinpresent <- c(TRUE, TRUE, FALSE, TRUE, FALSE)
> proteinpresent
[1] TRUE TRUE FALSE TRUE FALSE
```

If `TRUE` and `FALSE` are not enough we can create our own user-defined discrete states in R by using the `factor()` function. This can be used to simply represent the data in a more descriptive way, such as PRESENT/ABSENT instead of TRUE/FALSE. However, it is most useful when we have more than two states, for example when we want to flag the level of expression of a protein or gene as one of three states: over-expressed, unchanged, or under-expressed. In the example below an expression vector is created that captures the state of six genes.

```
expression <- factor(c("over","under","over","unchanged",
"under","under"))
```

R functions can be used to manipulate or derive information from such vectors. For example, `levels()` lists the discrete states present in the data, and the behaviour of the `summary()` function changes to show the occurrence of each level instead of the statistical metrics that would be returned if `expression` contained quantitative data.

```
> levels(expression)
[1] "over"      "unchanged" "under"
> summary(expression)
    over unchanged     under
       2         1         3
```

Lists

Lists are an extension of the vector concept in which the elements need not be of the same type, and may in fact be of sophisticated types, such as vectors, matrices, or lists. For example, a list could be used to store information about a protein, specifically the protein name, the PDB accession number, and the formula weight of the protein:

```
protein <- list("glucose oxidase", "1CF3", 63355)
```

This is a bit like a Perl array. Individual elements can be extracted from this list by specifying the number of the element, very much as would be done with a vector, so `protein[2]` would return the accession number 1CF3. Better still, the elements in a list can be named for easy reference, and accessed using the $ symbol. This is similar to a Perl hash.

```
> protein <- list(name="glucose oxidase", accession="1CF3",
weight=63355)
> protein$accession      # extract the accession number by
                         # name
[1] "1CF3"
```

It is also possible to add elements to a list using the $ symbol. In the example below, a vector of Gene Ontology (GO) IDs related to glucose oxidase is created and stored in x. This is added to the `protein` list defined above, and the contents of the list displayed.

```
> x <- c(16614, 50660, 6066, 6118)    # assign list of GO IDs
                                       # to x
> protein$GOIDs <- x                   # add x to list as new
                                       # GOIDs field
> protein                              # display content of
                                       # protein object

$name
[1] "glucose oxidase"
$accession
[1] "1CF3"
$weight
[1] 63355
$GOIDs
[1] 16614 50660 6066 6118
```

Lists are often used in R to pass parameters to a function, such as the parameters required when plotting a graph. Graphing options are necessarily of heterogeneous type, as they need to include the matrix of data to be plotted, textual labels, and logical settings that specify the plotting style. Similarly, lists are often used to return heterogeneous results from a function in a single object.

Data frames

Technically, a data frame is a specific type of list, but it is more convenient to think of data frames as an extension of the matrix concept. The main benefit over a matrix is that a single data frame can contain columns with different data types. A data frame object is therefore very similar to the spreadsheet object that is common in packages such as Microsoft Excel.

Mixing different data types in a single data frame is particularly useful when dealing with data sets that include DNA or protein sequences, accession numbers, and annotations. Consider the BLAST search results reproduced below.

Accession Number	Description	Score (bits)	E Value
CAA76841.1	albumin [Canis familiaris]	43.1	0.002
P02770	ALBU_RAT Serum albumin precursor	37.1	0.11
AAH85359.1	Albumin [Rattus norvegicus]	37.1	0.11
BAC34360.1	unnamed protein product [Mus musculus]	36.3	0.20
AAA37190.1	alpha-fetoprotein	36.3	0.20

The numerical data can be entered as vectors as in the earlier blood components example. The accession numbers are entered in exactly the same way, we just need to remember to enclose them in quotation marks (") as is common in other programming languages. An example of this in action is shown below.

```
> # concatenate numeric values into vectors as before
> score <- c(43.1,37.1,37.1,36.3,36.3)
> E <- c(0.002, 0.11, 0.11, 0.2, 0.2)

> # list of strings are brought together in the same way
> accn <- c("CAA76841.1","P02770","AAH85359.1","BAC34360.1","
  AAA37190.1")

> # now bring the vectors together to create data
> results <- data.frame(accession=accn, score=score,
  EValue=E)

> # check contents of the data frame
> results
  accession score EValue
1 CAA76841.1  43.1  0.002
2     P02770  37.1  0.110
3 AAH85359.1  37.1  0.110
4 BAC34360.1  36.3  0.200
5 AAA37190.1  36.3  0.200
```

Functions

The objects listed thus far are all designed for holding data. In R, functions are also defined as objects. User-defined functions are therefore stored in the workspace, and will be included in the list returned by objects(). Creating functions is explained in Section 4.1.8.

Other objects

So, most objects in R fall into one of the above classes. The class of a particular ob-ject can be determined using the `class()` function, so typing `class(protein)` after the example earlier would reveal that the protein object is a list. In most cases, R functions are written such that they adapt their behaviour to the class of object passed to them—they determine the class of the object before deciding exactly what to do. Other useful functions for finding out about objects are `length()` and `attributes()`, the behaviours of which are demonstrated below.

```
> class(protein)        # get the class of object
[1] "list"
> length(protein)       # get number of elements in object
[1] 4
> attributes(protein)   # get other attributes (element names
                        # in this case)
$names
[1] " name"     "accession"  "weight"     "GOIDs"
```

It is also possible to define new classes of object, with their own structures. These are sometimes referred to as S3 objects or S4 objects, reflecting which particular version (3 or 4) of the S language definition they relate to, as R sup-ports both versions. This is where R can start to get confusing but, as we will see later, this is a very valuable feature as classes can be created to capture data from complex biological applications. Objects contain individual components called *slots*, which can contain named elements. These slots and elements can then be accessed using the @ and $ operators respectively, but because the in-ternal structure of objects can be complex, and may be subject to change as software develops, the author of a particular class of object will usually produce a series of functions (often called *methods*) that extract data from within the depths of an object. Examples of working with complex objects can be found in Section 4.3.1.

4.1.6   Importing data into R

Bioinformatics is characterized by data sets that are usually large and often het-erogeneous. Standard formats to capture much of this data are emerging, but even so a lot of time is spent converting between file formats and getting data in and out of different software packages. R offers a great deal of functionality for importing data from a range of sources.

The built-in R function `read.table()` makes importing tabular data very simple, provided the data is available as a delimited text file (i.e. a text file con-taining one or more rows of data, each of which contains values separated by a specific character such as a comma). The example below loads data from the file "blood.csv" into the data frame `loadedresults`. The `sep` parameter is set to indicate the character used to separate columns—in this case a comma (,) and the `header=TRUE` option tells the function the first row of the file contains column headings.

```
loadedresults <- read.table("blood.csv", sep=",", header=TRUE)
```

For this to work, the file "`blood.csv`" will need to be in R's current working directory, because this is where R looks for files if a full path is not specified with the file name. Some people create a directory called `work` within the R program directory—a more resilient approach would be to work from a directory in your particular user area on a networked drive that is regularly backed up. You can find out the current working directory by using the `getwd()` function, set it using `setwd()`, and see which files are in that directory using the `dir()` function.

```
> getwd()                               # get working
                                        # directory

[1] "C:/Program Files/R/R-2.5.1"

> setwd("C:/Program Files/R/R-2.5.1/work")    # set working
                                        # directory

> dir()                                 # list files in
                                        # directory

[1] "blood.csv"
```

There is a range of optional arguments that can be used with `read.table()`, to cope with the many different ways in which tabular data can be represented in text files. These are thoroughly documented in the R help system (type `help(read.table)`). It is also possible to import data that has been stored in binary format, although this is never easy due to the many different ways in which binary files can be constructed. It is even possible to load files directly from remote servers on a network, or on the Internet, simply by specifying a full URL instead of just a file name, as in the example below which loads a file from our website.

```
filename <- "http://www.bixsolutions.net/blood.csv"
loadedresults <- read.table(filename, sep=",", header=TRUE)
```

It is also possible to import data directly from a relational database by issuing queries to a database server such as MySQL, which is clearly of great value in bioinformatics applications. This functionality is covered later in this chapter, in Section 4.3.2.

4.1.7   Data visualization in R

Visualizing data is useful in many bioinformatics applications, and R provides a number of built-in functions for graphing data, with many more elaborate graphing functions available in add-on packages (e.g. `scatterplot3d` and `ggplot2`). This is a distinct advantage of R over general purpose programming languages, as they do not have such native functionality. To illustrate R's visualization capabilities, we will use as an example a collection of protein fractionation profiles.

These were constructed by determining the abundance of several individual proteins in different fractions taken from a sample. The aim of the experiment was to determine which of the 12 proteins studied has the most similar properties to a particular protein of interest by comparing the fractionation profiles. Such studies are often carried out to determine previously unknown characteristics of proteins, for example to infer their subcellular location by association with proteins of known subcellular location (as in Sadowski, 2006). The data in our example comprises protein abundance data from six fractions for each of the proteins, and our basic aim is to find which protein has the most similar fractionation profile to the protein of interest over these six fractions. The data can be loaded in directly from a CSV file at `www.bixsolutions.net`:

```
X <- read.table("http://www.bixsolutions.net/profiles.csv",
sep=",", header=TRUE)
```

Looking at the data, we see that each column represents a protein (labelled p1 to p12), or the protein of interest (labelled x), and reading down a column gives us the abundance profile with respect to the six fractions. The magnitudes of the values are clearly different for the different proteins, but this is of little importance as we are only interested in identifying profiles of similar shape.

```
> X
    x  p1  p2  p3   p4   p5  p6    p7   p8   p9  p10  p11  p12
1   0 148   6   5  197    1  12     9    0    4    0   11    0
2   4 185   5   9  180   73   6     5    1    5   12   15    3
3  11 149 177 282  446  400   7     7    3    0    8  223    2
4  29 103 210 299 1264  912   3   599    2    2    6  865  387
5   7  72 131 197  520  171 181   301  411  864  561  266  763
6   1  75   7  11  125   34 241  1222  611 1175  216  133  511
```

The `matplot()` function is a powerful R function for plotting data contained in matrices and data frames. Using this, we can very quickly produce a graph of all the profiles together.

```
matplot(X,type="l")    # the letter "l" specifies a line plot
```

This will produce a basic plot, but to create a more professional and useful graph, it is necessary to utilize more of `matplot`'s many arguments—use `help(matplot)` to find out more about these. The example below uses the `col` argument to specify that R should cycle through six colours while plotting the lines, the `lty` argument to cycle through five different line types (solid, dotted, dashed, etc.), the `lwd` argument is used to boost the line width to 2, and axis labels are added to the plot using `xlab` and `ylab`. This results in the graph shown in Fig. 4.3.

```
matplot(X,type="l",xlab="fraction",ylab="quantity",col=1:6,lt
y=1:5,lwd=2)
```

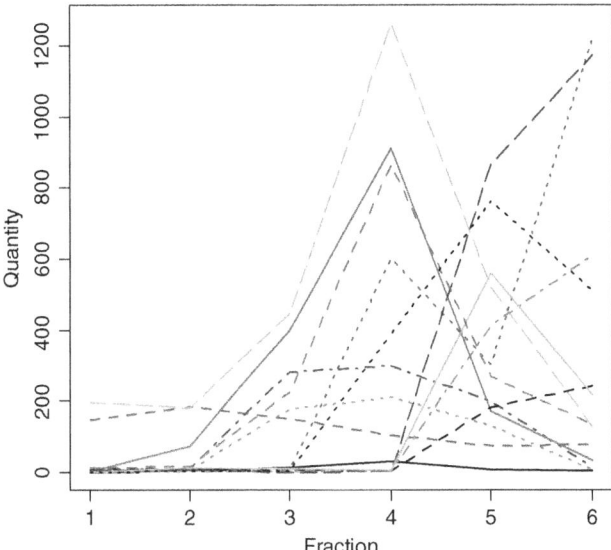

Fig. 4.3 The 13 individual protein profiles plotted together using the `matplot` function.

It is not really possible to see what we are looking for here, because the profile of interest is too low. One way to resolve this is to scale each profile so that it has a maximum value of 1. We can do this by dividing each column of the matrix by the maximum value in that column. To calculate the vector of these maximum values, we use the `apply()` function. Normally, functions such as `max` return a single value, but by using `apply()` we can apply such functions in a column-wise or row-wise manner. In this case, we apply the `max()` function to the columns of the X matrix. Setting the second argument to 2 indicates that we want to work on columns—if we wanted to work on rows this would be set to 1.

```
Xmax <- apply(X, 2, max)    # returns a vector containing the
                            # maximum value of each column
```

Another built-in function, `scale()`, can then be invoked in such a way that it divides each column by the maximum value of that column. By default, `scale()` also adds an offset to each column to centre the data, but in this case zero values have an important significance as these are the baselines of the profiles, so we switch centring off using `center=FALSE`.

```
Xscaled <- scale(X, scale=Xmax, center=FALSE)
```

The scaled version can be plotted using the `matplot()` function exactly as before—just replace X with `Xscaled`. However, the similarity between protein x and the rest of the proteins is still impossible to see because there is no indication of which line relates to which column. We can solve this problem by adding a legend using the `legend()` function. The key arguments here are x, which sets the horizontal position of the legend box in the graph's axes, and `legend`, which specifies the labels to apply to each line. In this case, we just take the names direct from the data frame by using `names(X)` as the legend text. The plotting agruments `col`, `lty`, `lwd` must be set to match the arguments passed to `matplot()` if the legend and plot are to match up. Note that a background colour has been specified for the legend box by setting the bg argument to `"snow"`—this is the name of one of many pre-defined colours in R, for a full list type `colours()` at the R prompt. So the legend command is:

```
legend(x=1,legend=names(X),col=1:6,lty=1:5,lwd=2,bg="snow")
```

We can now see from the resulting plot (Fig. 4.4) that the protein with the most similar profile to the protein of interest (labelled x) is p5. Reading off a graph like this is always going to be somewhat subjective, so we will look at more rigorous ways of quantifying these similarities later in the chapter, but for the moment this is a promising result, and a good indication of how graphs can be generated with just a few lines of R code.

The graphing capabilities we have seen so far are provided by R's built-in graphics package, which contains a range of other high level graphics functions for visualizing data. These include `barplot()` for plotting bar charts, `boxplot()` for producing *box and whisker* plots, `contour()` for contour plots, and `pie()` for pie charts. Each of these functions can be called using basic syntax, as in the examples below, which generate graphs from the protein profile data. Alternatively, these functions can be embellished by passing various other parameters, which you can learn about using the `help()` function.

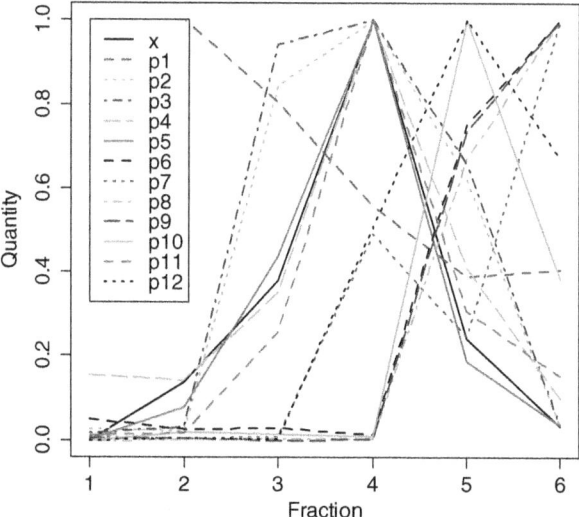

Fig. 4.4 The same data as in Figure 4.3, after range scaling and addition of a legend. Looking at the legend box, we can see the effect of the `col=1:6` and `lty=1:5` plotting arguments. As we look down the box on screen, we can see six colours (black, red, green, blue, cyan and magenta) repeating every six lines, and different line styles repeating every five lines, ultimately combining to produce a set of uniquely identifiable lines.

```
barplot(Xmax)          # bar chart showing max quantity of each
                       # protein
boxplot(X)             # distribution of quantities for each
                       # protein
pie(apply(X,2,sum))    # comparison of total quantity of each
                       # protein across all fractions
```

Primitive graphics functions

The above example shows how easy it is to generate graphs in R compared to using other programming languages such as Perl. The legend function is the epitome of this simplicity, in that it creates a complete legend box in a single line of code. Of course, other software can produce similar graphs, via a much more intuitive point-and-click interface, but what sets R apart is the ease with which these powerful commands can be combined with lower level primitive graphics functions to produce complex bespoke visualizations. This is crucial in bioinformatics, because we often need to generate complicated visualizations such as linkage maps, or unusual combinations of different graph types. R's `graphics` package facilitates such visualizations by supplementing its many high level graphic functions by a suite of more primitive functions for plotting basic shapes. For example, we can add a rectangle around the central part of the plot in Fig. 4.4 using the following call to R's `rect()` function. The minimum set of values passed to the function specifies respectively the left, bottom, right, and top bounds of the rectangle.

```
rect(3,-0.01,5,1.01)
```

Other parameters can be passed to specify other features of the rectangle, such as the fill colour and outline style. Similar functions include `arrows()`, `lines()`, `polygon()`, `points()`, `segments()`, and `symbols()`, all of which are reasonably self-explanatory. Information about these functions and all the parameters that can be passed to them can be found using the `help()` function. Textual annotations can be added at a specified point in a plot by using the `text` function, for example:

```
text(3.0,0.6,"similar \nprofiles")
```

These primitive graphics commands can be typed at the command line, but are more normally combined together in R programs to automatically generate bespoke plots (there is an example of this later, in Section 4.1.8). This functionality can be used directly within R, or can be used to generate graphical output as part of another tool, such as a web-based application (see Chapter 5). R uses vector graphics rather than the bitmapped graphics, so plots generated by R can be easily scaled up without loss of quality, which is ideal in situations where you might want to zoom in to a complex plot, or prepare figures for publication.

Creating interactive graphics

In some applications, it can be useful to interact with visualizations, to enable zooming in to regions or retrieval of annotations or data from another dimension.

R has built-in functions that make this possible. For example, the `locator()` function allows the user to select a specific position within a graph by clicking on it with the mouse. To see this in action, type `locator(n=1)` at the R command line while the graph in Fig. 4.4 is open. Then click somewhere in the graph. You may hear a sound (depending on how your computer is configured) and the *x* and *y* coordinates of the position you clicked will be returned in the R console. In an R program, this information could be used as the starting point for adding annotations to the plot or to retrieve some additional information about whatever is displayed at that point.

Devices

In the examples so far, we have issued plotting commands and graphs have appeared in a window. Working like this means that each new graph causes the old one to be lost, which is not helpful if we want to see multiple graphs simultaneously, to compare results for example. To accommodate this, R allows multiple windows to be open simultaneously. There are different functions for opening a new window, depending on which operating system you are using. These commands are shown below, but in practice we have found that `x11()` works in all three operating systems, so would recommend using that in your R programs to ensure that they are cross-platform.

```
x11()            # open a new window for plotting (in Linux)
windows()        # open a new window for plotting (in Windows)
quartz()         # open a new window for plotting (in Mac OS)
```

In R, a window is one particular instance of a *device* on which graphics can be displayed. Windows are therefore manipulated using a suite of device handling functions, whose names begin with `dev.`. Some examples are shown below.

```
> dev.list()      # list open devices
windows windows windows windows
      2       3       4       5
> dev.set(3)      # make window 3 the active device
```

The active device is where any graphics commands will be sent. If a window is currently the active device, the word ACTIVE will be shown in brackets in the title bar of that window. Windows are not the only type of R device. Another useful device is PDF, which can be created using the `pdf()` function. This creates a PDF (portable document format) file with the specified name, in the current working directory. For example:

```
pdf("figure1.pdf")   # create a PDF device called figure1.pdf
```

All graphical output will now be sent to that file, as long as it is the active device. This is very useful for capturing the output of an automated data analysis program, or for printing figures or sharing them with colleagues. A PDF reader such as Adobe Acrobat Reader is required to view the PDF, and you will probably need to close the device by issuing the command `dev.off(which = dev.cur())` before being able to access the file without throwing a sharing violation error. The functions `bmp()`, `jpeg()`, and `png()` are similar to `pdf()`, except that they create bitmapped image files (in BMP, JPEG, and PNG formats respectively) instead of PDFs. Again, these devices need to be closed before the file can be viewed.

Summary

This section has given an indication as to how R's built-in data visualization capabilities are used and, although this is just the tip of the iceberg in terms of what is possible, the general approaches hold regardless of the type of graph being drawn. There are more examples later in this chapter, but to see the full extent of the graphics functionality available in R's base graphics package, type `help(graphics)` at the R prompt.

4.1.8   Writing programs in R

At this point, you have probably realized that everything covered so far could have been done just as effectively, and probably much more quickly, in a spreadsheet application such as Microsoft Excel. You would be right of course, and there is no doubt that Excel is a fine tool for simple one-off manipulation of small data sets. However, bioinformatics is typified by large data sets, complex analysis algorithms, and the need for repetitive analysis. This is where R really comes into its own, as its command-line ethos is ideal for writing programs that automate data processing workflows. For example, producing an R program to automate the generation of the fractionation profile plots in the previous example is as easy as pasting all the relevant commands together in the correct order in a text file—the result of this is shown below. Any text editor can be used to do this, and the choice of editor to use with R is really down to personal preference. We would refer you to the survey of editors provided at the beginning of the previous chapter (Section 3.1.5), as the basic process of writing an R program is similar to writing Perl code. If you find a particular editor convenient for programming in Perl, then you may as well use the same editor with R.

Depending on your operating system, R may have some kind of internal text editor for writing programs, or will know about an editor installed on your computer. If so, you can initiate editing of a file either by opening a file using RGui's File menu, or by using the `edit()` function in the R Console.

```
edit(file="profiles.r")
```

If you feel uncomfortable with whatever editor R is presenting you with, you can tell it to use any other editor installed on your system by using the `options()` function. In the example below (for Windows), R is told to fire up Komodo Edit whenever we call the `edit()` function to edit a program.

```
options(editor = "komodo.exe")
```

Note that it is customary, but not compulsory, to give R programs file names with the extension .r. Hence the example below, which captures the sequence of commands issued in the graphing example from the previous section, should be saved in the working directory as profiles.r.

```
# PROFILES.R
#
# Simple R program to load a data matrix, scale it and plot
# the result.

# clear out the workspace first
rm(list = ls())

# load data frame from web site
X <- read.table("http://www.bixsolutions.net/profiles.csv",
sep=",", header=TRUE)

# rangescale data by dividing by the maximum value in each
# column
Xmax <- apply(X, 2, max)
Xscaled <- scale(X, scale=Xmax, center=FALSE)

# plot columns in matrix as lines on a single graph
matplot(Xscaled,type="l",xlab="fraction",ylab="quantity",col=
1:6,lty=1:5,lwd=2)

# add legend to graph
legend(x=1,legend=names(X),col=1:6,lty=1:5,lwd=2,bg="snow")
```

To execute a program we use the source() function. For this to work, the program file must be in the current working directory. We can then issue the command below.

```
source("profiles.r")
```

Alternatively, it is possible to run this program directly from www.bixsolutions.net by specifying the full URL. The command is:

```
source("http://www.bixsolutions.net/profiles.r")
```

So, now we have our program and we know how to execute it. If a laboratory colleague sends us a new data set, we can just change the filename in the read.table line and run the program on that new data, automatically generating the required graph in a matter of seconds. (Thanks to the use of matrices we do not even need to specify the number of proteins or fractions in the data set.) This type of scripting is a great time saver, and clearly very useful in its own right. However, R goes beyond simple scripting by allowing the creation of new functions, and fully fledged programs with structures similar to those covered in the Perl chapter.

Beyond scripting: loops and conditionals

To avoid repetition, we refer you to Chapter 3 for more detailed descriptions of control structures and their uses. In this section we just cover the basic syntax needed to define these structures in R. The syntax is in fact very similar to Perl.

Conditional statements are constructed using the `if` statement, a simple example of which is shown below:

```
> value <- -1
> if (value < 0) print("value is negative")
[1] "value is negative"
```

In this example, the less than (<) operator is used to check whether the number stored in value is less than zero (i.e. is it negative). Other comparison operators include greater than (>), equal to (==), combinations of these (<= and >=), and not equal to (!=). Multiple conditions can be combined using the and (&&) and or (||) operators, just like in Perl. If multiple commands need to be executed when the condition is met, it is necessary to group the commands into a code block defined by enclosing the commands between curly braces ({ }). Again, this is just like Perl, and an example is given below. It makes little sense to enter such complicated constructs at the command line, so such things are usually only used as part of programs.

```
if ((residue == "D") || (residue == "E")) {
    print("negatively charged amino acid")
    negativeresiduecount = negativeresiduecount + 1
}
```

R provides three statements for creating different types of `loop`: `for`, `repeat`, and `while`. A loop can be constructed using the `for` statement, in very much the same way as it would be in Perl. The generic format of the `for` statement is `for (i in range)`, where *i* is the loop variable, and *range* is the list of values that will be attributed to it on each pass through the loop. Typically, *range* is a series of integers, which in R is defined using the colon operator, for example `1:10` returns the integers from one to ten. However, `range` can in fact be any numerical R vector. As with `if` statements, multiple commands may be grouped together using curly braces such that they all execute on each cycle of the loop. The statements `next` and `break` may be used to exit the current cycle, or the whole loop, respectively.

The program below shows an example of a typical program combining a `for` `loop` and conditional statements with some of the primitive graphics functions introduced earlier, to produce a view of a protein sequence annotated with its secondary structure. A vector called `struct` is used to define the structure at each amino acid position, with 1 indicating that a residue is part of an alpha helix, 2 denoting a beta sheet, and 0 for other features such as turns and loops. The output of this program is shown in Fig. 4.5.

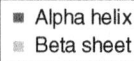

GARVHMDGARL MNA AVAL RI PPARL VEHCDSVSFCFSKG

Fig. 4.5 Annotated protein sequence produced using low level R graphics functions.

```
# SSSEQ.R
#
# Simple R program to display a sequence with structural
# annotation

# define sequence and secondary structure
seq <- "GARVHMDGARLMNAAVALRIPPARLVEHCDSVSFCFSKG"
struct <- c(0,0,2,2,2,2,2,1,1,1,1,0,0,0,0,0,0,0,0,0,0,
                 1,1,1,1,0,0,0,0,0,2,2,2,2,2,0,0,0,0)
residuecount <- 39;

# set up the window for plotting
x11() # may need quartz() for Mac or windows() for PC
plot.new()
plot.window(c(0,40),c(-20,20))

# plot a line representing the length of the sequence
segments(0.5,0,39.5,0)

# plot the sequence and features
for (i in 1:residuecount) {
      text(i,-2,substr(seq,i,i))   # write residue letter
      if (struct[i] != 0) {
            if (struct[i] == 1) boxcolour <- "firebrick"
            # alpha helix
            if (struct[i] == 2) boxcolour <- "yellow3"
            # beta sheet
            rect(i-0.5,-1,i+0.5,1,col = boxcolour,
            border = NA)
      }
}

# plot a legend
legend(x=0,y=8,legend=(c("alpha helix","beta sheet")),
    pch=15,col=c("firebrick","yellow3"),bg="snow")
```

Writing functions

Functions are essentially R programs that can be called just like the built-in functions we have been using so far in this chapter. They are analogous to functions/
subroutines in Perl, and as in Perl the benefit of an R function is that if we need
to perform something often, we can wrap the relevant code up into a function
that can then be called with just one line of code, either in a program or (in the
case of R) from the command line. This type of program code re-use increases efficiency of program development, not just by reducing the need to re-type things,
but also by reducing debugging time.

   R functions are defined using `function()`. Consider the example program
below, which defines a function called `rangescale`. This function scales all
columns in a matrix such that their maximum value is 1 by dividing by the
maximum value of each column, just as we did in Section 4.1.7. The R commands that make up the function are grouped together in the curly braces (`{ }`).
This group of commands is assigned to a function called `rangescale` with the
first line statement `rangescale <- function(X)`. Variables named between
the function brackets indicate that objects must be passed into the function,
and the variables listed in the brackets following the `return` command indicate variables that are passed out of the function on completion. In this case,
the function expects to see an object coming in, which is assigned to `X`, and
returns the range-scaled matrix, `Xscaled`. Note that any other variables used in
the function, (in this case just `Xmax`), are internal to the function and therefore
do not appear in the R workspace after the function is called. This concept of
the *scope* of variables is exactly the same as when the `strict` pragma is used
in Perl.

```
# RANGESCALE.R
#
# R program to define a function to rangescale columns of a
# matrix
rangescale <- function(X) {
      Xmax <- apply(X, 2, max)
      Xscaled <- scale(X, scale=Xmax, center=FALSE)
      return(Xscaled)
}
```

   Before we can use a function, we first have to save the program that defines
it (let's save the above program as `rangescale.r`), and then run that program
using `source()`. The newly defined function is treated by R just like any other
object, so once defined it is visible in the workspace. It can then be called just like
any built-in function.

```
source("rangescale.r")    # run the program to define the
                          # function
N <- rangescale(M)        # call the function to rangescale
                          # matrix M and place the result in N
```

4.1.9 Some essential R functions

We have already seen some of the most useful built-in R functions in action but, as mentioned at the start of the chapter, it is not possible for us to cover them all. However, there are a few generally useful functions that we wish we had known about when starting out in R, so we have gathered together brief introductions to those functions in this section. A common feature linking many of these functions is that they remove the need to use loops when dealing with data matrices. Avoiding loops makes R programs cleaner and faster, especially when dealing with large data sets.

apply(X, margin, fun, ...) and related functions

As we have already seen in Section 4.1.7, the benefit of the `apply()` function is that it can apply a function (specified in `fun`) to multiple values held in a matrix or list. The function can be applied to rows, columns, or both according to the value passed in `margin`. It can take a while to get in the habit of using `apply()` optimally, especially if you are used to programming in another language. To demonstrate `apply()` in use, let us again read in the protein profile data matrix from Section 4.1.7 and take a look at it:

```
X <- read.table("http://www.bixsolutions.net/profiles.csv",
sep=",", header=TRUE)
> X
    x   p1   p2   p3    p4   p5   p6    p7   p8   p9  p10 p11 p12
1   0  148    6    5   197    1   12     9    0    4    0  11   0
2   4  185    5    9   180   73    6     5    1    5   12  15   3
3  11  149  177  282   446  400    7     7    3    0    8 223   2
4  29  103  210  299  1264  912    3   599    2    2    6 865 387
5   7   72  131  197   520  171  181   301  411  864  561 266 763
6   1   75    7   11   125   34  241  1222  611 1175  216 133 511
```

If we perform the `sum()` function on this, we get a simple sum of all values in the matrix. By using `apply()`, we can sum over each individual column or row, as shown in the examples below.

```
> sum(X)
[1] 16099

> apply(X,1,sum)       # calculate sum for each row (margin=1)
[1]   393   503 1715 4681 4445 4362

> apply(X,2,sum)     # calculate sum for each column (margin=2)
    x    p1   p2   p3    p4    p5   p6    p7    p8   p9  p10  p11  p12
   52   732  536  803  2732  1591  450  2143  1028 2050  803 1513 1666
```

Because `apply()` can be used with any function, including user-defined functions, very complex processing can be carried out in this way. Even multi-argument functions may be applied—it is simply necessary to set all but one of the arguments to fixed values in the optional `apply()` arguments.

There are other functions such as `lapply()`, `sapply()`, `vapply()`, and `replicate()` that provide the same general functionality as `apply()`, but are optimized for slightly different use cases. The R help system provides the details of each variant.

sample(x, size, replace=FALSE, prob=NULL)

In statistical analyses, it is often necessary to randomly select a subset of items from a list, for example selecting a group of samples to use as a test set when doing pattern recognition. The `sample()` function fulfils this requirement by randomly selecting the specified number of items from a vector. This selection can be made according to a particular probability distribution (set using the `prob` argument) and can be made to pick items with (`replace=TRUE`) or without (`replace=FALSE`) the possibility of each sample being chosen more than once. In the following example, the names of three unique samples are randomly selected from a list of five.

```
> X <- c("GSM455115","GSM455121","GSM455118","GSM455120",
"GSM455125")
> sample(X,3)
[1] "GSM455125" "GSM455121" "GSM455120"
```

sort(x, decreasing = FALSE, ...)

The `sort()` function re-arranges values of a given vector or factor into numerical order (or alphanumerical order if the vector contains strings). The `decreasing` argument is used to specify whether the order should be ascending or descending.

```
> z <- c(0, 4, 11, 29, 7, 1)
> z
[1]   0   4  11  29   7   1
> sort(z)
[1]   0   1   4   7  11  29
> sort(z,decreasing = TRUE)
[1] 29 11  7  4  1  0
```

order(..., na.last = TRUE, decreasing = FALSE)

The `order()` function is used in similar situations to `sort()`, but works in a slightly different way. When passed a vector, instead of returning that vector with re-ordered elements, `order()` returns a new vector indicating the order of the elements. To demonstrate this, we can again use the protein profiles data set. Let's load it into R and remind ourselves what it looks like:

```
> X <- read.table("http://www.bixsolutions.net/profiles.csv",
sep=",", header=TRUE)
```

```
> X
      x   p1   p2    p3      p4   p5   p6     p7    p8    p9  p10   p11  p12
1     0  148    6     5     197    1   12      9     0     4    0    11    0
2     4  185    5     9     180   73    6      5     1     5   12    15    3
3    11  149  177   282     446  400    7      7     3     0    8   223    2
4    29  103  210   299    1264  912    3    599     2     2    6   865  387
5     7   72  131   197     520  171  181    301   411   864  561   266  763
6     1   75    7    11     125   34  241   1222   611  1175  216   133  511
```

Applying `order()` to the column for protein p1 returns a vector indicating the order of the values in that column:

```
> order(X$p1)
[1] 5 6 4 1 3 2
```

This tells us that the fifth item in the column has the lowest value (it is 72), the sixth item has the second lowest (75), then the fourth (103), and so on. This vector has many uses—one obvious example is to re-order the rows in the data table according to the values in the p1 column, like this:

```
> X[order(X$p1),]
      x   p1   p2    p3      p4   p5   p6     p7   p8    p9  p10   p11 p12
5     7   72  131   197     520  171  181    301  411   864  561   266 763
6     1   75    7    11     125   34  241   1222  611  1175  216   133 511
4    29  103  210   299    1264  912    3    599    2     2    6   865 387
1     0  148    6     5     197    1   12      9    0     4    0    11   0
3    11  149  177   282     446  400    7      7    3     0    8   223   2
2     4  185    5     9     180   73    6      5    1     5   12    15   3
```

rev(x)

Sticking with the theme of re-ordering, the `rev()` function reverses the order of the elements in the vector (or any other object for which reversal is defined) passed to it.

system.time(expr, gcFirst = TRUE)

Finding out how long R takes to carry out a given task is essential when optimizing program code, and can also be used to benchmark installations of R. The `system.time()` function provides a built-in method for timing the execution of any R activity. The command or function that you want to time is passed to `system.time()` in the expr argument. It is then executed and the elapsed time, in seconds, taken to execute that function will be returned. The example below demonstrates timing how long it takes R to produce a 10,000 by 10,000 matrix of normally distributed random numbers.

```
> system.time(replicate(1e4,x<-rnorm(1e4)))
   user   system elapsed
   8.64     0.22    8.86
```

t(x)

The `t()` function transposes the matrix or data frame passed to it, such that rows become columns and columns become rows. This is a very common operation in matrix maths and is often used when we need to present a data matrix to a function with a particular orientation. A trivial example of this in use is shown below. Examples of it being used in anger can be found later, in Sections 4.2.3 and 4.2.4.

```
> X <- matrix(1:9,3,3)
> X
      [,1] [,2] [,3]
[1,]    1    4    7
[2,]    2    5    8
[3,]    3    6    9
> t(X)
      [,1] [,2] [,3]
[1,]    1    2    3
[2,]    4    5    6
[3,]    7    8    9
```

table(...)

There are many situations in bioinformatics when we want to extract the frequency distribution for a given factor from a large data set. The `table()` function is a powerful tool for doing exactly that in a single line of code. To demonstrate this, we need some data with a reasonably large number of observations, so let's pull a built-in example R data set into the workspace and take a look at it:

```
> data(iris)
```

```
> View(iris)
```

This data set consists of 150 sets of morphological measurements from three species of iris flower. We can generate a table showing the frequency of occurrence of each petal width per species with the command below. This reveals a clear relationship between species and petal width that would have been very difficult to spot simply by looking at the original data.

```
> table(iris$Petal.Width, iris$Species)
```

	setosa	versicolor	virginica
0.1	5	0	0
0.2	29	0	0
0.3	7	0	0
0.4	7	0	0
0.5	1	0	0
0.6	1	0	0
1	0	7	0

```
1.1          0             3              0
1.2          0             5              0
1.3          0            13              0
1.4          0             7              1
1.5          0            10              2
1.6          0             3              1
1.7          0             1              1
1.8          0             1             11
1.9          0             0              5
2            0             0              6
2.1          0             0              6
2.2          0             0              3
2.3          0             0              8
2.4          0             0              3
2.5          0             0              3
```

The `table()` function provides a lot of additional functionality, so it is worth studying the documentation for it, and for the related `ftable()` function that generates multidimensional frequency tables in the case where three or more variables are investigated simultaneously.

which(x, arr.ind = FALSE, useNames = TRUE)

This is a useful function for returning the elements in an object that fulfil a particular criterion. Like `apply()`, it simplifies program code and speeds things up because we can avoid stepping though an object checking elements one by one. In the example below, a five element numeric array is created. When `which()` is called with the condition `seqlength>20` it returns the indices of the elements that meet that condition.

```
> seqlength <- c(16,47,35,12,45)
> which(seqlength>20)
[1] 2 3 5
```

This is a quick and easy way to filter items out of large data structures according to a particular quantitative characteristic, for example pulling out genes that have a fold change over a particular threshold in a transcriptomics experiment.

4.1.10   The RStudio integrated development environment

Thus far, we have been interacting with R directly, using the very basic interface that it provides. This is fine for grasping the concepts underpinning R using simple examples, but if you are planning more substantial interactions with R we strongly recommend RStudio, an open source integrated development environment (IDE) designed specifically for R. RStudio provides an advanced R console augmented with useful features, including a view of the workspace, a command history, file editor, file browser, package manager, plotting tools, and integrated documentation. This IDE makes the process of working in R much more comfortable, and particularly helps when developing and debugging R programs.

An example of RStudio in action is shown in Fig. 4.6. The program window is split into four panes, the lower left of which is the familiar R console into which commands can be entered and results returned. The pane in the top right summarizes the content of the current workspace, providing a more convenient alternative to the `objects()` function. As well as showing the current workspace objects, it also allows you to view the content of these objects simply by clicking on them. Tools at the top of this pane provide the ability to load and save the workspace, and load text-based data sets (e.g. CSV) into R as a data frame. The top left pane contains a text editor for developing R programs. This editor provides a number of useful R-specific features, including automatically coloured text and auto-completion (activated using the Tab key). Finally, at the bottom right is a multi-function pane that provides a file browser, package manager, online help, and plotting window. The plotting window has an Export button that makes it easy to export plots to files in a user-specified format and resolution.

The beauty of RStudio's IDE is approach is that all these panels work together. So, for example, when a program is loaded it appears in the file editor pane. This program can then be run simply by clicking the Source button at the top of the pane. This sends a `source` command to the R console to execute the program.

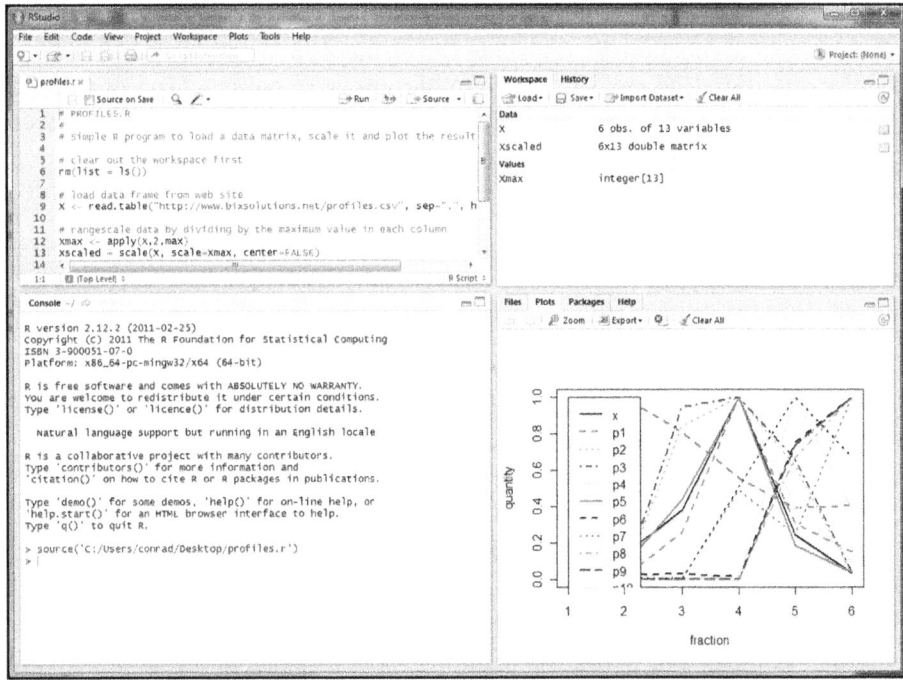

Fig. 4.6 The RStudio IDE running under Windows. The four panes provide (clockwise from top left) a text editor for writing R programs, a list of objects in the current workspace, plotting output, and finally the R console. Most of the panes have multiple functionality, accessed via tabs at the top – for example the workspace pane doubles as a command line history viewer.

Any plots produced by the program automatically appear in the plots pane, and the workspace pane is updated to show objects created by the program. RStudio also has integrated version control, a topic which we cover in more detail in Chapter 6.

RStudio can be downloaded from `www.rstudio.org`, where there are desktop versions for Windows, Mac OS, and Linux. There is also a server version for Linux-based servers, which makes RStudio functionality available remotely via a web browser. This latter option is useful if you set up a dedicated server for computationally intensive R jobs and want to log in from time to time to start analyses and check results.

4.2 Multivariate data analysis

Much has been made of the relatively large size of data sets emanating from modern bioanalytical techniques, such as high-throughput sequencing and gene expression analysis. As mentioned previously, another key characteristic of such data is that it is multivariate, by which we mean that multiple values are acquired from each sample or time point we analyse. These values might be gene expression ratios, protein quantities, or metabolite concentrations. Multivariate analysis techniques have been developed specifically for the investigation of this type of data. Typical tasks include exploratory analysis, where we simply want to visualize a data set in some meaningful way, through to classification, where we seek to assign each sample to a defined class according to the characteristic pattern of the measured variables. In rare cases we may seek to perform multivariate calibration, where multiple measured variables are reduced to a continuous value relating to something of interest. The built-in support for matrix mathematics makes R an ideal platform for multivariate data analysis.

4.2.1 Exploratory data analysis

The aim of data exploration techniques is to provide a way of visualizing variation within large multivariate data sets. This is sometimes an end in itself, but it is also a useful way of evaluating whether the data is of sufficient quality or sufficient information content to warrant further study. For example, there is clearly no point in expending effort attempting to classify samples into different groups according their gene expression profiles if initial exploration of the data shows that there is no sign of correlation between the data acquired and the sample types analysed.

4.2.2 Scatter plots

One of the simplest, yet most effective, forms of exploratory analysis is the construction of scatter plots. A typical application of this in bioinformatics is the identification of differentially expressed genes from microarray data. By plotting, for each gene, a point on a graph at coordinates (a_j, b_j), where a_j is the expression level of gene j in sample A and b_j is the expression level in sample B, genes which show substantially different expression levels between the two samples can be clearly seen. Typically, the expression values are plotted on log scales to provide more clarity to the figure. Genes with similar expression levels fall along a diagonal line across the plot. Genes that fall more than a specified distance from this line can be considered to exhibit a significant difference in expression between the two samples. The definition of a significant difference varies depending on the application, and according to the general level of noise in the data, but typically a two-fold change in expression would be considered significant. Lines marked on the scatter plot representing a two-fold change can be superimposed so that the genes of interest can be clearly seen.

4.2.3 Principal components analysis

A significant limitation of the scatter plot approach is that it is limited to pairwise comparisons, with just two samples in any one plot. If we want to compare the data from more than two samples, or compare the actual expression profiles of multiple genes, then more advanced techniques are required. One such technique is principal components analysis (PCA).

PCA is a way of reducing a large multivariate data matrix into a matrix with a much smaller number of variables (called principal components, or PCs), without losing important information within the data. In mathematical terms, PCA is the reduction of the original data matrix, X, into two smaller matrices, the scores, T, and loadings, P. The product of the scores and the transposed loadings, P', plus a residual matrix, E, gives the original data matrix (Equation 4.1).

$$X = T.P' + E \qquad (4.1)$$

There are a number of algorithms for calculating T and P, the most common being singular value decomposition (SVD)—one of the standard matrix manipulations alluded to in Section 4.1.4. We will not delve further into the maths here, suffice to say that the way that the PCs are calculated means that they are delivered in the order of largest variance first, hence the first PC (PC1) captures the most information in the data, PC2 the second most information, and so on. The scores

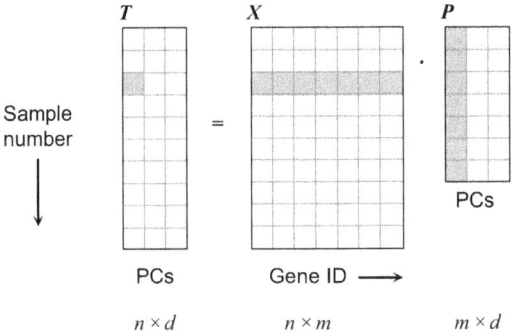

Fig. 4.7 Relationship between the data matrix (X), scores matrix (T) and loadings matrix (P) in principal components analysis. In this simple example, the number of samples, n, is 10, the number of measured variables (e.g. genes), m, is 7, and the number, d, of PCs considered is 3. The highlighted row in X and column in P show what is required to generate the first PC score for sample 3.

matrix T is determined by multiplying X by the matrix of loadings, P, as shown in Fig. 4.7. In simple terms, this means that the scores for a particular sample are weighted sums of the original variables. For example, the first PC score for the third sample in the data matrix shown in Fig. 4.7 would be calculated as:

$$t_{3,1} = x_{3,1}p_{1,1} + x_{3,2}p_{2,1} + x_{3,3}p_{3,1} + x_{3,4}p_{4,1} + x_{3,5}p_{5,1} + x_{3,6}p_{6,1} + x_{3,7}p_{7,1} \qquad (4.2)$$

In many cases just the first two or three components are sufficient to capture the bulk of the variance (hence the bulk of the information) in a given data set. Each sample can then be plotted on a simple two- or three-dimensional graph at the position dictated by its first two or three PCA scores. The relative positions of the samples in this plot indicate the relative similarities between samples, with similar samples appearing at similar positions within the graph. Variance in the higher PCs is often due to experimental noise, so plotting only the first two or three PCs not only simplifies interpretation of the data, it also reduces the noise.

To do PCA in R, we can use the `prcomp()` function, which is part of the `stats` package included in the basic installation of R. A program to perform PCA on the protein profiles data from Section 4.1.7 is shown below. Much of this program will be familiar from the previous examples, indeed it makes use of the `rangescale()` function defined previously. (So, for the program to work, `rangescale.r` should be available in the working directory.) The new material is in the latter part of the program, which performs PCA and then generates a labelled scatter plot in which the position of the point representing each profile is defined by its PCA scores.

```
# PCA_EXAMPLE.R
#
# Program to load in data matrix, calculate principal
# components and plot resulting scores.

rm(list=ls())              # clear workspace

source("rangescale.r")     # define our rangescale function

# load data matrix from file
X <- read.table("http://www.bixsolutions.net/profiles.csv",
sep=",", header=TRUE)

Xscaled = rangescale(X)     # scale the profiles

result = prcomp(t(Xscaled), center=FALSE)   # perform PCA on
                                            # transpose

# extract the scores matrix from the result
scores=result$x

# plot PC1 against PC2
plot(scores[,1], scores[,2], xlab="PC1",ylab="PC2")
```

```
# add labels to points (note 0.005,0.003 offset to avoid
# obscuring points)
text(scores[,1]+0.005, scores[,2]+0.003, names(X))
```

Note that `prcomp()` returns the results in an object of the specially defined class `prcomp`. The PCA scores are contained in the x component of this object, and are extracted using the $ operator. Note also that we had to transpose the data matrix using `t()` to get it into the correct orientation for `prcomp()`. The result of running this program is shown in Fig. 4.8.

4.2.4 Hierarchical cluster analysis

Hierarchical cluster analysis (HCA) is another exploratory data analysis technique which, like PCA, is designed to reveal relationships between samples, or between the molecular entities (e.g. genes) being studied. The result of HCA is a tree diagram, or dendrogram, in which each sample is represented by a branch, and the distance between branch tips indicates the level of similarity between samples. Such diagrams are used in many areas of bioinformatics, due to their ability to represent large multivariate data sets in a reasonably intuitive way.

The dendrograms are created by a recursive process in which the pairwise similarity between every sample and every other sample is calculated. The samples representing the two most similar samples are then joined using branches whose

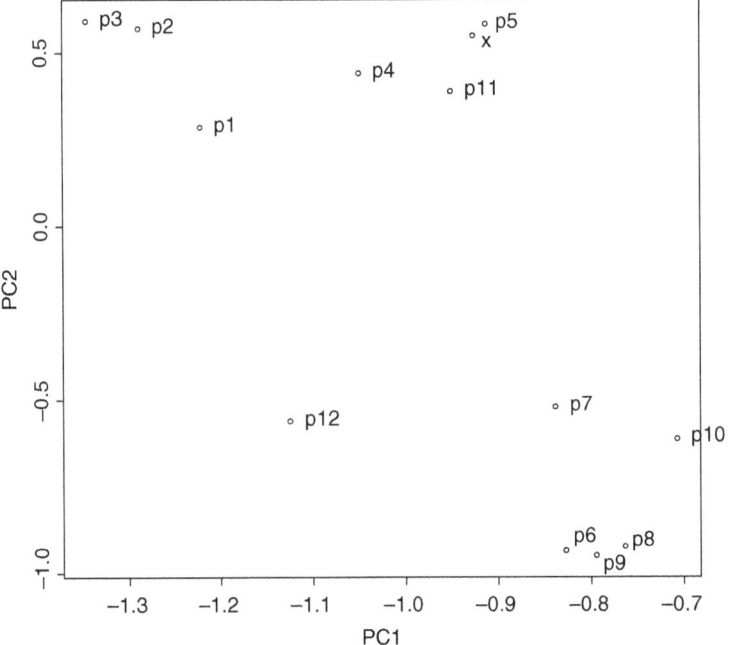

Fig. 4.8 PCA scores plot generated from the profiles data set. The protein of interest (x) and protein p5 appear very close in the plot, indicating that their profiles are similar, suggesting that the two have similar properties. Other groupings, such as p6, p8 and p9 are also eminently sensible if we refer back to the profiles in Figure 4.4.

length is related to the level of similarity between the samples. The process is then repeated, with the two samples already accounted for being agglomerated in such a way that they can be considered as a single sample. This process is repeated until all samples have been joined together. This method is capable of displaying the relationship between entities in a data set, and unlike PCA it is easily extended to very large data sets without cluttering the plot or losing information.

All hierarchical clustering follows the general approach set out above, but there are a lot of variations in how the similarity between samples is calculated, and how samples are joined together. The primary method of determining the level of similarity between two samples is by calculating the distance between them in the multidimensional space of the measured variables (e.g. the quantity of protein measured in the fractionation example). The process is easy to understand for two measured variables, as shown in Fig. 4.9, but is equally applicable to any number of variables. Taking the two-dimensional case in the figure as an example, the most intuitive distance measure is the Euclidean distance—the shortest distance between the two points. This distance, d, is trivially calculated using Pythagoras' theorem:

$$d_{A,B} = \sqrt{(A_1 - B_1)^2 + (A_2 - B_2)^2} \tag{4.3}$$

Extending this to further variables simply involves adding the squared differences for the other variables within the square root. For the case of N variables, the calculation for each sample would be:

$$d_{A,B} = \sqrt{\sum_{n=1}^{N} (A_n - B_n)^2} \tag{4.4}$$

However, the Euclidean distance is not the only measure. If we want to particularly emphasize samples which are markedly different from others, we can amplify the distance by squaring it. For the two-dimensional example, the squared Euclidean distance is simply Equation 4.3 with the square root removed.

If we want to emphasize the difference between samples according to the value of the largest difference between values of a single variable, regardless of

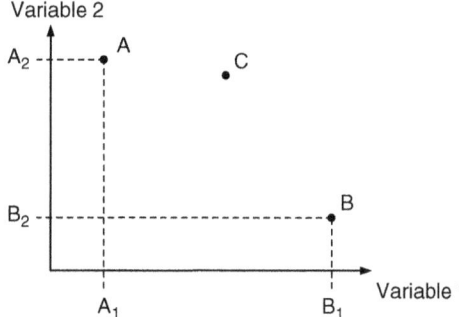

Fig. 4.9 Illustration of distance between samples in variable space. In this case, we consider two samples, A and B, with two measured variables. There are many ways in which the distance between the samples can be calculated.

what that variable is, we can use the maximum distance (sometimes called the Chebychev distance):

$$d_{A,B} = \max|A_n \quad B_n| \tag{4.5}$$

R has a built-in function, `dist()`, for calculating the distance between objects described in a multivariate data matrix. The distance measure used is selected by setting the `method` parameter to one of the following methods: `"euclidean"`, `"maximum"`, `"manhattan"`, `"canberra"`, `"binary"`, or `"minkowski"`. The command for calculating the Euclidean distance matrix is shown below. Note that the data frame, `Xscaled`, is transposed, using the `t()` function, because the distance function expects each row of data to represent an object.

```
d <- dist(t(Xscaled), method = "euclidean")
```

A dendrogram object can then be created from this distance matrix using R's `hclust()` function. This object can then be plotted using the `plot()` function, resulting in the dendrogram shown in Fig. 4.10. The commands to do this are shown below. Note that the `plot()` function detects that an `hclust()` dendrogram object has been set to it, and deals with it accordingly.

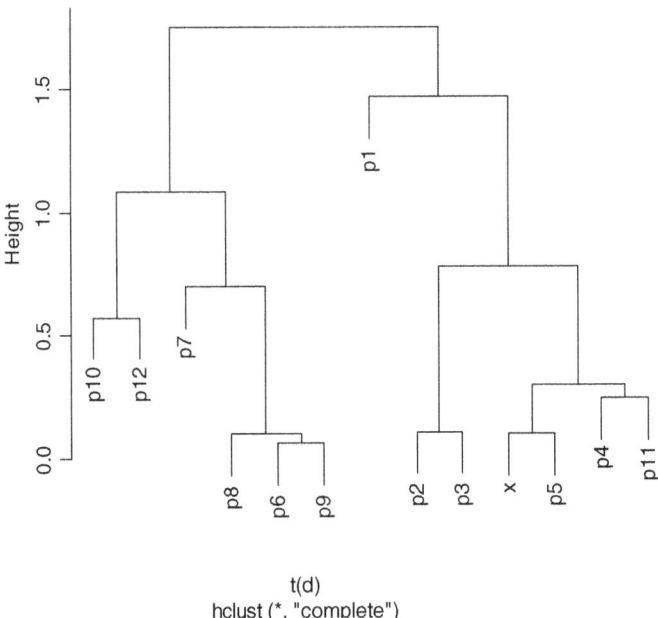

Cluster dendrogram

Fig. 4.10 Dendrogram generated from the protein profiles dataset. Note that the profile of protein p1 is shown as very different to the other profiles. Looking at Figure 4.4, this is not surprising as the profile of p1 is clearly different from the others, effectively peaking much earlier. In the PCA plot in Figure 4.8, p1 did not appear as such an outlier because, taking a global view of the dataset across all samples, the variance associated with the first few points in the profile was not particularly significant, so p1 would only appear as a outlier if less significant PCs were plotted (i.e. PCs other than 1 and 2).

```
dendrogram <- hclust(t(d), method = "complete", members = NULL)
plot(dendrogram)
```

Just as there is a choice of method for calculating the distance, or similarity, between two samples, so there is a range of linkage algorithms for joining clusters together as the clustering process progresses. Essentially, a linkage algorithm defines which point in a cluster is used to represent that cluster when the distances are calculated. The most obvious approach is the weighted average, where each cluster is represented by the average position in the variable space of the samples that make up the cluster—this essentially represents the centre of gravity of the cluster. Other popular methods include complete and single linkage. Using complete linkage, the distance between two clusters is calculated using the largest distance between individual points in those clusters—this promotes tight clusters over those with more variance. Single linkage is the opposite, where the distance is measured according to the closest two points in the two clusters—this allows clusters to be joined on the basis of just two similar samples, regardless of the spread across the variable space that each cluster exhibits.

A more advanced linkage algorithm, called Ward's method, moves away from simple geometric solutions, and joins clusters not just on simple distance measures but according to which of the agglomerated clusters will have the least variance. This approach has the benefit of promoting tight clusters, but doesn't suffer the sensitivity to outliers found in complete linkage. For this reason, it is often used as the linkage algorithm of choice. The `hclust()` function supports all the above linkage methods, and these can be selected by setting the `method` parameter to `"average"`, `"complete"`, `"single"`, or `"ward"`.

Clearly, there is a wide range of possible combinations of parameters for performing HCA, and experience shows that these can result in markedly different dendrograms, leading to potentially different interpretations of the data set. It is therefore very important to ensure that the particular distance measures and linkage algorithms used are appropriate, either by considering in detail how each approach works and how this relates to the particular data set being analysed, or by following best practice described in the literature for similar data sets. It is also important to consider the *robustness* of the results obtained—if a particular clustering behaviour is observed only in the dendrogram created by a very specific set of HCA parameters, then it may not be wise to assume that the clusters genuinely represent the relationships between the samples.

In most applications, the purpose of hierarchical clustering is to reveal relationships between samples according the multiple measured variables for each sample. Sometimes, however, we may instead (or also) want to reveal the relationships between the measured variables. A good example of this is gene expression data, where instead of clustering the samples, we often want to cluster the genes according to the similarity of their behaviour across those samples. Doing this in R is simply a matter of using the `t()` function to transpose the data prior to the process of creating a dendrogram, so that the samples become variables and vice versa.

In situations where it is beneficial to get an indication of the detailed content of the data matrix, as well as both the relationship between the samples and between the variables, R has an excellent built-in function called `heatmap()`, which uses `dist()` and `hclust()` to perform hierarchical clustering on the data matrix and its transpose and then appends dendrograms generated from this to the two sides of a square image representing the original data matrix. The rows and columns of the data matrix are re-arranged such that they line up with their respective dendrogram branches. The results of using this function with microarray data are shown in a later example.

4.2.5   Pattern recognition

In many bioinformatics applications, we are particularly interested in being able to classify objects (be they samples, patients, or molecules) according to some measured characteristics. For example, many papers have been published showing how genomic, metabolomic, or proteomic profiles can be used to classify biological samples according to phenotype, for example to differentiate between *healthy* and *diseased* states for particular diseases. This is important because it raises the possibility of detecting diseases according to the behaviour of multiple biomarkers, rather than a single biomarker as has traditionally been the case. This has the potential to improve accuracy of diagnosis, simply because it takes into account more biological factors. Indeed, it is also the first step towards discovering diagnostic biomarkers, which is a valuable activity in its own right. Such classification could be done by looking at the output of an exploratory technique such as HCA or PCA, but we really want an automated computational method if we are to ensure objectivity and high data throughput.

Pattern recognition is one name given to the data analysis approach used to achieve this. It involves building a classification model from a data matrix acquired from samples of known class. The model is effectively a mathematical transformation relating the measured variables to a number indicating the class of sample (e.g. 0 for healthy, 1 for diseased). Crucially, a separate matrix of data from samples of known class is collected and used to test the resulting classification model. The performance of the classification model can therefore be quoted using easily understood quantitative measures such as the proportion of test samples that are correctly identified by the model. Alternatively, the performance can be specified in terms of the specificity and sensitivity of the model, which are derived individually from the proportion of correctly identified positive samples and correctly identified negative samples. Some pointers for starting to build classification models in R are provided in Section 4.3.3.

4.3   R packages

So far in this chapter, we have limited ourselves to using the basic installation of R. This is useful in its own right, thanks to a number of powerful built-in packages, including `graphics` and `stats`, functions from both of which we have already used in this chapter. It is possible to find out which packages are present in your particular installation of R by typing `library()` at the command prompt. Although R clearly has a lot of functionality with just the basic packages, what makes R particularly popular among the bioinformatics community is the vast number of high quality user-contributed packages that are available. The number of packages available is already impressive, and more are being released all the time. Indeed, for many bioinformatics applications, there is probably an R package out there that does at least part of what you want—it is just a case of finding it. This clearly saves a lot of coding effort, but you need to be prepared to spend a substantial amount of time searching for packages, finding out how to use them, and incorporating them into your own analysis pipeline. At the time of writing there is no easy way to find the right package for a specific task, other than browsing through the CRAN archive or searching the web with relevant queries, for example 'R SVM package' for a package that implements support vector machines.

If you know the name of the package you want, the process of downloading and integrating it into R is actually very straightforward. You simply select a package from the CRAN archive and install it from within R, either via the Packages menu in the R GUI if that is available in your operating system (Windows or Mac OS), through the Linux package manager if you are using Linux (like Perl modules, R packages are often available through the regular Linux package managers), through the Tools menu or Packages tab in RStudio, or by typing the following at the command line and following the on-screen instructions:

```
install.packages()
```

Alternatively, packages can be added from local archive files (.zip or .tar.gz), which you might have acquired from the web, from a colleague, or perhaps even written yourself (see Section 4.3.4). As with the CRAN packages, this can be done via the R GUI in Windows and Mac OS, or under Linux with a command issued from the Linux command line (not the R console):

```
R CMD INSTALL name_of_package
```

Some of the best known, and most useful, bioinformatics packages are part of the Bioconductor Project (`www.bioconductor.org`). Bioconductor packages fall roughly into two groups. Packages in the first group are designed to provide basic infrastructure support for doing routine tasks—such as fetching data from repositories and manipulating data for later analysis—these can save a huge amount of development time as you can avoid re-inventing the wheel when it comes to tedious things like parsing specific data formats. The second type of Bioconductor package provides R implementations of innovative techniques for the analysis of biological data. It is gradually becoming the norm that someone publishing a new data processing algorithm will make the algorithm available as an R package for the benefit of the community, possibly as part of Bioconductor. For historical reasons, many of the Bioconductor packages are oriented around microarray data analysis, but many of these packages are being generalized or expanded to cover other data types.

4.3.1   Installing and using Bioconductor packages

Bioconductor packages can be installed like any other R packages, for example Bioconductor's popular `limma` package for microarray analysis can be seen in the list of CRAN packages shown by `install.packages()`. However, as Bioconductor is made up of a number of individual packages, many of which are dependent on one another, it is recommended that newcomers start by installing the core packages using the `biocLite.R` script provided at `www.bioconductor.org`. This can be done by typing the following at the command line:

```
source("http://www.bioconductor.org/biocLite.R")
biocLite()
```

This can take a few minutes as the packages are downloaded, unpacked, and installed. Typing `library()` will confirm that R now has a number of additional libraries installed, with names like `Biobase` and `Biogenerics`. Note that although the installation of the package copies the relevant files, and makes the package available to R, to actually use the functions within that package it is necessary to load the package into the current R session using the `library()` function. Once a package has been loaded in that way, it is possible to find descriptions of the functions within that package using the `help()` function as described earlier. This obviously requires you to know the names of functions in the packages. Larger packages often have general information about the package, which can be accessed by passing the name of the package to the `help()` function. For example, to get started with Bioconductor's Biobase package, we would type:

```
library(Biobase)    # load package
help(Biobase)       # tell us something about the package
```

After loading Bioconductor packages, PDF documentation covering key topics can be accessed from within R by typing `openVignette()` at the R prompt, and selecting from the list of options that appear. You can also access the PDF documentation directly via the web, which is useful for researching which packages to use without having to install them.

Getting data from GEO using the GEOquery package

Other elements of Bioconductor can be added at any time, simply by passing the name of the desired package to the `biocLite()` function. In the example that follows, we are going to see how a Bioconductor package called `GEOquery` can be used to load data from the NCBI's Gene Expression Omnibus (GEO) into R. The `GEOquery` package can be added to R using the command below.

```
biocLite("GEOquery")    # add GEOquery package to R
```

To make the functions from the package available for use in the current R session, we must load it using the `library()` function:

```
library(GEOquery)       # load package
```

The `GEOquery` package contains a function called `getGEO()`, which provides an excellent example of the powerful capabilities of Bioconductor packages. `getGEO()` provides a one-line solution to loading data from GEO straight into R. For example, to load the data set with accession number GDS2577 into an object called `dset`, the command is:

```
dset <- getGEO("GDS2577")
```

The GDS2577 data set was collected as part of a study into tissue repair mechanisms (Otu *et al.*, 2007). The data set contains gene expression data collected from two very different mouse tissue types: developing embryonic liver and regenerating liver. Furthermore, each tissue type was analysed at multiple time points.

It may take a minute for `getGEO()` to download and parse the data, due to the amount of data involved, and during the parsing stage R may become unresponsive. Patience is a virtue here—eventually control will be passed back to the R command line. The whole data set is now stored in the object `dset`. To avoid having to go through the download and parsing process again, you might want to save this object locally using R's `save()` function. This function saves the specified object, or list of objects, in a file with the specified name. In the example below, we save the `dset` object in the file "GDS2577".

```
save(dset, file="GDS2577")    # save dataset to the file
                              # "GDS2577"
```

The object can then be quickly loaded back into workspace in a future session using the load function:

```
load("GDS2577")               # load dataset from file "GDS2577"
```

The `dset` object created by `getGEO()` is an instance of a fairly complex data class, the structure of which mirrors that of the GEO database (specifically GEO's GDS class in this case). It is possible to extract information from the GDS object using the operators discussed towards the end of Section 4.1.5. For example, typing `dset@header$description` at the R prompt will return the description of the data set from the header slot in the GDS object, which contains metadata from the GEO record. However, this approach is not recommended as it depends on the internal structure of the object, which may be different in future versions of `GEOquery`. Instead, you are encouraged to use *methods* included in the `GEOquery` package that extract information from the data structure for you. One such method is `Meta()`, which is used to return metadata associated with the data set, including the description of the study, which would be retrieved by typing `Meta(dset)$description`. The `Columns()` method returns a data frame containing information about the individual samples. The first three columns returned from the GDS2577 data set are shown below (a fourth column called *description* has been omitted as the entries within it are rather detailed descriptions of each sample that would not fit on the page).

```
> Columns(dset)
          sample        specimen              time
1         GSM161128     developing liver      10.5 dpc
2         GSM161129     developing liver      10.5 dpc
3         GSM161130     developing liver      11.5 dpc
4         GSM161131     developing liver      11.5 dpc
5         GSM161132     developing liver      12.5 dpc
6         GSM161133     developing liver      12.5 dpc
7         GSM161134     developing liver      13.5 dpc
8         GSM161135     developing liver      13.5 dpc
9         GSM161136     developing liver      14.5 dpc
10        GSM161137     developing liver      14.5 dpc
11        GSM161138     developing liver      16.5 dpc
12        GSM161139     developing liver      16.5 dpc
13        GSM161108     regenerating liver    0 h
14        GSM161109     regenerating liver    0 h
15        GSM161110     regenerating liver    1 h
16        GSM161111     regenerating liver    1 h
17        GSM161112     regenerating liver    2 h
18        GSM161113     regenerating liver    2 h
19        GSM161114     regenerating liver    6 h
20        GSM161115     regenerating liver    6 h
21        GSM161116     regenerating liver    12 h
22        GSM161117     regenerating liver    12 h
23        GSM161118     regenerating liver    18 h
24        GSM161119     regenerating liver    18 h
25        GSM161120     regenerating liver    24 h
26        GSM161121     regenerating liver    24 h
27        GSM161122     regenerating liver    30 h
28        GSM161123     regenerating liver    30 h
29        GSM161124     regenerating liver    48 h
30        GSM161125     regenerating liver    48 h
31        GSM161126     regenerating liver    72 h
32        GSM161127     regenerating liver    72 h
```

This output shows the two types of sample in the data set, regenerating liver and developing liver, and shows which sample is which. For each liver type, samples were taken for analysis in duplicate at different time points. A data frame containing the actual expression data from these samples can be extracted from the dset data set object using GEOquery's Table() method. One way to inspect the data is by assigning it to another object and then using the edit() function on that object.

```
X <- Table(dset)
X <- edit(X)
```

Looking at the table in the data editor provides a detailed view of the whole data set. Each row is associated with a particular gene probe, which is identified by the spot ID_REF (e.g. 1415672_at) and the gene name (e.g. Golga7). Each column is associated with a particular sample, identified by its unique GEO ID, for example GSM161129. The data clearly takes the form of a matrix, which means it is suited to many different types of analysis, as explained earlier in this chapter. Indeed, since the object returned by `Table(X)` is a data frame, we can deal with it exactly as we dealt with data frames earlier. However, to make full use of the many functions that Bioconductor provides for microarray analysis, it is necessary to convert the data set object from the GEO-specific GDS class into a more generic object based on the Biobase *ExpressionSet* class. `GEOquery` provides a function, `GDS2Set()` which does this conversion for us. If the `do.log2` parameter is set to TRUE, the expression values are transformed by taking logs to base two during the conversion process, something that is commonly done in microarray data analysis.

```
eset <- GDS2eSet(dset, do.log2 = TRUE)    # covert data to
                                          # experiment set
```

Again, this may take some time to complete, due to the size of the data set, but now that we have the data in this generic Bioconductor format we can analyse it using a wide range of functions from different Bioconductor packages. For example, we can extract a matrix containing the actual expression data from the data set using Bioconductor's `exprs()` method, and use the familiar `summary()` function to get some headline statistics on the gene expression for each sample.

```
Y <- exprs(eset)    # extract the expression values
summary(Y)          # show some statistics for each sample
```

Among other things, this reveals that there is a large number of missing values associated with each sample, as evidenced by the number of NA entries. About 3% of 45,101 values are missing from most of these samples, which is not uncommon in such data. We can also have a cursory look at the relationship between the samples by using R's `heatmap()` function to plot a heatmap and associated dendrograms for a subset of the data (using just the first 50 probes in this case, labelled with the identifiers from the GEO data set).

```
heatmap(Y[1:50,],labRow=X[1:50,2])
```

The result is shown in Fig. 4.11. A difference between the sample types can clearly been seen, even with this very small arbitrary selection of gene probes. This is not unexpected given the very different biology of regenerating and developing tissue, and the clustering is very similar to that seen in the heatmap for this experiment at the GEO website, which takes into account data from all the probes. This kind of quick and dirty analysis of data is not recommended! Normally, we would use proper gene selection criteria and take into account factors such as data scaling, but the aim of this example is simply to give a taste of how—with very little code—data can be imported into R and visualized. Indeed, the danger of using powerful tools such as R and Bioconductor is that complex

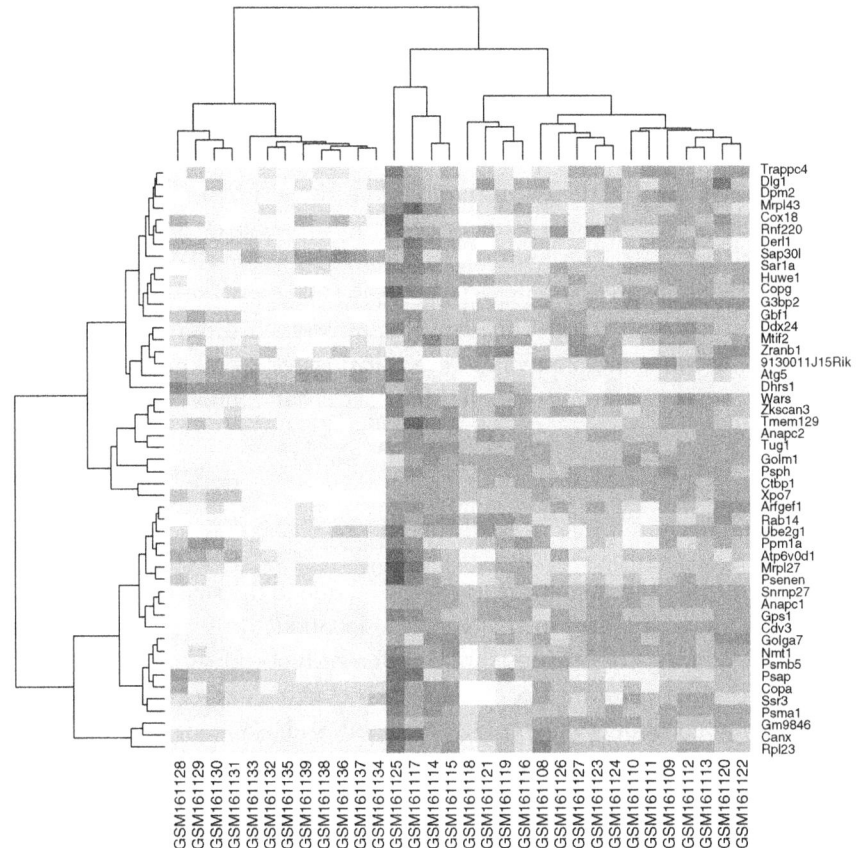

Fig. 4.11 Heatmap and associated dendrograms generated from the first 50 probes from GEO dataset GDS2577. The expression of individual genes is clearly different between the two sample types (developing and regenerating liver).

functions can be used without any knowledge of their underlying statistical algorithms, potentially leading to bogus results. We therefore recommend that you always gain a good understanding of algorithms that you are using by reading relevant background material, or by consulting experts in the subject.

Getting data from GEO is quite handy for a tutorial like this, because the pre-processing has already been done before the data is deposited in GEO. In many applications, the starting point is raw data from an instrument (e.g. .CEL files from Affymetrix microarray analysis), and pre-processing is required before any meaningful analysis can be done, but Bioconductor includes a number of packages which include functions to make the transformation from raw data to *ExpressionSet* object fairly painless. These packages tend to be platform dependent, and include `affy`, `arrayMagic`, and `oligo`.

Bioconductor is a truly massive resource, and is one of the reasons for the popularity of R among bioinformaticians, particularly those dealing with microarray data. In this section, we have only scratched the surface of its functionality. The main reason for not going further is the fact that the particular Bioconductor functionality of interest is naturally application dependent. Indeed, if you are working on something other than microarray data you may well have to look elsewhere to find useful packages. To learn more about Bioconductor we thoroughly recommend the book *Bioinformatics and Computational Biology Solutions Using R and Bioconductor* (Gentleman *et al.*, 2005). Such books should be quite accessible now that this chapter has helped you overcome the hurdle of getting started with R.

4.3.2 The **RMySQL** package for database connectivity

In the previous chapter we saw how easy it is to access MySQL databases in Perl programs using the Perl DBI modules. Similar connectivity is provided in R by a package called RMySQL, and the DBI package on which it is built. All we need do to enable this functionality is to install the RMySQL package from CRAN and load it into R. The associated DBI package will be installed and loaded automatically. (If you are using Linux, it may be far more straightforward to search your package manager first as this can solve problems with unmet dependencies.) Together, these packages allow us to connect to a MySQL database, find out information about it, input data, extract data directly into R objects, and even issue standard SQL queries like those described Chapter 2. Similar packages are available for other RDBMSs, but we focus on MySQL here since it is the RDBMS of choice in this book. After downloading from CRAN, to load the RMySQL packages into R ready for use, we must not forget the command:

```
library(RMySQL)
```

The examples below show how the available functions can be used to interact with the example database from Chapter 2.[1] If you have that database to hand, you can issue these commands (inserting the correct username and password) and you will get results back from the database. The process is very similar to using Perl DBI, in that the first step is to establish a connection to the database. This is done with the command below (entered on one line).

```
dbh <- dbConnect(dbDriver("MySQL"), dbname = "PCR_experiment",
user="conrad", password="donuts")
```

If the connection is successful, a database connection handle is returned (in this case into dbh), and this is then used to refer to the database in subsequent commands. For example, we can get a list of tables in the database by passing the handle to the dbListTables() function. This is very similar to the MySQL SHOW tables command, except that the tables are returned as a list within R instead of just being printed to the screen.

```
> dbListTables(dbh)
[1] "experiment" "kit" "kit_order" "scientist" "supplier"
```

[1]If you don't already have it, you can create and populate this database using the PCR_database_ create.txt and PCR_database_populate.txt source files available at www.bixsolutions.net.

Where this functionality gets really useful is when we use much more powerful functions such as `dbReadTable()`, which allows us to read data directly into a data frame, and `dbWriteTable()`, which is used to populate a table with data from R. An example of an R session in which the contents of a whole table is extracted from the `PCR_experiment` database (previously opened with handle in dbh) is shown below.

```
> orders <- dbReadTable(dbh,"Kit_order")  # copy data to data frame
> orders                                  # display content of data frame
  order_number manufacturer            kit_name           supplier
1  1           The Epsilon Kit Company Basic PCR Kit 1    Epsilon Chemicals
2  115         The Epsilon Kit Company Basic PCR Kit 2    Epsilon Chemicals
3  121         The Epsilon Kit Company Basic PCR Kit 1    Epsilon Chemicals
4  380         Simply Solutions        PCR Visual Dye Kit Experiments_R_US
```

Reading whole tables into R somewhat defeats the purpose of having the data in a database—normally it is desirable to only copy out the data you need, to minimize memory usage and maximize performance. This can be achieved by issuing queries using `dbGetQuery()`. In the example below, this function is used to import just the costs of PCR kits from the `Kit` table of the example database.

```
> dbGetQuery(dbh,"SELECT kit_cost FROM Kit")
  kit_cost
1   49.99
2   19.99
3   29.99
```

Indeed, any valid SELECT query can be executed in this way, and the results will be returned to a data frame in R, allowing you to process those results using all the power of R and its associated packages. This ability to harness the complementary strengths of R and MySQL makes for a very powerful combination. In the example interaction below, one of the more complex queries from Chapter 2 is issued and the output placed in a data frame called `result`.

```
> result = dbGetQuery(dbh,"SELECT scientist_email,
COUNT(scientist_email) FROM Experiment GROUP BY scientist_
email;")
> result # check the results
  scientist_email COUNT(scientist_email)
1   c.bessant@bixsolutions.net         1
2   d.oakley@bixsolutions.net          2
3   i.shadforth@bixsolutions.net       2
```

Even SQL commands that make major changes to the database, such as CREATE and DROP statements, can be issued by calling `dbGetQuery()`.

Finally, as with Perl, it is good practice to disconnect from the database once you have finished with it, by calling the `dbDisconnect()` function.

```
dbDisconnect(dbh)
```

4.3.3 Packages for multivariate classification

There exists a plethora of methods for constructing a classification model—far more than can be dealt with in detail here, but the details can be found in chemometrics textbooks (Brereton, 2009; Otto, 2007). Provided that samples from similar classes cluster well in a PC scores plot, one of the easiest solutions is to divide up the scores plot into sections using a collection of linear boundaries. New samples are then identified according to which side of the boundaries they fall on. This approach is referred to as linear discriminant analysis (LDA). LDA is capable of automatically generating the boundaries using fairly simple mathematics, and the technique can be extended to multiple dimensions—in three dimensions the boundary becomes a two-dimensional plane, and in higher dimensions it's a hyper plane. This means that LDA can be used on the original data matrix as well as on PCA scores, regardless of the number of variables measured. An R implementation of LDA is provided as the function `lda()` in the MASS package included in the basic R installation.

In more complex data sets, where there are many classes of sample or classes of sample which cluster in an unusual shape or with a lot of variance, it is not always possible to separate classes using simple linear features defined by LDA. There are various approaches that can be used for these tougher problems, including support vector machines (of which there is an implementation in a CRAN package cryptically called `e1071`), random forests (see the `randomForest` package), and neural networks (with the `nnet` package).

4.3.4 Writing your own R packages

In the future, after acquiring a wealth of R experience, you might end up producing some really useful functions that you would like to share with the community. The best way to do this is by making these functions available as a package. Anyone can create a package, using commands available within R, and anyone may submit a package for inclusion in CRAN.

An R package is not just a collection of functions—it must also include documentation and various other bits of information about the package. A package may also include data, demos, and examples. When putting a package together, these different components need to be placed in specifically named files and subdirectories so that R knows where to find the relevant information when someone installs your package on their computer. If you explore the directory into which you have installed packages (typically the `library` subdirectory of wherever you installed R) you will see that each package has its own directory, each containing files and directories with standard names, such as CONTENTS, `help`, and so on. The whole process of creating a package is necessarily prescriptive, with a particular data format being required for the documentation, and special attention being paid to filenames to ensure that the package functions on all the operating systems that R supports. Perhaps most importantly, you also need to make sure your package is thoroughly tested, and is optimized for speed and memory usage. For these reasons, production of an R package is really only a task for a very experienced R programmer so it is therefore not sensible to go into details in this introductory chapter. Suffice to say, producing your own packages is possible, and if you want to know more you should check out the *Writing R Extensions* PDF manual that comes with R.

4.4 Integrating Perl and R

As we have emphasized throughout this book, increasingly few bioinformatics tasks can be solved efficiently using a single tool. Integration of tools is therefore crucial. We have already seen above how easy it is to hook up R to MySQL to enable R's sophisticated analysis and visualization capabilities to be applied to data stored in a relational database. What about integrating Perl with R? One of the great benefits of R's command-line interface is that it is very easy to construct commands outside of R and pass them to R for processing. This means that we are able to make use of the visualization capabilities of R, and the analysis capabilities provided by R and the plethora of available packages, from within Perl programs. Furthermore, the fact that R is open source means that such hybrid software solutions can be distributed without concerns about licensing. An example of integrating Perl and R, for the purpose to generating graphs within web pages, is discussed further in Chapter 5.

4.5 Alternatives to R

There are many software packages available for the analysis of biological data. Such software may be freely available via the web, it may come bundled with an instrument (for example microarray scanners are likely to ship with image processing software), or the software may be sold commercially (for example, for protein identification from mass spectrometry data). Almost without exception, these packages feature graphical user interfaces to make their full range of functionality available to the average biologist. As mentioned earlier, such user interfaces actually make it more difficult to build bioinformatics solutions, particularly where there is a need to integrate with other tools. In considering genuine alternatives to R, we therefore restrict ourselves to packages that are primarily command-line based, and which have native support for storage and manipulation of matrices, something that we have seen is key in many bioinformatics applications. It should be noted that this section is not exhaustive, but it captures the main packages that we have seen in use across the bioinformatics community.

4.5.1 S+

S+ is a commercial statistics platform from TIBCO software (`www.insightful.com`). In many ways, it is the obvious alternative to R, as both R and S+ are based on the same underlying statistical language—S. Most code is therefore interchangeable between the two packages, and some R packages can be used with S+. TIBCO also offer their own S+ specific add-ons, called *modules*. Having said that, although R and S+ share the same core language, there are differences in the S+ implementations of additional features, such as devices, certain graphics

functions, and Internet connectivity. So, for example, the main example programs in this chapter will not work in S+ without modification.

The main benefits that S+ offers are a more developed user interface, and more
consistent documentation. This makes the learning curve of S a little less challenging than that of R, but the downside is that there are fewer S users than R users,
so free help and code are harder to find on the web. Due to the similarity with R,
there is little to add about S+, except to say that TIBCO tend to offer free trials of
the software, particularly for students, and that is by far the best way to find out
whether it is the tool for you.

4.5.2 Matlab

Matlab from Mathworks (`www.mathworks.com`) is another commercial alternative to R, currently available for a range of operating systems including
Windows, Mac OS, and Linux. It has a strong pedigree in numerical data analysis
in the engineering sector and its use has more recently spread to the analysis of
chemical and biological data. Being a commercial package, it has many of the
same benefits as S-Plus, particularly excellent and consistent documentation.
There is a lot of common functionality between Matlab and R, but the syntax
is different. A sample Matlab session is shown below. The double arrow (>>) is
the Matlab command prompt, and the % symbol is used to denote the start of a
comment.

```
>> x = [2.0, 2.8, 3.9, 4.0, 4.8, 6.5]    % assign vector to x
x =
    2.0000    2.8000    3.9000    4.0000    4.8000    6.5000
>> y = round(x.^2) % square elements of x and round
y =
    4       8       15      16      23      42
>> plot(x, y, 'b:') % plot y vs x as dotted blue line
```

If you are moving from Matlab to R, or vice versa, there is a very useful document on CRAN called *R for Octave Users* (`cran.r-project.org/doc/contrib/`
`R-and-octave.txt`), which catalogues direct relationships between Matlab
syntax and the R equivalent. Alternatively, if you are moving from Matlab to R,
there is an R package called matlab available via CRAN that provides a subset
of popular Matlab functions for use in R, such as `imagesc()`, `ones()`, and `re-
shape()`.

Matlab offers similar programming capabilities to R—programs can be written in text-based `.M` files, and can include loops, conditional statements, and
user-defined functions. Matlab also has a reasonable object-oriented programming model, which can be easier to get to grips with than R objects. Matlab
comes complete with a fully integrated IDE, which is very much like RStudio.
The IDE includes a graphical user interface designer called GUIDE, which allows
for the creation of front ends for Matlab code, such as the one shown in Fig. 4.12.
Although GUIs can be created in R, the process is not currently as slick as in
Matlab.

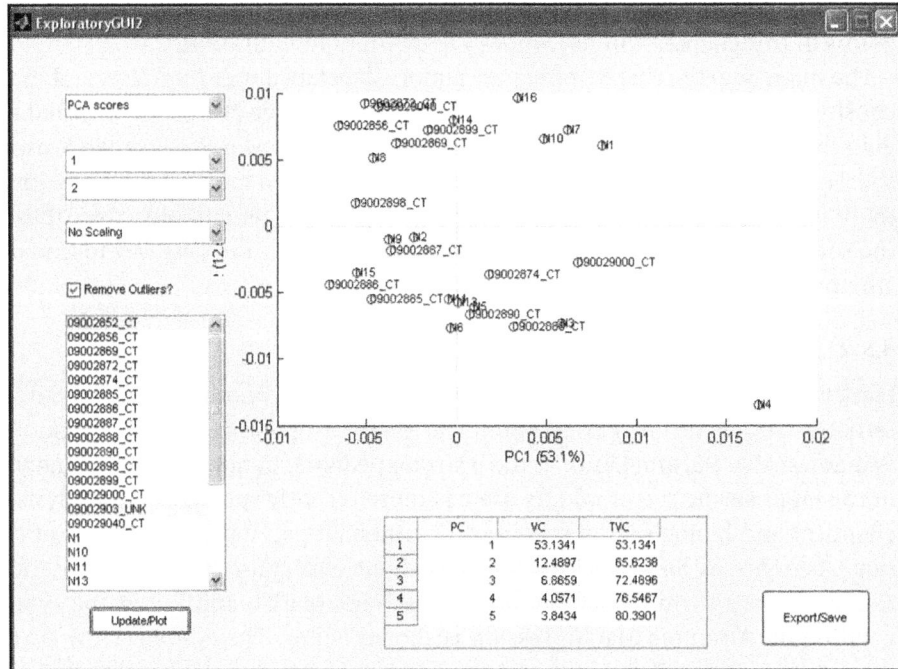

Fig. 4.12 A graphical front end for an exploratory data analysis application, created in Matlab using GUIDE.

Visualization is another area where Matlab challenges R. Matlab's graphs look smarter than R base graphics (though the best R graphics packages can produce output to rival Matlab) and as well as being able to manipulate the figures using low-level graphics functions like those in R, all figures are interactive and their detailed appearance can be tweaked via the graphical user interface. This is particularly useful when preparing figures for publication.

Toolboxes are the Matlab equivalent of R packages. Many toolboxes are sold by Mathworks themselves, but there are several third-party commercial toolboxes on the market, and even more freely available efforts. There is a dedicated bioinformatics toolbox, which provides a reasonable core of functionality, but does not have the breadth of coverage provided by Bioconductor.

The great disadvantage of Matlab is that it costs money, and people with whom you may want to share your programs may not have access to it. Mathworks do sell a compiler that allows Matlab programs to be made available as standalone entities that don't require the user to have a Matlab license, but this is not quite the same as being able to pass someone your source code.

4.5.3   Octave

Octave, also referred to as GNU Octave, is to Matlab was R is to S-Plus. It is a free package that has very similar syntax to Matlab. Indeed, with sufficient care, programs can be written that run on both Matlab and Octave without modification. Octave's website is `www.octave.org`, and the software can be freely downloaded from there. Octave is primarily developed for Linux-based systems. At the time of writing, a Mac OS version can be downloaded from the website, and Octave can be made to run under Windows with some effort. Interactions with Octave are via a command-line interface very similar to R and Matlab. The similarity with Matlab means that Matlab syntax can be typed in right away, with familiar results. For example, the interaction below mimics the Matlab example given previously.

```
octave:1> x = [2.0, 2.8, 3.9, 4.0 4.8 6.5] % assign vector to x
x =
   2.0000    2.8000    3.9000    4.0000    4.8000    6.5000
octave:2> y = round(x.^2) % square elements of x and round
y =
    4       8      15      16      23      42
octave:3> plot(x, y, 'b:')
```

Like R and Matlab, Octave offers various visualization capabilities. Visual results are actually rendered by a separate program, Gnuplot, to which Octave sends commands to produce the required graphics. In general, the visual output is inferior to both R and Matlab.

Programming in Octave is very similar to programming in Matlab, thanks to the similarity of syntax between the two. Like R and Matlab, Octave's functionality can be increased through the addition of packages, many of which can be found on Octave-Forge (`octave.sourceforge.net`). Generally, the Octave community tends to be focused more on engineering than biology, so there are slim pickings for the bioinformatician in the packages available.

4.6 Summary

The ability to reliably perform advanced data analysis using the latest algorithms is a key requirement in many bioinformatics applications, and R allows us to fulfil this requirement in most cases. R's key features are the inherent support for importing and manipulating data matrices, the ease with which graphs can be constructed, and the dynamic ecosystem that has delivered a plethora of packages covering the majority of routine bioinformatics tasks. We have seen in this chapter how R can provide a wide range of analyses, from simple statistics through to microarray pre-processing and sophisticated data visualization and analysis, all with very little actual coding. In the next chapter, we see how this power can be harnessed for use in web-based tools through integration with Perl and web frameworks.

References

Brereton, R.G. (2009). *Chemometrics for Pattern Recognition*. Wiley: Chichester, UK.

Gentleman, R., Carey, V.P., Huber, W., & Irizarry, R.A. (2005). *Bioinformatics and Computational Biology Solutions Using R and Bioconductor*. Springer-Verlag: Berlin.

Otto, M. (2007). *Chemometrics: Statistics and Computer Application in Analytical Chemistry*. Wiley: Chichester, UK.

Otu, H.H., NaxerovaK., Ho K., Can H., Nesbitt N., Libermann T.A., & Karp S.J. (2007) Restoration of liver mass after injury requires proliferative and not embryonic transcriptional patterns. *Journal of Biological Chemistry*, **282**: 11197–204.

Sadowski, P., Dunkley, T. P. J., Shadforth, I. P., Dupree, P., Bessant, C., Griffin, J.L., & Lilley, K.S. (2006) Quantitative proteomic approach to study subcellular localization of membrane proteins. *Nature Protocols*, **1**: 1778–89.

CHAPTER 5

Developing web resources

In this chapter we show how to bring together the skills covered in the previous chapters to make data and analysis tools available via the web. This moves us onto the subject of web development, a skillset that is frequently required in bioinformatics and elsewhere. It is no coincidence that all the major biology databases, and many of the key bioinformatics tools, are accessed via the web. The reason is that everyone has a web browser, regardless of their operating system, and there are plenty of helpful tools available for developing web interfaces to Perl programs and MySQL databases. You might think that web interfaces are only important in situations where we want to make our work accessible via the Internet, but even for local applications used by a single organization or group, a web interface accessible over a local network is often the most painless way to interface users with your software.

5.1 Web servers

Before we get onto the task of web development, we first need a basic understanding of the tools responsible for serving content to web users—*web servers*. A web server is a software package that runs on a computer (typically also called a server when used for this, although it may be just a humble PC), and serves web pages and related files to other users over a network. The Apache HTTP Server (commonly referred to as *Apache Web Server* or just *Apache*) is the most popular web server used in the world today, and is the server solution of choice in bioinformatics due to being free, open source, fast, highly configurable, and available on most operating systems. Installing Apache is not a prerequisite for working through the examples in this chapter because you can view static web pages simply by loading them into a browser, and for dynamic pages we will be using a web framework that comes with its own development server. However, for more serious web development you will ultimately need to have your own server, so we explain how to go about this (with a focus on Apache) in Appendix B.

5.2 Introduction to HTML

HTML (short for HyperText Mark-up Language) is the text-based data format that is the basic building block of the web that we know and love today. It is a mark-up language based on the concepts of *tags* and content—very similar

Building Bioinformatics Solutions. Second Edition. Conrad Bessant, Darren Oakley and Ian Shadforth.
© Conrad Bessant, Darren Oakley, and Ian Shadforth 2014. Published 2014 by Oxford University Press.

to XML described in Chapter 2. Since it is a text-based format, it is possible to generate and edit HTML documents in a standard text editor, such as those reviewed at the start of Chapter 3. Just be sure to save the file with the extension .html. One main difference between HTML and XML is that in XML the tags are freeform—you get to decide what they are, as the tags are directly relevant to the data that they contain. In HTML the tags are already defined in a vocabulary. The reason for this is that XML is a data storage and description language, so it needs to be flexible. HTML, on the other hand, is more concerned with data presentation; therefore most of the tags used in HTML dictate how your data is presented in a web page, so the definitions and syntax of these tags needs to be established up front. Here are some simple examples of HTML tags:

```
<p>This is a paragraph</p>
<p>This is a paragraph with some <strong>bold</strong> text
    </p>
```

As you can see, this is very similar to XML. Textual data is simply surrounded by tags that indicate how the text will be formatted on the final web page. You may also remember *attributes* in XML, these are options that can be added to a tag—these are also present in HTML, as are nested tags. For example:

```
<table border="1" cellpadding="5" cellspacing="5">
    <tr>
        <td>Column 1</td>
        <td>Column 2</td>
    </tr>
</table>
```

As with XML, the attributes confer extra properties on the data that the tags encase—in the above example we are dictating the size of the border and the spacing between the cells and text for a table of data.

5.2.1   Creating and editing HTML documents

Before we move on to discussing more details of HTML, we need to decide on a tool to use to create our web pages. As with Perl and R programming, the writing of HTML documents (as well as CSS and JavaScript, which we'll briefly cover later) can be done in any good text editor. There are many editors available that are suitable for the job; in fact most of the tools listed in Chapter 3 can also be used as editors for HTML, CSS, and JavaScript. Whatever you used for writing Perl programs should be fine for working through this chapter.

5.2.2   The structure of a web page

The basic structure of a web page is shown below:

```
<doctype>
<html>
```

```
<head>

</head>
<body>

</body>
</html>
```

The entire page is surrounded in `<html>` tags; and within the page there are two main sections: the `<head>` and `<body>` tags. In the head section we place metadata about our web page, such as the title, the author, any copyright messages, keywords, and we also place links to external resources that can be used in our web page, such as CSS or JavaScript files, which we will come to later. The body section of the web page is home to all of the content of the web page.

The `doctype` declaration is the way of telling a web browser how the web page is formatted (be it either HTML5—the current standard—or an earlier variant of HTML), so it knows how to interpret the mark-up, and which rendering mode it should use (web browsers have different rendering modes for different situations). Without this declaration at the top of the page, the browser would have to guess which form of mark-up you are using, resulting in the page not displaying in the way that was intended.

Here is an example of a doctype declaration for an older standard of HTML (version 4.01, Transitional):

```
<!DOCTYPE HTML PUBLIC "-//W3C//DTD HTML 4.01 Transitional//EN"
        "http://www.w3.org/TR/html4/loose.dtd">
```

And here is an example for the latest standard, HTML5:

```
<!DOCTYPE html>
```

You will notice that the latter example is vastly simplified and does not even mention that it is for HTML version 5. The reason for this is that in previous versions of the HTML standard there was a DTD (Document Type Definition) specification that defined exactly how the standard should be implemented. In HTML5 there is, at the time of writing, no complete and exhaustive specification (it is still being actively developed and enhanced), so there is no DTD to link to. It is still necessary to include the HTML5 doctype as it lets the web browser know to use a standards-compliant rendering mode.

5.2.3   HTML tags and general formatting

Tables 5.1 and 5.2 list some of the most common HTML tags available for use within web pages.

For more information on the tags available in HTML we recommend visiting the excellent W3Schools tutorial (`www.w3schools.com/html/html5_intro.asp`)—this is particularly important because, although it is already in widespread use, the HTML5 specification is still under development, so details may have changed

Table 5.1 Tags commonly used in the `<head>` section of web pages

HTML	Description
`<title>`	This defines the title of the webpage.
`<meta>`	This is used for metadata such as author or keywords. i.e.: `<meta name="`**`keywords`**`" content="genes, genome" />` `<meta name="`**`description`**`" content="..." />` `<meta name="`**`author`**`" content=" name or email" />`
`<style>`	Used for in-line CSS style declarations (discussed later).
`<link>`	Used to link an external file into our webpage i.e. a CSS stylesheet file: `<link rel="stylesheet" href="style.css" type="text/css" />`
`<script>`	Used to add in-line JavaScript code to your webpages or link out to an external JavaScript file: `<script src="code.js" type="text/JavaScript" />` `<script type="text/javascript"> code ... </script>`

Table 5.2 Tags commonly used in the `<body>` section of web pages

HTML	Description
`<p>`	A paragraph tag – used to denote a paragraph of text. `<p>This is a paragraph of text.</p>`
` `	A line break.
`<hr />`	A horizontal rule.
`<strong>`	Used to write important text (i.e. in **bold**). `<p>This is some <strong>bold</strong> text.</p>`
`<em>`	Used to write emphasised text (i.e. in *italics*). `<p>This is text in <em>italics</em>.</p>`
`<h1>,` `<h2>,` `...,` `<hx>`	Headings – the numbers indicate the level of heading. `<h1>Page Title</h1>` `<h2>Sub-Title</h2>` `<h3>3rd Level Title</h3>`
`<a>`	Links – these are used to create links between different pages within a website, and also to link out to other websites. `<a href="/about">internal link</a>` `<a href="http://www.google.co.uk">external link</a>` `<a href="mailto:help@bixsolutions.net">email link</a>`

HTML	Description

`<ul>`,
 	
`<ol>`
	Unordered (bulleted) and ordered (numbered) lists. These tags work in the same way and share the same tag to define a 'list item': `<li>`
	
	`<ul>`
	    `<li>First on list</li>`
	    `<li>Next on list</li>`
	`</ul>`

`<table>`	These are used to make up tables and require the use of the extra tags `<tr>` for a table row, `<th>` for a table heading column, and `<td>` for a normal table column.
	
	`<table>`
	    `<tr>`
	        `<th>Heading 1</th>`
	        `<th>Heading 2</th>`
	    `</tr>`
	    `<tr>`
	        `<td>Content 1</td>`
	        `<td>Content 2</td>`
	    `</tr>`
	`</table>`

`<img />`	Used to insert images into webpages. The `src` attribute indicates the image filename and its position in the filesystem relative to the HTML file. The `alt` attribute is a given name for your image that may be displayed under certain circumstances. `<img src="/images/image1.png" alt="Image 1" />`

`<code>`	This is used to indicate programming code and will be printed to the screen exactly as it is typed in the document.
	
	`<code>`
	    `#! /usr/bin/perl`
	    `print "Hello World!\n";`
	    `exit;`
	`</code>`

`<pre>`	This is used for pre-formatted text and will be printed to the screen exactly as it typed in the document.
	
	`<pre>`
	    `This is some pre-formatted text.`
	    `and it will be printed to screen as is,`
	    `leading spaces and all.`
	`</pre>`

continued

Table 5.2 (*continued*)

HTML	Description
`<!-- -->`	This is how we put comments in our document. These comments will not be shown when the page is viewed in a browser. `<!-- This will not be printed to screen -->`
`<form>`	This creates a form for user input. A form can contain numerous other elements such as text fields, check boxes, radio-buttons and more. We shall go into forms and the elements that they contain in more detail shortly.

since this book went to press. Before we move on, it is worth noting three rules of good HTML writing practice that sometimes get overlooked:

- In HTML, single (often known as *empty*) tags such as br, which produces a line break, need to be self-closing: that is <br> is incorrect, and should be <br />, where the forward slash before the greater than symbol indicates that the tag is self-closing. Note however that there is one single exception to this rule, the `doctype` tag. This does not need to be, and should not be, closed.

- Nested tags, that is tags opened inside another set of tags, must be closed within the first set of tags, as tags are not allowed to overlap. Here is an example: Hello is allowed, Hello is not allowed as the tags overlap.

- You must use lowercase letters for all tags and attributes. In addition to this, attributes must be surrounded by quotes.

5.2.4 An example web page

Now that we have given you an overview of some of the more basic HTML tags, we can put that all together to create an example web page that we can view in a web browser. Create a file called test.html with your text editor and paste in the code below (or, as usual, download it from www.bixsolutions.net).

```
<!DOCTYPE html>
<html>
    <head>
        <title>An Example Web page</title>
        <meta name="author" content="Joe Bloggs" />
        <meta name="description" content="My First Web
        page!" />
    </head>
    <body>
        <h1>My First Web page</h1>
        <p>This is a quick test of some of the tags from the
        book ... </p>
```

```
<p>This is a <a href="http://www.bixsolutions.
net">link to the book's web site</a>.</p>
<p>This is an unordered list:</p>
<ul>
    <li>unordered element 1</li>
    <li>unordered element 2</li>
</ul>
<p>This is an ordered list:</p>
<ol>
    <li>element 1</li>
    <li>element 2</li>
</ol>
<h2>More Stuff</h2>
<p>This is an image:</p>
<!-- We're using an external image here for
convenience -->
<!-- Oh, by the way - this is a comment! -->
<img src="http://www.rcsb.org/pdb/images/1cf3_
bio_r_250.jpg" alt="Glucose Oxidase" />
<p>This is a table:</p>
<table border="1" cellspacing="5" cellpadding="5">
    <tr>
        <th>Header 1</th>
        <th>Header 2</th>
    </tr>
    <tr>
        <td>Data 1</td>
        <td>Data 2</td>
    </tr>
</table>
<p>Finally, this is a code example:</p>
<code>
    #! /usr/bin/env perl<br />
    print "Hello World!\n";<br />
    exit;
</code>
    </body>
</html>
```

Save this file and then open it up in a web browser. You should now have a simple web page in front of you, demonstrating some of the structures and tags that we have just described. This is all we are going to cover on basic HTML. A good way to get more comfortable with it is to try to modify test.html and see the results of your changes. If you want to go further, we recommend reading through the tutorials on the W3Schools website and making a few more web pages that link together using hyper-links (the <a> tag) to form a mini website.

5.2.5 Web standards and browser compatibility

As with all of the other subjects in this book, HTML is a form of programming language, and as such it can be programmed correctly or incorrectly. The main difference is that if you deviate from the HTML standard or make some other mistake while writing a web page, nothing tells you that you have made a mistake. In Perl, the program would refuse to run if there was a mistake in the code; a web browser, on the other hand, will try to render a web page with many errors in its code, and often does quite a good job at covering up the mistakes. The ability of web browsers to deal with poorly formatted HTML code is both a blessing and a curse. It is a blessing in the fact that you can get away with small mistakes without ever realizing that you made them, and it is also a curse when something very subtle goes wrong with your web page on a certain web browser and you have to then spend hours trying to find what is causing the problem. This is where following and adhering to the standards in the first place can help because, if you have standards compliant HTML, all browsers should render it nearly identically. We say *nearly* for a reason—unfortunately the rendering engines used in the different web browsers available today are quite different to each other in the way that they handle fonts and spacing, and even in their default behaviour. This is why you should always test your web pages on different web browsers, even when your site has perfectly valid HTML, just to make sure the page renders as you want it to.

To ensure that we are following the standards correctly we can check the code using a validation tool. There are numerous different validation tools available on the Internet either in the form of web based services or plugins for web browsers. The one we recommend is the W3C validation service found at `validator.w3.org`. This is built and run by the World Wide Web Consortium (W3C), the people who write the specifications for HTML, so it is safe to say that if your page passes as valid there, it really is valid. To use the validation service you can simply give the tool a URL to check, upload a file, or cut and paste in some HTML code, it will then check your code for correctness.

That concludes the basics of creating *static* web pages: pages that always display the same information. If we want to use the web as an interface for users to interact with databases and analysis tools that we have written, then we need to combine HTML with a programming language such as Perl.

5.3 Programming for the web using Perl

Perl has been used to create dynamic websites on the Internet for a very long time, and is still in heavy use today. In the early days, a common approach to creating web applications in Perl was to use a protocol (and module of the same name) called CGI (Common Gateway Interface) that allows Perl code to run within a web server. These days, however, CGI is not the recommended way of making a web application because, although it makes it easy to get started on a simple application, as the application becomes more complex it gets far harder to maintain within the realms of CGI. The preferred method now is to use a *web framework*.

There are many web frameworks available for Perl, a simple web search for 'perl web framework' will point you to the most popular options of the moment. For the purpose of this chapter, we have chosen to use Mojolicious (`mojolicio.us`) as we have found it gives a good balance between power, flexibility, and ease of learning.

5.3.1 `Mojolicious::Lite`

The recommended way to start working with the Mojolicious framework is to begin with the `Mojolicious::Lite` package. This is a thin wrapper around the full web framework, so everything you learn about `Mojolicious::Lite` will also apply to full Mojolicious applications if you wish to pursue more advanced web development in the future, but it has the advantage that your entire web application is contained within a single file so it is easy to manage the code.

Before we get started, we must first install Mojolicious:

* *Windows:* Search for and install the package 'Mojolicious' within PPM.

* *Linux:* Search for and install Mojolicious via your package manager.

* *Mac OS X:* Simply install via cpanm in the terminal (`cpanm Mojolicious`).

If you have any issues with the above, we would recommend consulting the installation instructions on the Mojolicious wiki: `github.com/kraih/mojo/wiki/Installation`.

Hello, world

Let's get started with `Mojolicious::Light` by building the most simple web application possible:

```perl
#!/usr/bin/env perl

use Mojolicious::Lite;

get '/' => sub {
  my $self = shift;
  $self->render(text => 'Hello World!');
};

app->start;
```

Save this code into a file called `hello_world.pl` and enter the following command in the console:

```
morbo hello_world.pl
```

This starts up a web server called Morbo (bundled with Mojolicious for ease of development and debugging), and you can now visit your application (in a web browser) using the URL `localhost:3000` and see that it simply prints the text, 'Hello World'. You can stop Morbo at any time be pressing Ctrl-C within the command-line window, though it may take a few seconds for Morbo to respond to this. It is normal at this point to ask who else might be able to log into your computer and see this page—unless you have specifically configured your firewall to allow access to Morbo, the answer, reassuringly, is nobody.

Let's have a look at what we just did in more detail. The first obvious thing in the code is that we use `Mojolicious::Lite`. You should also notice that we have not used `strict` or `warnings`; these are not required as Mojolicious adds them in automatically for us. The next piece of code starts with a Mojolicious supplied function: `get` (this is what is known as a routing function). This command tells the application, for all GET requests (a HTTP verb—the default request made by web browsers to a web server), to the / address (the homepage), run the following subroutine. When invoked, the subroutine receives a `Mojolicious::Controller` object (`$self`) that we call the `render` function on and pass our 'Hello World' text to. It is `render` that sends data to be presented in the web browser.

The final part of the application is the `app->start` command. This is required at the end of all `Mojolicious::Lite` applications (instead of an exit statement) to tell the server to start listening for connections. Quite a lot happened in this first example, but it is all relatively straightforward and is boilerplate setup code that you will need in all `Mojolicious::Lite` applications. The only real piece of functionality in the above application was the render command.

Templates

In our first example we cheated slightly, in that we only output raw text to the web browser; normally we would expect web applications to output HTML. The way in which we generate HTML in Mojolicious applications is through the use of templates. Let's start with an example and expand upon this:

```perl
#!/usr/bin/env perl

use Mojolicious::Lite;

get '/' => sub {
  my $self = shift;
  $self->render('index');
};

app->start;
__DATA__


@@ index.html.ep
<!DOCTYPE html>
<html>
  <head>
    <title>Hello, World!</title>
  </head>
  <body>
```

```
    <p>Hello, World!</p>
  </body>
</html>
```

This example takes our initial application and adds in a few new concepts. The first thing you will notice is that the call to render has been modified; instead of passing through a text string, (denoted by `text =>` in the previous example), we are now passing the name of a template to call (`index`—found at the base of the code).

Below the main body of the application, and after the `app->start` call, we have a `__DATA__` literal; this tells Perl that the remainder of the text in the file is not Perl code, but data for use in the above code. The way in which this is used in `Mojolicious::Lite` is that we put our templates here, and separate them with lines beginning with `@@`, where we also give a 'filename' to each template. In our example we have a single template called `index.html.ep`, indicating that it is a template called index, to represent HTML (`.html`)—the `.ep` part is the template language in use (Embedded Perl). In a full (non-lite) Mojolicious application, these templates would be separate files, but in a Lite application we can keep things simple by having them within the same file as the code.

If you run this version of the application with Morbo, you can again visit it at `localhost:3000`. You will see little difference in the appearance of the page compared to the previous version, but at least you can now be comfortable in the knowledge that the page is correctly rendered HTML, and it should be displayed with a proper title in your browser.

Embedded Perl

The template language we use within this chapter is known as Embedded Perl (EP). This is a simple template language bundled with Mojolicious that allows you to write plain HTML and include small snippets of Perl code when needed. We shall explain a little bit more about the use of EP in the next section and will be using it throughout the rest of this chapter. We will not cover every detail. If you want a complete explanation of EP and what it can do, please consult the documentation that we link to in the Going Further section a bit later.

Layouts

Most of the time when creating a modern web application you will have more than one page within your site. If this is the case, and you are using templates, you do not want to have to repeat the basic HTML skeleton for each and every template that you create. This is where layouts are useful. Below is an example of our previous application, modified to use a layout:

```
#!/usr/bin/env perl

use Mojolicious::Lite;

get '/' => sub {
  my $self = shift;
```

```
    $self->render('index');
};

app->start;
__DATA__

@@ layouts/default.html.ep
<!DOCTYPE html>
<html>
    <head>
      <title><%= title %></title>
    </head>
    <body>
      <%= content %>
    </body>
</html>

@@ index.html.ep
% layout 'default';
% title 'Hello, World!';
<p>Hello, World!</p>
```

As you can see, not a huge amount has changed, but we now have two template files, one called `layouts/default.html.ep` and the other `index.html.ep`; the first being our new layout file, and the second being our template for the index action. The naming scheme for the layout file is significant. In full Mojolicious applications, you may want multiple layouts, so there would be a layouts directory and different templates can use different layouts. In `Mojolicious::Lite` we have a name like `layouts/xxx` to mimic this structure.

The next thing to bring to your attention is the `<%= title %>` and `<%=` `content %>` tags within the layout, these are EP tags (well, the `<%= %>` part is— see Table 5.3 for details) that are used to denote the beginning and end of a block of Perl code. In this case we are calling `title` and `content` attributes within the context of the template (or the `stash` as it is known). Basically, the content returns the HTML (and EP) of the template that is calling the layout, and `title` is a variable that is defined in the template that is calling the layout, and if you look at the `index` template you will see that this says to use the 'default' layout (via the layout method), and sets a `title` via the title method.

5.3.2 Debugging Mojolicious applications

In the past, debugging web applications was a real pain, but with modern web frameworks, debugging and fixing errors is a whole lot easier. To demonstrate this, we are going to introduce a number of errors into our example code. First off, let's invoke a syntax error in our `hello_world.pl` Perl code so the code cannot even run. Simply add some text in the space below the shebang line that is

Table 5.3 Embedded Perl tags and line-start characters

Tag	Description
`<% code %>`	Perl code that is executed, but no output is generated.
`% code;`	`<% %>` variant can be anywhere within the line/text, but the % variant must be at the start of a line – i.e. this line can only be Perl code.
`<%= code %>`	Perl code that is executed, and output is XML escaped before being placed into the template.
`%= code;`	
`<%== code %>`	Perl code that is executed, and output is placed directly into the template (no escaping).
`%== code;`	
`<%# code (commented) %>`	This is a code comment – no output will be placed into the generated template.
`%# code (commented);`	

not a valid Perl statement and then try and run the program again with Morbo. You will immediately notice that the web server has not started up correctly and details of the syntax error are printed to your console; this is exactly like catching syntax errors with regular Perl programs.

Now let's look at other types of errors (non-syntax related). First off, fix the error you introduced into the `hello_world.pl` code and save the file. If you have not already stopped Morbo you should notice that it has automatically restarted itself and the corrected web page is being served, because Morbo watches for changes to your files. This is a very handy feature for developers as it means that you don't have to keep stopping and starting the server to see the effect of changes to your code. Keep Morbo running, but now add a `die` statement in the middle of the index action as so, then save your changes:

```
get '/' => sub {
  my $self = shift;
  die;
  $self->render('index');
};
```

This time all will appear normal in the terminal; but if you now visit `local-host:3000` again in your web browser; you should be greeted by the Mojolicious error page, a screenshot of which is shown in Fig. 5.1. Error pages such as this are a great development of modern web frameworks; in the top pane we have a snippet of code, showing exactly where the program died, and below that we have a table detailing the state and environment of the application when it crashed. This is usually all the information you need to catch and fix most errors.

5.3.3   Routes

Routes in web applications mean matching the URL address (or path) requested by the web browser to an action within your application. So far, in all our examples we have defined a single route; the `'/'` route for our index/home page,

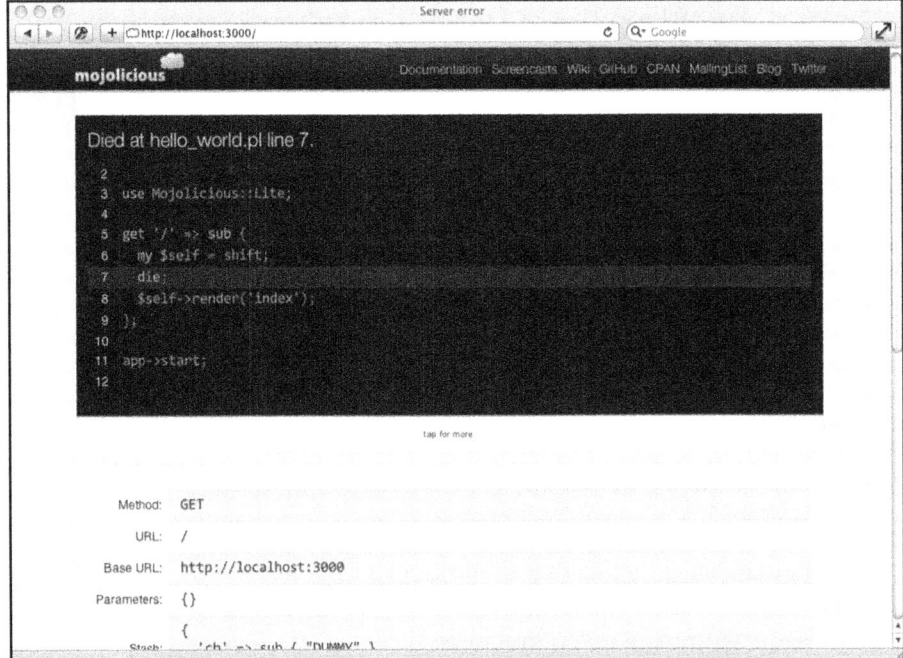

Fig. 5.1 A screenshot of the Mojolicious error page. This shows exactly where errors are in the code (in the top panel) and the current state/environment of the application (in the bottom table).

which is found when there is no path or just a trailing slash following a website's address (or domain name as it is more typically known). However, we can also associate actions or web pages with any other path you may wish, as demonstrated in the example below.

```perl
#!/usr/bin/env perl

use Mojolicious::Lite;

get '/' => sub {
  my $self = shift;
  $self->render('index');
};

get '/foo' => sub {
  my $self = shift;
  $self->render('foo');
};

get '/protein/show/:id' => sub {
  my $self = shift;
  $self->render('protein_show');
};
```

```
app->start;
__DATA__

@@ layouts/default.html.ep
<!DOCTYPE html>
<html>
  <head>
    <title><%= title %></title>
  </head>
  <body>
    <%= content %>
  </body>
</html>

@@ index.html.ep
% layout 'default';
% title 'Home Page';
<h1>Home Page</h1>

@@ foo.html.ep
% layout 'default';
% title 'Foo';
<h1>Bar</h1>

@@ protein_show.html.ep
% layout 'default';
% title "PDB Structure: $id";
<img src="http://www.rcsb.org/pdb/images/<%= $id %>
_bio_r_500.jpg"></img>
```

In this example application we have three routes defined. The first two are quite straightforward; our home (/) route, as seen before and a simple route that matches (/foo) –that is, www.example.com/foo would match this route if our app was running at www.example.com. The third route is a dynamic route that captures addresses that begin with /protein/show/, and have a simple string after the /show/ part, for example /protein/show/3QVP will match this route. A special feature of this route is that the last value within the URL will be captured by Mojolicious and placed within the stash, (with the variable name $id— matching the :id we put in the route) so it is available within the controller (the subroutine that is handling the route) and template. In this example, we insert $id into a PBD URL that, provided $id is a valid PDB protein accession number, returns an image of the protein for display within the page. A more common use case for a feature like this might be to use the $id as a lookup key on a database record.

5.3.4 Interfacing with databases within a web application

Let us now look at a more useful example of a web application—displaying information from a database. This is a common use of web applications in bioinformatics and is a good way to combine the techniques of Perl DBI covered in Chapter 3 with the power of Mojolicious.

In the following example, we connect to the public Ensembl database (www. ensembl.org), query the database for a short list of genes and other information, and then print these to a web page. Here is the complete code:

```perl
#!/usr/bin/env perl

use Mojolicious::Lite;
use DBI;
use DBD::mysql;

# Create a database connection as an application attribute
app->attr(dbh => sub {
  my $self = shift;

  my $dbh = DBI->connect(
    'DBI:mysql:homo_sapiens_core_47_36i:ensembldb.sanger.
    ac.uk',
    'anonymous'
  );

  return $dbh;
});

get '/' => sub {
  my $self = shift;

  # Prepare our query
  # NOTE: q() is the same as surrounding our text in single
  # quotes.
  my $query = q(
    SELECT es.synonym, sr.name, g.seq_region_start, g.seq_
    region_end
    FROM seq_region sr, gene g, external_synonym es
    WHERE es.xref_id = g.display_xref_id
    AND sr.seq_region_id = g.seq_region_id
    AND es.synonym IS NOT NULL
    ORDER BY g.seq_region_start ASC
    LIMIT 500
  );
  my $dbh = $self->app->dbh;
  my $sth = $dbh->prepare($query);

  # Run the query
  $sth->execute();
```

```
  # Stash the query string and the results
  $self->stash(
    query => $query,
    results => $sth->fetchall_arrayref()
  );

  $self->render('index');
};

app->start;
__DATA__

@@ layouts/default.html.ep
<!DOCTYPE html>
<html>
    <head>
      <title><%= title %></title>
    </head>
    <body>
      <%= content %>
    </body>
</html>

@@ index.html.ep
% layout 'default';
% title "Example Database Query";
<h1>Example Database Query</h1>
<p>
  The table below was created by running this query on the
  'homo_sapiens_core' database at
  <a href="http://www.ensembl.org">Ensembl</a>
</p>
<pre>
  <%= $query %>
</pre>
<table border="1" cellpadding="3" cellspacing="3">
  <thead>
    <tr>
      <th>External Synonym</th>
      <th>Seq Region Name</th>
      <th>Seq Region Start</th>
      <th>Seq Region End</th>
    </tr>
  </thead>
  <tbody>
    <% foreach my $result (@{$results}) { %>
```

```
    <tr>
      <% foreach my $item (@{$result}) { %>
        <td><%= $item %></td>
      <% } %>
    </tr>
  <% } %>
  </tbody>
</table>
```

The above program is quite large, but when we break it down into smaller sections, very little of the program is new to us—it is just a combination of the techniques we have learned with Mojolicious and DBI. Below we define the basic steps taken within the program:

1 We declare the modules we are going to use (Morbo will return a compilation error if these have not already been installed).

2 We set up a connection to the database as an attribute on the Mojolicious application itself. You will see that this is called in the index route with the code `$self->app->dbh`. This is the preferred method of connecting to resources such as databases in Mojolicious, as the connection code itself is not run until it is requested within a route (this concept is often called *lazy evaluation*).

3 We declare the index route and, within that, we prepare a SQL query. Note that this SQL query is not as straightforward as some of the examples that you have seen thus far. Unfortunately, this is the reality of us using the Ensembl database for our query—the Ensembl database has a large and complicated schema. That said, it is only a combination of three table joins to gain access to the information that we need—given time working with databases, queries such as this become second nature.

4 Following this we execute the SQL statement against the Ensembl database, retrieve the results and store this data, and the SQL query itself, within the stash:

```
# Stash the query string and the results
$self->stash(
  query     => $query,
  results   => $sth->fetchall_arrayref()
);
```

All this does is make these two variables (`$query` and `$results`) available for use within the templates.

5 Finally we render our template—you can clearly see how the stashed variables are used in the code.

It is from these simple concepts that we can build larger and more powerful web applications with Perl, DBI, and Mojolicious and it represents the core functionality of many bioinformatics database front ends available on the web.

5.3.5 Getting user input via forms

The most common way of getting information from users on the Internet is via forms. They can be found on almost any website that allows user interaction and should be instantly familiar to anyone using your web applications. Forms are a standard part of HTML and are therefore very easy to implement. Here is the basic structure of a HTML form:

```
<form action="URL for form handling" method="get/post">
    contents of our form - standard HTML and form elements
    allowed
</form>
```

Basically it is a `<form>` tag, with two attributes that dictate how the form, and the information contained within it, are handled:

• The first attribute, `action`, details where the information entered into the form should be sent to be processed when it is submitted via the user clicking a *submit* button. You set this attribute to a URL that points to a web application route to handle the input from the form.

• The second attribute is the method that the web browser should use to pass these details on to our program. This is explained in more detail later.

The actual content and appearance of the form is defined by elements that are placed between the two `<form>` tags. Table 5.4 lists the various input elements available, but take note of the fact that most use the same tag (`<input>`), it is only the `type` attribute that is used to differentiate them.

You will notice that there are three common attributes shared between the form elements described. These are `type`, `name`, and `value`. Here is a brief description of their uses:

• `type`—this attribute defines the type of input element.

• `name`—this is used as an identifier for the information that a user enters into a given form element. When we describe passing data from the form on to a program to process the data, the information that is entered into each form element will be accessed via the name attribute. A good way to think of this would be that you are passing a data structure similar to a Perl hash, so the information contained within the form can be accessed via the name values of each element. Note that things such as check boxes and radio buttons should share the same name attribute when they are together in a group (for a single question in your form). If they all have different names, the form would treat them as if you were asking a separate question with each button or box.

• `value`—in the case of check boxes, radio buttons, and selection drop downs, (where we have pre-defined content) the content of the value attribute is what is passed to your form parsing program (via the `name` attributes). Note that the value attributes of buttons (`submit` and `reset`) are not passed onto our form handling programs—these are used purely to tell the browser what text to place on the buttons.

Table 5.4 Commonly used HTML form elements

Input Type	Description

Text Fields	Text fields are used wherever you want your users to type letters or numbers into a form.
	
	``First name: <input type="text" name="firstname" />
	``
	
	```Last name: <input type="text" name="lastname" />```

Password Fields	Password Fields are used whenever you want a user to input sensitive information such as a password. They are basically text fields, but the text in the box is only visible as a series of asterisks.
	
	``Username: <input type="text" name="user" />
	``
	
	```Password: <input type="password" name="password" />```

Text Areas	Text areas are used when you want your users to enter a large chunk of text (that could not reasonably be entered into a text field). Note that you must define the size of the text area that is to be displayed with the `rows` and `cols` attributes. ```<textarea rows="10" cols="30"> Please enter your text here. </textarea>```

Radio Buttons	Radio buttons are used when you want your users to select exactly one item from a number of choices that you offer them.
	
	```<input type="radio" name="sex" value="male">male </input>```
	
	``
	``
	
	```<input type="radio" name="sex" value="female"> female</input>```

Checkboxes	Checkboxes are used when you want users to select zero or more options from a number of choices.
	
	```I have a bike:```
	
	```<input type="checkbox" name="vehicle" value="Bike" />```
	
	``
	``
	
	```I have a car:```
	
	```<input type="checkbox" name="vehicle" value="Car" />```
	
	``
	``
	
	```I have an airplane:```
	
	```<input type="checkbox" name="vehicle" value="Airplane" />```

Input Type	Description
Drop Down Boxes	Drop down boxes are used as alternatives to radio buttons or check boxes. In their default mode (shown below), the user can only select one option from a defined list.

```
<select name="go_ontologies">
    <option value="BP">Biological Process
    </option>
    <option value="MC">Cellular Component
    </option>
    <option value="MF">Molecular Function
    </option>
</select>
```

| Buttons | Buttons are used to submit or reset forms. Reset buttons clear all user entered information from a form, whereas a submit button will pass all of the information held within the form to the handling code that is defined in the form's `action` attribute. Note, unlike other form elements the value attribute on submit and reset buttons simply defines the text that is shown on the button. |

```
<input type="submit" value="Submit"/>

<input type="reset" value="Reset" />
```

Now, what happens once a form has been filled out by a user and they hit the submit button? The information contained within the form is passed on to a handling program, but before we show an example form and its associated form handling code, we first need to describe the two methods of form submission.

Submitting forms via POST and GET

With HTML forms there are two ways to submit data, these are via POST and GET. These two methods are introduced below.

◆ **GET**—This method of form submission can be used whenever users are not submitting sensitive data (i.e. a search form). It is not normally used when data from a form is to be used to submit data to a database linked to a website (as this can inadvertently reveal too much about the underlying structure of your site and database and potentially leave you open to security breaches), and it is most definitely NOT used for submissions that involve things like passwords. The reason for this is that the information contained within your form becomes part of the URL that the form gets submitted to, that is if you had a text box with the name `search_query`, the URL constructed to deal with the form would be something like:

```
/search?search_query=user_entered_string
```

where every entry in the form would follow the question mark (`?`) and multiple name/value pairs would be linked with an ampersand (`&`).

◆ **POST**—This method of form submission is used whenever you would like to keep submitted information in the background, so it is not clear for anyone to see once the form has been submitted. POST is suitable for forms that are used to enter information into a database and password entries. The information submitted from a form using POST is not appended onto the end of the processing URL.

So, the basic rule of thumb is for simple things such as searches, use GET. One of the benefits of this is that it allows people to link to predefined searches, as the search parameter forms part of the URL and saves them submitting a form each time they want to repeat an old search. One example of this is YouTube links, for example www.youtube.com/watch?v=RVo7jCXwVKE. For anything that involves entry to a database, or things such as password handling, use POST.[1]

Now that we have covered this, let's look at an example of a form. The example web application below defines a form for submitting a gene name to another route for processing (we will complete the application and add the other route shortly).

```perl
#! /usr/bin/env perl

use Mojolicious::Lite;

get '/' => sub {
  my $self = shift;
  $self->render('index');
};

app->start;
__DATA__

@@ layouts/default.html.ep
<!DOCTYPE html>
<html>
  <head>
    <title><%= title %></title>
  </head>
  <body>
    <%= content %>
  </body>
</html>

@@ index.html.ep
% layout 'default';
```

---

1 Although using POST hides submitted data from view, it will not stop someone intercepting the data on its way to the server. Sensitive data should therefore also be encrypted—something we don't have space to cover here.

```
% title 'Basic Web Form';
<h1>Basic Web Form Example</h1>
<p>
  Please enter a gene name and click 'Search' to get a
  report of information for a gene.
</p>
<p>
  (If you are at loss for a something to search for,
  try <strong>p53</strong> or <strong>ATP%</strong>).
</p>
<form method="get" action="/results">
  <p>
    Gene:
    <input type="text" name="gene" size="15" />
    <input type="submit" value="Search" />
    <input type="reset" value="Clear" />
  </p>
</form>
```

As you can see, this is a pretty standard Mojolicious web application, we have a single route and this page is generating a HTML form that is going to submit data to a route called /results using the GET method. Now let's add in the /results route and its accompanying code and template.

```
#! /usr/bin/env perl

use Mojolicious::Lite;
use DBI;
use DBD::mysql;

app->attr(dbh => sub {
  my $self = shift;
  my $dbh = DBI->connect(
    'DBI:mysql:homo_sapiens_core_47_36i:ensembldb.sanger.
    ac.uk',
    'anonymous'
  );
  return $dbh;
});

get '/' => sub {
  my $self = shift;
  $self->render('index');
};

get '/results' => sub {
```

```perl
    my $self = shift;
    my $gene = $self->param('gene');

    my $query = q(
        SELECT es.synonym, sr.name, g.seq_region_start, g.seq_
region_end
        FROM seq_region sr, gene g, external_synonym es
        WHERE es.xref_id = g.display_xref_id
        AND sr.seq_region_id = g.seq_region_id
        AND es.synonym LIKE ?
        ORDER BY g.seq_region_start ASC
        LIMIT 500
    );
    my $dbh = $self->app->dbh;
    my $sth = $dbh->prepare($query);

    $sth->execute($gene);

    $self->stash(
        gene => $gene,
        query => $query,
        results => $sth->fetchall_arrayref()
    );

    $self->render('results');
};

app->start;
__DATA__

@@ layouts/default.html.ep
<!DOCTYPE html>
<html>
  <head>
    <title><%= title %></title>
  </head>
  <body>
    <%= content %>
  </body>
</html>

@@ index.html.ep
% layout 'default';
% title 'Basic Web Form';
<h1>Basic Web Form Example</h1>
<p>
```

```
    Please enter a gene name and click 'Search' to get a
    report of information for a gene.
</p>
<p>
    (If you are at loss for a something to search for,
    try <strong>p53</strong> or <strong>ATP%</strong>).
</p>
<form method="get" action="<%= url_for '/results' %>">
  <p>
    Gene:
    <input type="text" name="gene" size="15" />
    <input type="submit" value="Search" />
    <input type="reset" value="Clear" />
  </p>
</form>

@@ results.html.ep
% layout 'default';
% title "Search Results For: '$gene'";
<h1>Search Results For: '<%= $gene %>'</h1>
<p>
  The table below was created by running this query on the
  'homo_sapiens_core' database at
  <a href="http://www.ensembl.org">Ensembl</a>
</p>
<pre><%= $query %></pre>
<table border="1" cellpadding="3" cellspacing="3">
  <thead>
    <tr>
      <th>External Synonym</th>
      <th>Seq Region Name</th>
      <th>Seq Region Start</th>
      <th>Seq Region End</th>
    </tr>
  </thead>
  <tbody>
    <% foreach my $result (@{$results}) { %>
      <tr>
        <% foreach my $item (@{$result}) { %>
          <td><%= $item %></td>
        <% } %>
      </tr>
    <% } %>
  </tbody>
</table>
```

The vast majority of the new additions to our program should already be familiar. It is essentially the same as the program from earlier in this section when we talked to the Ensembl database via our web application. The only addition here is that it is no longer a static query—we are now accepting input from the user via a form (or via adding parameters to the URL). The way that we accept input parameters within a Mojolicious application is through the use of the `$self->param` function, which is used in the above program as follows:

```
my $gene = $self->param('gene');
```

This returns the content of the element with the nam `gene` on the form that we set to forward to this program, and we use it as the argument in our query to the database defined earlier in the program. This query is then run against the Ensembl database, and the results returned as a web page exactly as before. This is how we pass parameters from a web form to a program for processing, something that is widely used in online bioinformatics tools.

5.3.6   Deploying a Mojolicious application

So far we have only used Mojolicious with the local development server, Morbo. This is a great way of developing your web application, but it is not suitable for supporting end users as it is only capable of handling a low volume of connections and is not designed for production use (e.g. re-loading all of your Perl code on each request is not very efficient). Therefore, when we deploy Mojolicious applications we use a different application server and setup—this process is commonly known as *deployment*.

There are several options available for deploying your Mojolicious web applications. If you are looking for a hosted solution where you do not have to take care of the server setup and configuration, we would recommend Heroku (`www.heroku.com`). Deploying an application to Heroku is very simple, and your first basic application is free.

If, on the other hand, you already have access to your own Apache web server (see Appendix B for details of how to do this) and would like to host your web applications yourself, your options are much greater. Our recommended method of deployment would be to use the Mojolicious application server[2] `Mojo::Server::Daemon` (if your server runs Windows), or the Hypnotoad application server (if your server runs Linux, Mac OS, or any other flavour of Unix) and *proxy* requests to this via Apache. This deployment/application architecture is described in Fig. 5.2. The reason we recommend this approach is that:

- It allows you to host multiple web applications from the same server (using Apache virtual hosts to proxy to many different Perl applications).

- As a performance improvement, it allows you to use the Apache web server to serve static files (images, CSS, and JavaScript files from the /`public` directory in your application) directly, without going via the application server.

---

2 An application server is a term used to describe the server process that runs your Perl code. Morbo is another example of an application server.

- It separates your main web server (i.e. Apache, the thing that recognizes that the incoming request is for `www.example.com`) from your application server (your Perl application). This has a number of benefits:

  - Security: you can run your Perl applications as simple, non-privileged user accounts that are completely separate from Apache and the rest of your system. If the worst happens and your application is compromised, intruders would only have access to files and data available to the user you run your Perl processes as, nothing more.

  - You can update your Perl code without needing to re-start Apache. This is most beneficial if you run multiple sites through a single Apache server as only one of your websites potentially goes offline during deployment.

  - If your web service becomes popular, or too resource intensive for your single machine (hosting both Apache and your application servers), it is quite simple to move your application server (Perl) onto another machine (or multiple machines) and update your Apache proxy to point to the newer machine(s).

For details of how to deploy a Mojolicious application using the methods above, we refer you to `mojolicio.us/perldoc/Mojolicious/Guides/Cookbook#DEPLOYMENT` (for `Mojo::Server::Daemon`), `github.com/kraih/mojo/wiki/Hypnotoad-prefork-web-server` (for Hypnotoad) and `github.com/kraih/mojo/wiki/Apache-deployment` for information on setting up Apache as a proxy to your application.

5.3.7 Going further with Mojolicious

This brings us to the end of our basic introduction to Perl programming for the web with Mojolicious. If you would like to learn more about Mojolicious, we recommend the screencast tutorials (`mojocasts.com`) and the Mojolicious documentation (`mojolicio.us/perldoc`) and wiki (`github.com/kraih/mojo/wiki`).

5.4 Advanced web techniques and languages

The basics of HTML and web programming described so far are acceptable for developing basic in-house bioinformatics tools; but if you aspire to become an accomplished bioinformatics web developer, you will need to understand the more advanced technologies that build on this basic foundation. In particular, it is worth having an awareness of cascading stylesheets (CSS) and JavaScript—two technologies found on almost all modern websites.

5.4.1 Cascading stylesheets

Cascading stylesheets (CSS) help make web applications look nicer—taking us beyond that 1990s look of plain white backgrounds and standard black text. This is done by defining a *style* for how the various different aspects of your HTML

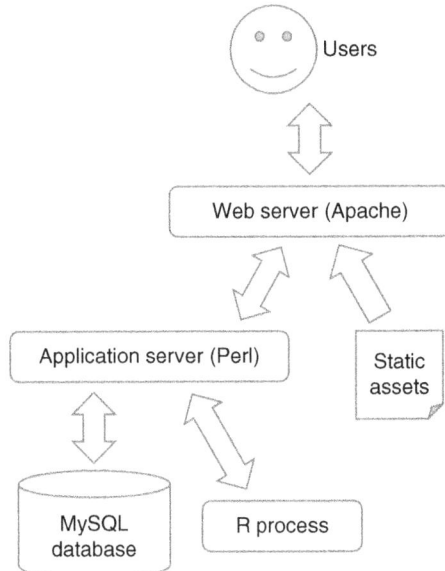

Fig. 5.2 Our recommendation for the basic architecture of a deployed web application. In this diagram we see the relationship between the different components of a web application and the way in which data flows between them. The R process is optional – it is only necessary if data analysis within R is required within the web application.

pages are displayed—you can change almost anything: the font, colour, and size of your text, as well as where different sections of your content are positioned. The things that you can do purely with a good understanding of CSS are vast. In this section we will look at the core concepts of CSS and point you in the direction of further learning resources.

The syntax of CSS is quite simple—it is just a list of style declarations that can be stored in a separate file (known as a CSS stylesheet) or placed at the top (within the `<head>` section) of a HTML page. A web browser will use these styles when rendering the page. Here we concentrate on the latter (embedded within a HTML page) use of CSS, which is fine for simple pages.

To demonstrate the effect of CSS on a HTML document, open up a HTML file that you have created (for example `test.html` which we used to show off our knowledge of HTML tags), and add the segment of code below somewhere between the `<head>` tags (e.g. just below the last `<meta>` tag), then open the file in your browser to see how it affects the visual appearance of the page.

```
<style type="text/css">
    body {
        font-family: 'Lucida Grande', Verdana, Arial, Sans-
        Serif;
        color: #444;
    }
```

```
    code {
        color: blue;
    }

    table th {
        background-color: black;
        color: white;
    }

    a:hover {
        text-decoration: underline;
        color: red;
    }
</style>
```

The CSS is the code enclosed within the `<style>` tags, which are used to inform the web browser that the code within is CSS and should be treated as such. The CSS is made up of a list of declarations; these start with the name of the element to which the declarations are to be applied (e.g. the first declaration here is `body`) then all styles that are to be applied to that element are enclosed within curly braces.

In the example above, we apply two style declarations to the contents of the `<body>` tag within the HTML (everything that we see on our web page is housed within the body tag so this is a good place to set default styles for the entire page to follow)—these declarations are `font-family` and `color`. These two declarations affect the font that is used to display all text within our web page, and the colour that is used for the text. The font colour is defined by a hexadecimal code #444—this translates to a light grey colour text; all colours in web pages can be defined in this notation—for further information, consult the Wikipedia entry on web colours (`en.wikipedia.org/wiki/Web_colors`).

In the next declaration, we alter the way in which code blocks appear in our web page—we have changed the font colour to blue (notice how we can also use the word blue to define the colour we wish to use—for simple colours we can use the names, for more exotic shades we have to revert to using the hexadecimal codes).

The third declaration defines new rules for table heading cells (`<th>` tags). We colour the text white and make the background colour of the cell black.

Finally, we demonstrate one last concept of CSS: states. In this last declaration we define what happens to links when they are *hovered* over (when a user puts their mouse pointer over a hyperlink)—this is known as the hover state. In this instance we define that the text that makes up the link should become underlined and red in colour—this is an effect that you would have no doubt noticed in use all over the Internet to signify links. Other available states for use with links and CSS are as follows:

* `link`—this is a link that has not been used, nor is a mouse pointer hovering over it,

- `visited`—this is a link that has been used before, but has no mouse on it,

- `active`—this is a link that is in the process of being clicked,

- `hover`—this is a link that currently has a mouse pointer hovering over it/on it (as used above).

As you can see from viewing your altered web page in a browser, just a small amount of CSS can considerably improve the appearance of a web page. To find out more about CSS, we recommend looking at the W3Schools website again—the section covering CSS techniques (`www.w3schools.com/css`) is very clear and concise. If you prefer a book, *CSS3: The Missing Manual* (McFarland 2013) is a very good text to introduce you to the subject and will help you progress into more advanced aspects of CSS quickly.

Using CSS stylesheets with Mojolicious

There are two simple ways to use CSS stylesheets within a Mojolicious application. You can either include the CSS within the head tags of your template in much the same way as we have just done in our above example, or you can store the CSS as a static asset and then refer to it.

To use CSS as static assets you will need to create a `public` directory in the same directory as your Mojolicious application, and within that we would recommend creating another directory called `stylesheets` where you can house all of your CSS stylesheets. Now, if you created a stylesheet called `app.css` within that directory, you would simply have to add the following line of code within the head section of your template in order to have it included in all your pages that use that template:

```
<link href="/stylesheets/app.css" rel="stylesheet"
type="text/css" />
```

5.4.2 JavaScript, JavaScript libraries, and Ajax

What we have looked at so far with Perl and Mojolicious are server-side technologies—all of the data processing happens on the web server, and only HTML is sent out to the user's browser. JavaScript, on the other hand, is a client-side programming language—it runs within the browser of the person viewing your web page and can only act on the content of the web page that the user is viewing at the time. This sounds a lot more restrictive than Perl and other server-side technologies, but the great benefit of JavaScript is the immediacy of feedback, which can make web pages seem much more dynamic and interactive. Before we move on, we should clear up one common misconception: JavaScript has nothing to do with the Java programming language covered in the next chapter—it is just a shared love of caffeinated beverages that caused the name Java to be used for both languages. Some common uses of JavaScript are:

- Validation—you can write JavaScript functions to check the content of web forms; for example, when you ask a user for an email address you can immediately check that they have actually entered something that looks like a valid e-mail address.

- Dynamically changing the content of a web page—you can write JavaScript functions that adapt the content of your web page based on a user's actions. Examples include multi-part forms, where the latter parts of the forms are dependent on what was entered in the earlier part of the form, and collapsible lists.

- Animation—recent advances in JavaScript techniques make it possible to produce detailed and very slick animations on almost any element of a HTML page. It is possible to hide a section of content, and upon the click of a button (or even just hovering over a defined item on your page), have another section slide, fade, or just appear, ready for use.

- Ajax—Ajax (Asynchronous JavaScript and XML) is a term used to describe one of the more complex uses of JavaScript. The traditional method of sending data to and from the server in web technologies is before and after each page *refresh* (i.e. when you submit a form, data is sent to the server, and the web page does a complete refresh whilst the form data is processed). Ajax changes this concept slightly in that it allows communication of data in small pieces between the web page and server without page refreshes, so you can submit forms or perform searches on databases without having to do a full page refresh to get your desired results to the screen. This allows web developers to produce a much *richer* experience for their users and can make web-based applications behave much more like desktop applications.

Given that JavaScript is not a core technology in bioinformatics, we cannot justify going into details of the language here. If you want to find out more, the W3Schools site is a great place to get started (`www.w3schools.com/js`). In terms of books, we recommend *JavaScript: The Definitive Guide* (Flanagan 2011) as this covers the basics of the language and goes all the way up to more advanced techniques. For Ajax development specifically, the *Getting Started* guide at the Mozilla Development Centre (`developer.mozilla.org/en-US/docs/AJAX/Getting_Started`) is a reasonable starting point. One thing to be wary of from the very beginning is that JavaScript can behave differently in different web browsers, so you need to test your code on the browsers that you think your users are likely to be using.

Remarkably, it is possible to harness much of JavaScript's client-side power within your web pages with only the vaguest understanding of the language, thanks to some mature JavaScript libraries that implement a range of popular functionality. Importantly, these libraries have been written with maximum browser compatibility in mind. The most popular libraries at the time of writing are:

- jQuery (`jquery.com`)—The most widely used JavaScript library of those listed here. jQuery aims to be 'a fast, small, and feature-rich JavaScript library. It makes things like HTML document traversal and manipulation, event handling, animation, and Ajax much simpler'.

- jQuery UI (`jqueryui.com`)—Governed by the jQuery Foundation, jQuery UI is a set of user interface controls that can be used to give your web pages a professional look and feel that makes your web applications easier to interact with. The controls include things like Accordian, to support expanding content on web pages, a date picker, and a progress bar. These controls can be added to a HTML page by inserting just a few simple lines of code.

- Yahoo! User Interface Library (`yuilibrary.com`)—This is a set of JavaScript utilities and controls, for building richly interactive web applications. YUI is not as straightforward to learn and start working with as jQuery, but some web developers prefer the approach that YUI adopts.

- Dojo (`dojotoolkit.org`)—Like YUI, Dojo is a set of JavaScript utilities and controls for building rich web applications. In addition to the basics, Dojo also includes tools for optimizing web interfaces for mobile devices and has charting and graphics capabilities.

5.5   Data visualization on the web

One of the more challenging tasks that you might need to perform in bioinformatics is that of producing graphical output derived from a given data set. We saw in Chapter 4 that R is a very powerful tool to use for this, but what if you want to make such figures available to people dynamically via the web? In this section we explain some common techniques that you can use via Perl to produce graphical displays dynamically on the web.

5.5.1   Using R graphics in Perl

In Chapter 4, we saw how visualizations such as plots, bar charts, and heatmaps can be created using R's powerful graphics capabilities. Generating these types of sophisticated visualizations on the fly and serving them up within R would clearly be very useful. The `Statistics::R` Perl module provides a basic bridge between Perl and R, allowing you to do just that. The module can be installed from CPAN, PPM, or your package manager as described previously.

Using Statistics::R

Using `Statistics::R` within your Perl programs is really rather simple and best described with an example. Here is some code that sends commands to R to produce a PNG output file within the current directory:

```
#! /usr/bin/env perl

use Statistics::R;

my $R = Statistics::R->new();
$R->startR;

$R->send('data <- c(1, 3, 6, 4, 9)');
```

```
$R->send('png("barplot.png", width=500, height=500)');
$R->send('barplot(data)');

$R->stopR;
exit;
```

As you would expect, we start off with a `use` statement for `Statistics::R`; we then create a `Statistics::R` object (`$R`) and call the method `startR` on it—this starts an R process that we can then talk to via our `$R` object. We then send through R commands as simple text strings via the send method (again on our `$R` object), before finally closing the R session with `stopR`.

In the above example we have only been communicating with R in one direction—sending data from Perl into R—now what if we want to go the other way and get the result of some computation from R into our Perl program? The good news is that this is possible with `Statistics::R`, but the bad news is that this is not friendly or graceful. Let's show you what we mean with an example:

```
#! /usr/bin/env perl

use Statistics::R;

my $R = Statistics::R->new();
$R->startR;

$R->send('x = 123');
$R->send('print(x)');
my $ret = $R->read;

print "The returned value was: '$ret'\n";

$R->stopR;
exit;
```

This code works in the same way as the previous example; starting an R session and sending some commands, but the main difference here is that we call the `read` function on the `Statistics::R` object. This function reads the last output from the R session and returns it. The above code will generate the following output:

```
The returned value was: '[1] 123'
```

As you can see this is taking the output from R that would otherwise be printed to the terminal and returning it in full as a string to Perl—nothing more. This is not ideal as it would be much more preferable to get a parsed output of this result (i.e. the number 123 as an integer in the above example), but we can understand why this is the proposed solution as the output from R could be a vector, a matrix, a string, an error message, or pretty much anything, so writing a generic solution for all the possible combinations would be near impossible. At least, because it is your program sending the R commands, you should know what form of output you expect them to produce and can therefore write an output handling subroutine to convert that to a Perl variable for further use.

Using Statistics::R in web applications

Now that we have an idea of what we can do with `Statistics::R`, we can use this in the context of a web application. What we would like to do is demonstrate a method for getting R generated images on the web using Mojolicious. This is not entirely straightforward as we cannot simply get the resulting image from an R program as a variable within our Perl program and show this to the user—we need to save the images to disk (in a defined location) and serve them automatically via Mojolicious.

Before we get started with any code, we need to do a small amount of preparation. In the directory where you are going to write your web application, create a directory called `public`, and within that directory, create another one called `r-images`. Mojolicious has a useful feature where it serves static files automatically for you, and the default location for static files is the `public` directory we have just created. We created the `r-images` directory within `public` as it is not normal practice to have files stored directly within `public`; usually we have a series of directories for storing images, CSS, and JavaScript files separately, and in our case `r-images`.

A complete program that produces a graph in R and displays it within a web page is shown below.

```perl
#! /usr/bin/env perl

use Mojolicious::Lite;
use Statistics::R;
use DateTime;

app->attr(r => sub {
  my $self = shift;
  my $R = Statistics::R->new();
  return $R;
});

get '/' => sub {
  my $self = shift;

  # Build our R Script in a multi-line string
  my $timestamp = DateTime->now();
  my $img_name = "r-image-$timestamp.png";

  my $rcmd = <<RCMD;
rangescale <- function(X) {
  Xmax <- apply(X, 2, max)
```

```
  Xscaled <- scale(X, scale=Xmax, center=FALSE)
  return(Xscaled)
}

X <- read.table("http://www.bixsolutions.net/profiles.csv",
sep=",", header=TRUE)
Xscaled <- rangescale(X)
d <- dist(t(Xscaled), method="euclidean")
bitmap(file="$img_name", type="png256", width=6, height=6,
res=96)
dendrogram <- hclust(t(d), method = "complete", members =
NULL)
plot(dendrogram)
dev.off()
RCMD

  chdir('public/r-images');
  my $R = $self->app->r;
  $R->startR;
  $R->send($rcmd);
  $R->stopR;

  $self->stash(img_name => $img_name);
  $self->render('index');
};

app->start;
__DATA__

@@ layouts/default.html.ep
<!DOCTYPE html>
<html>
  <head>
    <title><%= title %></title>
  </head>
  <body>
    <%= content %>
  </body>
</html>

@@ index.html.ep
% layout 'default';
% title 'Using R in a Perl Web App';
<h1>Using R in a Perl Web App</h1>
<p>Here is the resulting image generated by R</p>
<img src="/r-images/<%= $img_name %>" alt="R and Perl" />
```

If you save this as a Perl file (in the `public` directory made earlier), run it with Morbo, and visit `localhost:3000` you should see an R-generated image in the web page, a screenshot of which can be seen in Fig. 5.3. This image has been generated by R code called via Perl and then inserted into a HTML page by Mojolicious.

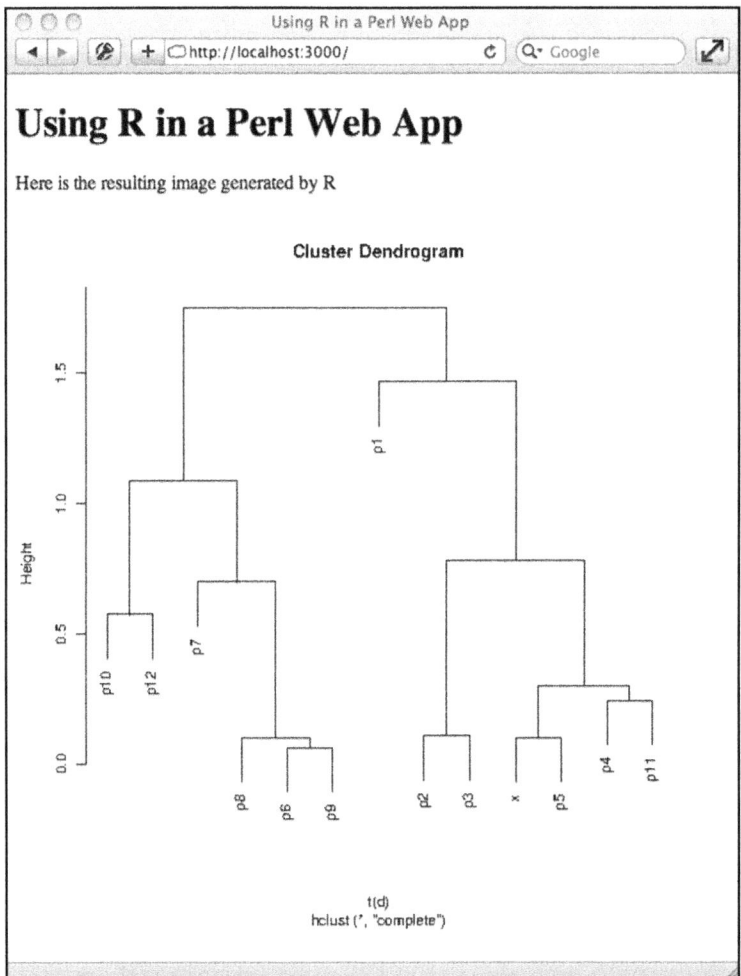

Fig. 5.3 A screenshot of a dendrogram automatically generated by R being displayed on a Perl driven web page.

No individual component of this program is particularly new to us, but the innovation here is in bringing everything together. The key steps in the program are as follows:

1 Before we declare any routes, we first set up the code for building a `Statistics::R` object as an attribute on the Mojolicious application—just as we did with our database connection code earlier.

2 Then within the index route:

◆ We first define a unique filename for our image with a timestamp as part of the filename (using `DateTime`). We do this because we are programming for a web page that could potentially be accessed multiple times by many different people—if we only had a single file name to use, we could be halfway through writing one file when the next person comes along and starts overwriting it. This way, we try to ensure that these sorts of file complications are avoided.

◆ We then declare our complete R program in a single string variable through the use of a document (as described in Chapter 3). You will notice that we are recycling one of the R programs from Chapter 4 (clustering performed on the protein profiles data set). The data set in this instance is hard coded into the program (well, a link to the CSV file on the Internet is). We only did this for brevity so that the program is easier to read—this data could just as easily have come from a database query, or from user input via forms.

◆ We then change our working directory to `public/r-images` and then run the R script via `Statistics::R`, saving the image to disk before stashing the image name (so we can use this in the template) and then rendering the index template.

3 Finally, in the `index` template we present a standard web page, but we also insert a `img` tag with the `src` attribute pointing to our newly created image file—this is all we need to do to get Mojolicious to serve the file correctly as it is located within the `public` directory.

This is all we need to do to be able to use R generated images within our web pages. It is not as elegant as we might like, because we have to first save the images to disk before we can utilize them, but it is a simple approach that generally works well. If we were to make this available as a public service we would need to periodically remove all of the images within the `public/r-images` directory, otherwise we risk using up all the server's storage.

Shiny: Serving R web output directly from R

As an alternative to the `Statistics::R` approach, a relatively new R package called `shiny`, available from CRAN and documented at `www.rstudio.com/shiny`, provides a particularly elegant way to make interactive R graphics and tables available via the web. It runs a web server within R, serving up interactive pages each of which can be defined by just two R programs: one to specify the user interface to be displayed on the web page, and the second to perform the data analysis and visualization that needs to take place on the server. At the time of writing, `shiny` is still in beta, but it looks like it could become a very efficient tool for visualizing data within websites.

In many applications, however, we might not need the sophistication of processing and visualisation provided by R—we may just need simple graphs. For these applications, there are many alternatives in Perl, one of which is `Chart::Clicker`.

5.5.2 Plotting graphs with `Chart::Clicker`

`Chart::Clicker` is a graph drawing library built upon the `Cairo` graphics library that can be used to produce very pleasant and modern looking graphs.

Installing Chart::Clicker

`Chart::Clicker` is not a standard part of the Perl distribution, so we first need to install it.

Windows

Unfortunately we were unable to find a good build of `Chart::Clicker` for the Windows platform, owing to `Chart::Clicker`'s strong dependence on Unix tools. If you particularly want to experiment with this, we recommend setting up a Linux virtual machine as described in Appendix C. Alternatively, you can skip ahead to the next section, which shows similar functionality from `SVG::TT::Graph`.

Linux

You should find pre-built packages for `Chart::Clicker` within your package manager.

Mac OS

To get `Chart::Clicker` to work on Mac OS you have to jump through a few hoops and install numerous dependency libraries. This is nothing too taxing, it just takes a little while to complete. Below you will find all of the required commands needed to get it installed; we have also posted these on the bixsolutions.net forum (`www.bixsolutions.net/forum/thread-98.html`) for your convenience.

```
brew update
brew install fontconfig
brew install freetype
brew install libpng
brew install cairo

export ULOPT=/usr/local/opt
export LIBPKG=lib/pkgconfig

export PKG_CONFIG_PATH=/usr/X11/$LIBPKG:$PKG_CONFIG_PATH
export PKG_CONFIG_PATH=$ULOPT/pixman/$LIBPKG:$PKG_CONFIG_PATH
export PKG_CONFIG_PATH=$ULOPT/fontconfig/$LIBPKG:$PKG_CONFIG_
PATH
export PKG_CONFIG_PATH=$ULOPT/freetype/$LIBPKG:$PKG_CONFIG_
PATH
export PKG_CONFIG_PATH=$ULOPT/libpng/$LIBPKG:$PKG_CONFIG_PATH
export PKG_CONFIG_PATH=$ULOPT/cairo/$LIBPKG:$PKG_CONFIG_PATH

cpanm Chart::Clicker
```

Using Chart::Clicker

With `Chart::Clicker` installed correctly we can get started writing our first program to demonstrate how it works:

```
#!/usr/bin/perl

use strict;
use warnings;

use Chart::Clicker;

# Create a Chart::Clicker object.
my $cc = Chart::Clicker->new(
  width => 600,
  height => 400,
  format => 'png'
);

# Add data to our chart.
$cc->add_data('Set 1', [5.8, 5.0, 4.9, 4.8, 4.5, 4.2]);
$cc->add_data('Set 2', [0.7, 1.1, 1.7, 2.5, 3.0, 4.5]);

# Set a title for the chart.
$cc->title->text('Line Chart');
$cc->title->padding->bottom(5);

# Finally, save our image.
$cc->write_output('chart_clicker_line.png');

exit;
```

The above example produces the graph shown in Fig. 5.4. The main steps in the program are as follows:

1 Make Perl aware that we need `Chart::Clicker`, via the use statement.

2 Create a new `Chart::Clicker` object—to this we pass options for the size and format of image that we want to output.

3 Add our data sets to the `Chart::Clicker` object via the `add_data` method; to this we pass a label for the data set, and the data set itself within an array reference.

4 Set any further rendering options for our chart; that is we set a title and a small amount of padding below it.

5 Write the resulting image into a file.

`Chart::Clicker` is very flexible in the types of graphs it produces, here is an extension of the above code that produces a bar chart instead of a line graph:

```
#!/usr/bin/perl

use strict;
use warnings;
```

```perl
use Chart::Clicker;
use Chart::Clicker::Renderer::Bar;

# Create a Chart::Clicker object.
my $cc = Chart::Clicker->new(
  width => 600,
  height => 400,
  format => 'png'
);

# Add data to our chart. We use a hash this time to represent
# data as we're drawing a bar chart and we want to say which
# group the values belong to.
$cc->add_data('Set 1', { 1 => 5.8, 2 => 5.0, 3 => 4.9, 4 =>
4.8 });
$cc->add_data('Set 2', { 1 => 0.7, 2 => 1.1, 3 => 1.7, 4 =>
2.5 });

# Set a title for the chart.
$cc->title->text('Bar Chart');
$cc->title->padding->bottom(5);

# Replace the standard (Line) renderer.
my $renderer = Chart::Clicker::Renderer::Bar->new(opacity =>
.6);
$cc->set_renderer($renderer);

# Get the image 'context' and set some values
my $def = $cc->get_context('default');
$def->range_axis->baseline(0); # Make the y-axis start at
$def->domain_axis->tick_values([qw(1 2 3 4)]); # x labels
$def->domain_axis->format('%d'); # x-axis label formatting

# Ask clicker to "fudge" the edges with some padding so the
# bars show up properly - this is a bug that we unfortunately
# have to work around.
$def->domain_axis->fudge_amount(.25);

# Finally, save our image.
$cc->write_output('chart_clicker_bar.png');

exit;
```

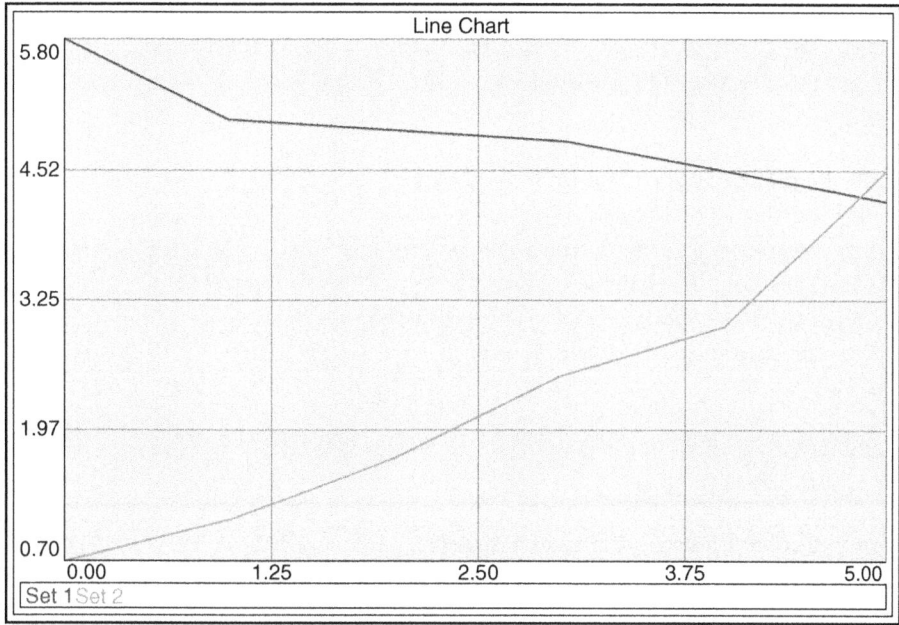

Fig. 5.4 A line graph produced using `Chart::Clicker`.

As you can see, this is a little more complex than the original example, but the main thing to understand is that to produce a different type of chart, all you have to do is change the *renderer* from the default—the rest of the code is really just formatting to make the resulting image look nice.

There are many more renderers available for use with `Chart::Clicker` for drawing other standard graph types and, if you need it, there is enough flexibility to produce completely custom images. For more information on what you can do with `Chart::Clicker` we recommend the CPAN documentation (`metacpan.org/module/Chart::Clicker`) and tutorial (`metacpan.org/module/Chart::Clicker::Tutorial`).

Using Chart::Clicker in web applications

You are probably already thinking that images produced by `Chart::Clicker` could quite nicely be incorporated in dynamically generated web pages. We are happy to report that it is almost as simple as cutting and pasting your `Chart::Clicker` code into a Mojolicious program, and it is also conceptually simpler than our previous example of using R within a web application. Here is the full code (there is no setup or directories needed):

```
#! /usr/bin/env perl

use Mojolicious::Lite;
use Chart::Clicker;
use Chart::Clicker::Renderer::Area;
```

```perl
get '/' => sub {
  my $self = shift;
  $self->render('index');
};

get '/graph/:type' => sub {
  my $self = shift;
  my $type = $self->param('type');
  $self->res->headers->content_type('image/png');
  $self->res->body(draw_graph($type));
  $self->rendered(200);
};

sub draw_graph {
  my ($type) = @_;

  my $cc = Chart::Clicker->new(
    width => 400,
    height => 300,
    format => 'png'
  );

  # Add data to our chart.
  $cc->add_data('Set 1', [5.8, 5.0, 4.9, 4.8, 4.5, 4.2]);
  $cc->add_data('Set 2', [0.7, 1.1, 1.7, 2.5, 3.0, 4.5]);

  set_renderer($cc, $type);

  $cc->draw;

  # return raw PNG binary data
  return $cc->rendered_data;
}

sub set_renderer {
  my ($cc, $type) = @_;

  if ($type eq "area") {
    my $renderer = Chart::Clicker::Renderer::Area->new(
      opacity => 0.75
    );
    $cc->set_renderer($renderer);
  }
}

app->start;
```

```
__DATA__

@@ layouts/default.html.ep
<!DOCTYPE html>
<html>
  <head>
    <title><%= title %></title>
  </head>
  <body>
    <%= content %>
  </body>
</html>


@@ index.html.ep
% layout 'default';
% title 'Using Chart::Clicker in a Web App';
<h1>Using Chart::Clicker in a Web App</h1>
<p>Here are the resulting images generated by
Chart::Clicker</p>
<img src="/graph/line" alt="Line Chart" />
<img src="/graph/area" alt="Area Chart" />
```

To see this code in action, save it in a Perl file and run it with Morbo. Now, if you visit `localhost:3000` in your web browser you should have a correctly rendered web page with two images (the dynamically created graphs) displayed, a screenshot of which you can see in Fig. 5.5.

This example gives us a full web application that dynamically generates images for display on a web page. To achieve this we have simply put together our knowledge of `Chart::Clicker` and Mojolicious from before. The main difference between this example and the `Statistics::R` example is that we do not have to save an image to disk when using `Chart::Clicker`, instead we can serve image data directly from within Perl, so in order to do this we have created a second route for serving the image data (as a single route can only serve one thing at a time—text or binary (image) data, not both). Here's the code for that route once again:

```
get '/graph/:type' => sub {
  my $self = shift;
  my $type = $self->param('type');
  $self->res->headers->content_type('image/png');
  $self->res->body(draw_graph($type));
  $self->rendered(200);
};
```

This route does the following:

1 Captures the type of graph that we want to generate from the URL being used to request our route (look at the `index` template to see how we've called this route).

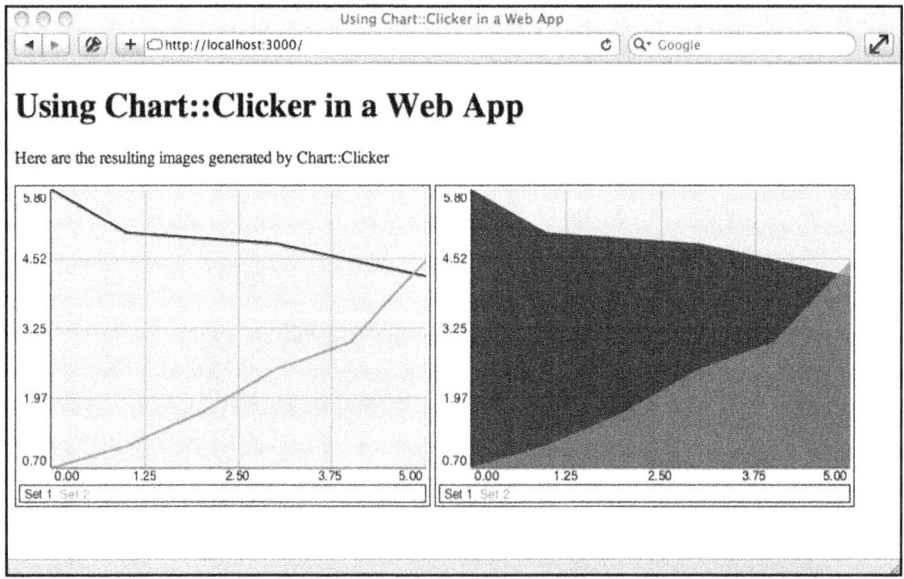

Fig. 5.5 `Chart::Clicker` generated images in use within a Mojolicious web application.

2 Sets the `content_type` of the response to `image/png`; this tells the web browser to treat this response as a PNG formatted image.

3 Places within the body of the response (what would normally be a template) the result of our `draw_graph` function—this is the binary image data as we are simply returning `$cc->rendered_data` from our function rather than writing it to a file.

4 Sets the status code of the response to 200—in HTTP, this means that our response is 'OK' (i.e. this is normal and nothing has gone wrong).

These are the only new things—we are returning bespoke image data from a Mojolicious route rather than text, or textual data from a template. The content of the `draw_graph` function is almost identical to our previous `Chart::Clicker` example, just in subroutine form.

This is a simple example of how you can use `Chart::Clicker` within a web application. Admittedly it would not be this simple in a real bioinformatics web application, as you would not usually hard code all of your graph data within the program. More typically, you would take data from the user, from a data analysis process, from a database query, or maybe a mixture of the three.

5.5.3 Plotting graphs with `SVG::TT::Graph`

The graphs generated by `Chart::Clicker` are perfectly functional and very nice to look at, but `Chart::Clicker` is not the only chart drawing library available for Perl; one good alternative is `SVG::TT::Graph`.

Functionally `SVG::TT::Graph` is very similar to `Chart::Clicker`, the main difference is the final output format of the images. With `Chart::Clicker` we have been producing bitmapped image files (JPEG, GIF, and PNG). SVG::TT::Graph on the other hand produces SVG (Scalable Vector Graphics) images. SVG is a graphics standard that allows the production of infinitely scalable high-quality graphics, by creating graphic objects that are rendered by the browser rather than bitmapped images rendered on the server. SVG is an XML-based format and is recognized and rendered well by the majority of modern web browsers.

Installing SVG::TT::Graph

Installing SVG::TT::Graph and all of its dependencies is a fairly trivial process.

- *Windows:* Search for and install the package 'SVG-TT-Graph' within PPM.
- *Linux:* Search for and install SVG::TT::Graph via your package manager.
- *Mac OS X:* Install via cpanm: cpanm SVG::TT::Graph.

Using SVG::TT::Graph

Using SVG::TT::Graph is very much like using Chart::Clicker. Let's start with our first example program.

```perl
#! /usr/bin/perl

use strict;
use warnings;

use SVG::TT::Graph::Bar;

# Field names for the x-axis
my $fields = ["1st","2nd","3rd","4th","5th","6th","7th","8th",
"9th"];

# Our data set
my $data1 = [1, 2, 5, 6, 3, 1.5, 1, 3, 4];

# Create our new graph object
my $graph = SVG::TT::Graph::Bar->new({
  height             => '300',
  width              => '400',
  fields             => $fields,
  x_title            => 'X Label',
  show_x_title       => 1,
  y_label            => 'Y Label',
  show_y_title       => 1,
  scale_integers     => 1,
  stagger_y_labels   => 2,
  show_graph_title   => 1,
  graph_title        => 'A simple graph'
});
```

```
# Add data to our graph
$graph->add_data({
  data => $data1,
  title => 'Dataset 1',
});

# Print our image to file
open(IMG, '>barchart.svg') or die $!;
print IMG $graph->burn();
close IMG;

exit;
```

The above code generates an output file called `barchart.svg`, which can be viewed in any web browser that can handle SVG graphics. An example of the output can be seen in Fig. 5.6.

As you can see this image is just as pleasing on the eye as the `Chart::Clicker` graphs. Let's go through the steps in the above example code to generate our image:

1 As with `Chart::Clicker`, we have a `use` statement that declares what type of graph (renderer in the case of `Chart::Clicker`) we are going to produce: `use SVG::TT::Graph::`*graph_type*`;`. The options are `Bar`, `BarHorizontal`, `Line`, `Pie`, and `TimeSeries`. In our examples, we are going to show you the `Bar` and `Line` options—the others are just as straightforward.

2 We establish the fields for our data to fit into, and then we establish a data set. Note that these are both array references.

3 We create a new `SVG::TT::Graph` object, passing it various attributes that dictate how the resulting image will look. Note the differences here between

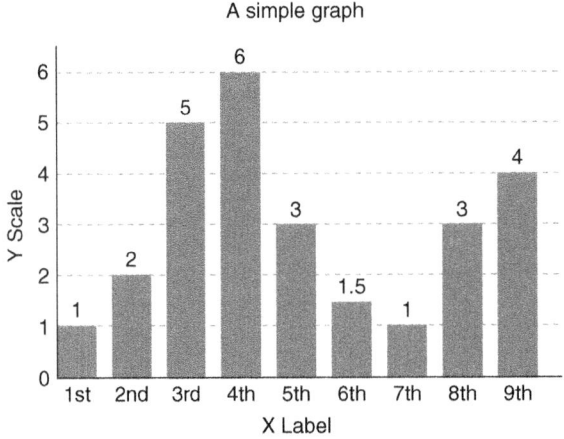

Fig. 5.6 A bar chart produced through the use of `SVG::TT::Graph`.

SVG::TT::Graph and Chart::Clicker—all of the attributes are set upfront in the new() method with SVG::TT::Graph. Another difference is that SVG::TT::Graph allows you to pass in a fields attribute—this lets you specify which fields your data points belong to, thus producing a more controlled final chart with a nicely labelled x-axis.

4 We load the data set into our graph with the use of the method add_data, also passing a title for the data set. If we are adding more than one data set (see below), this section of code is simply repeated with the new data set.

5 Finally we print out our image using the ->burn function. In the above example we simply print this to a file. Note that we do not need to set our print mode to binary—because SVG is XML-based the file consists only of text.

Now that we have explained the basics of using SVG::TT::Graph, we can show you one of its major differences from Chart::Clicker, how bar charts handle multiple data sets. Here is some code to demonstrate this:

```perl
#! /usr/bin/perl

use strict;
use warnings;

use SVG::TT::Graph::Bar;

# Field names for the x-axis
my $fields = ["1st","2nd","3rd","4th","5th","6th","7th","8th",
"9th"];

# Our data sets
my $data1 = [1, 2, 5, 6, 3, 1.5, 1, 3, 4];
my $data2 = [1, 1, 4, 7, 2, 3, 7, 4, 6];

# Create our new graph object
my $graph = SVG::TT::Graph::Bar->new({
  height           => '300',
  width            => '400',
  fields           => $fields,
  x_title          => 'X Label',
  show_x_title     => 1,
  y_label          => 'Y Label',
  show_y_title     => 1,
  scale_integers   => 1,
  stagger_y_labels => 2,
  show_graph_title => 1,
  graph_title      => 'A simple graph'
});
```

```
# Add data to our graph
$graph->add_data({
  data => $data1,
  title => 'Dataset 1',
});

$graph->add_data({
  data => $data2,
  title => 'Dataset 2',
});

# Print our image to file
open(IMG, '>barchart2.svg') or die $!;
print IMG $graph->burn();
close IMG;

exit;
```

If you run this program and look at the output, you will see that the way SVG::TT::Graph handles multiple data sets in bar graphs is to produce compound bar graphs where each data set is overlaid in front of the other. This contrasts with the bar chart renderer from Chart::Clicker which produces charts in which the data sets are represented in separate bars (although Chart::Clicker can also make compound bar charts). These types of compound bar graphs can often be useful, but can easily become cluttered and hard to understand if many data sets are represented. One workaround for this is to use the line graph function of SVG::TT::Graph. To do this, just replace the two instances of the word Bar in the above program with the word Line and run the program again.

Using SVG::TT::Graph in web applications

Like Chart::Clicker, it is very easy to use SVG::TT::Graph within a web application—in fact, the approach we take is identical (one route for serving the index page, and another for serving dynamically generated graphs). Below is a full example of a working web application that generates two images on a web page—a screenshot of which can be seen in Fig. 5.7:

```
#! /usr/bin/env perl

use Mojolicious::Lite;
use SVG::TT::Graph::Bar;
use SVG::TT::Graph::Line;

get '/' => sub {
```

```perl
  my $self = shift;
  $self->render('index');
};

get '/graph/:type' => sub {
  my $self = shift;
  my $type = $self->param('type');
  $self->res->headers->content_type('image/svg+xml');
  $self->res->body(draw_graph($type));
  $self->rendered(200);
};

sub draw_graph {
  my ($type) = @_;

  my $fields = ["1st","2nd","3rd","4th","5th","6th","7th",
"8th","9th"];
  my $graph = svg_tt_graph_obj($type, $fields);

  $graph->add_data({
    data => [1, 2, 5, 6, 3, 1.5, 1, 3, 4],
    title => 'Dataset 1'
  });

  $graph->add_data({
    data => [1, 1, 4, 7, 2, 3, 7, 4, 6],
    title => 'Dataset 2'
  });

  return $graph->burn();
}

sub svg_tt_graph_obj {
  my ($type, $fields) = @_;

  my $graph_options = {
    height          => '300',
    width           => '400',
    fields          => $fields,
    x_title         => 'X Label',
    show_x_title    => 1,
    y_label         => 'Y Label',
    show_y_title    => 1,
    scale_integers  => 1,
```

```
    stagger_y_labels => 2,
    show_graph_title => 1,
    graph_title => 'A simple graph'
  };

  if ($type eq 'line') {
    return SVG::TT::Graph::Line->new($graph_options);
  } else {
    return SVG::TT::Graph::Bar->new($graph_options);
  }
}

app->start;
__DATA__


@@ layouts/default.html.ep
<!DOCTYPE html>
<html>
  <head>
    <title><%= title %></title>
  </head>
  <body>
    <%= content %>
  </body>
</html>


@@ index.html.ep
% layout 'default';
% title 'Using SVG::TT::Graph in a Web App';
<h1>Using SVG::TT::Graph in a Web App</h1>
<p>Here are the resulting images generated by
SVG::TT::Graph</p>
<embed src="/graph/line" type="image/svg+xml" height="300"
width="400" />
<embed src="/graph/area" type="image/svg+xml" height="300"
width="400" />
```

The way in which the above code works is virtually identical to our
`Chart::Clicker` example except for three small differences:

• The use of `SVG::TT::Graph` within the `draw_graph` subroutine.

• The returned content type from the graph route is `image/svg+xml`.

• Instead of using `img` tags, we use `embed` tags to embed SVG images within HTML.

It is also worth noting at this point that you can also produce SVG graphics
with `Chart::Clicker`. The changes are very minimal, we simply change the

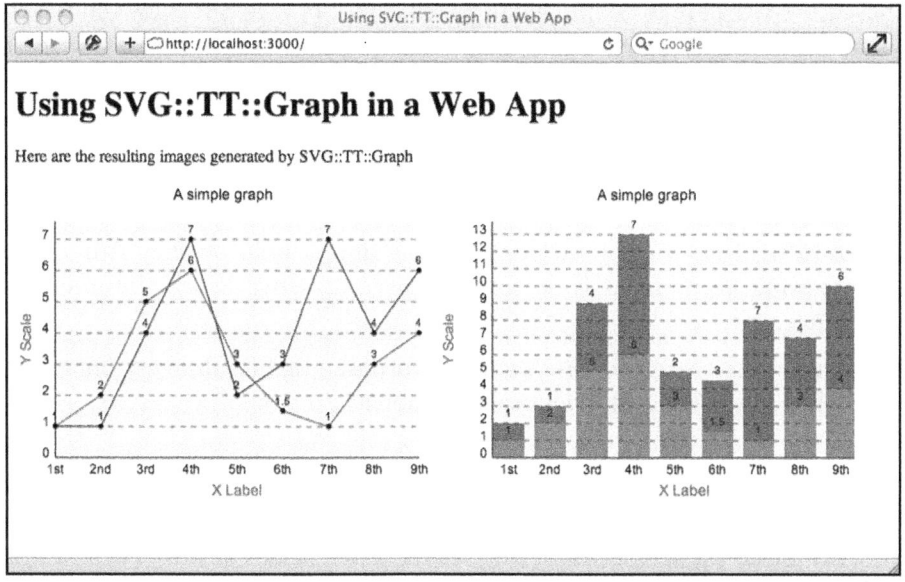

Fig. 5.7 `SVG::TT::Graph` generated images in use within a Mojolicious web application.

returned content type (in the image route), used `embed` tags (in the outputted HTML) and tell `Chart::Clicker` to output as SVG instead of PNG, so if you preferred the graphs produced by `Chart::Clicker`, but really wanted SVG output, you can use this as you wish.

5.5.4   Primitive graphics with Perl

As in R, there are low-level functions for drawing primitive shapes in both the `Cairo` and `SVG` graphics libraries available from CPAN (`Cairo` is the library that `Chart::Clicker` builds upon). Although using these low-level functions can be hard work, they are essential if you need to create the kind of very advanced bespoke visualizations that can be extremely valuable in bioinformatics. For example, they can be used to display gene structure, mass spectra, and interaction networks. Much of the data visualization on bioinformatics websites therefore employs these techniques. For more information about this topic, we refer you to the documentation for the `Cairo` (`cairographics.org`) and `SVG` (search CPAN for SVG) modules.

5.5.5   Drawing graphs and graphics using JavaScript

So far, we have talked about producing graphs and graphics on the server via R and Perl. Another option is to produce graphics *within the browser* using JavaScript. This can be appealing for two reasons: there are many JavaScript libraries for producing graphs and other graphics and it also takes the burden of generating images off your server and onto the user's computer. Unfortunately, we do not have the space here to go into detail on this topic, but we can point you to a few of our favourite JavaScript libraries for drawing web graphics:

- D3.js (d3js.org)—A general-purpose data visualization framework capable of producing rich and interactive visuals as well as basic graphs and plots.

- Dojo (dojotoolkit.org)—One of the rich JavaScript toolkits mentioned earlier in this chapter. Dojo includes a suite of graphics and charting capabilities.

- Flot (flotcharts.org)—A graph-plotting library built on top of the jQuery framework. Flot's focus is on simple usage, attractive looks, and interactive features.

- Raphaël (raphaeljs.com)—A general-purpose graphics drawing library that also has a charting component called gRaphaël (g.raphaeljs.com).

5.6   Summary

The main message to take from this chapter is that the extra work needed to add a web interface to a software tool or database produced in Perl, R, or MySQL is nothing to fear, so you can now start to implement fully fledged bioinformatics solutions as conceptualized way back in Chapter 1 (Fig. 1.1). However, it should be apparent that any serious project of this nature is likely to require a number of different individual elements, such as HTML templates, CSS stylesheets, multiple Perl programs, JavaScript, R code, and possibly an underlying relational database. To develop such complex projects effectively requires a certain amount of discipline and organization—these traits are at the heart of *software engineering*, the subject of the next chapter.

References

McFarland, D (2013). *CSS3: The Missing Manual*. O'Reilly: Sebastapol, California, USA.
Flanagan, D. (2011). *JavaScript: The Definitive Guide*. O'Reilly: Sebastapol, California, USA.

CHAPTER 6

Software engineering for bioinformatics

With this book, a computer, and an Internet connection, you have everything you need to start producing powerful bioinformatics software. This is mainly thanks to the open-source software movement making incredibly powerful software development tools freely available via the web, complete with detailed reference documentation and support forums where you can get answers to even the most specialized questions. However, this turns out to be both a blessing and a curse.

It is a blessing in that people with very little software development experience are able to pull various pieces of software together, pasting in program code from here and there, adding in a couple of powerful Perl modules and R packages, and release the result back into the community. The curse is that this approach (often called *hacking*) has some very real limitations in terms of the size and maintainability of the system you can build, and it is not always obvious where these limits lie. Consequently, it is easy to spend a few days knocking together a perfectly respectable little program but ultimately frittering away years struggling unsuccessfully to build it into something more substantial.

The problem is that hacking allows you to make very rapid progress at the start of the project, but if that project starts to grow you may unwittingly cross some boundary that starts to make your approach unproductive. The more boundaries you cross, the worse your problems become, until eventually you stop making any form of real progress. You may be working very hard and writing lots of code, but your effort is entirely spent fixing problems and every problem you solve creates several more to take its place. Over the past 30 or so years, the discipline of software engineering has developed specifically to ease the development of large and maintainable systems. Many books have been written on the subject, among which *Software Engineering* (Sommerville, 2010) is a good place to start.

Many of today's bioinformaticians have no reason to become experts in software engineering. They simply want to spend a few hours writing a self-contained program for their own use, and may discard it in a matter of days or weeks. However, as the bioinformatics field grows and matures, software projects are gradually growing larger and more complex and are intended to serve considerably larger user bases for many years or even decades. Of particular note is a

Building Bioinformatics Solutions. Second Edition. Conrad Bessant, Darren Oakley and Ian Shadforth.
© Conrad Bessant, Darren Oakley, and Ian Shadforth 2014. Published 2014 by Oxford University Press.

move towards multi-developer projects—many of today's bioinformatics projects are team efforts. Maintaining efficiency when working on something with other people is never easy and it becomes a particular challenge when working on complex software, especially if team members are distributed around the globe or come and go as the project progresses. Software engineering skills are therefore becoming increasingly important in bioinformatics.

This chapter provides a practical introduction to some of the fundamental areas of software engineering from which we feel you and your software would gain the most immediate benefit. These areas are:

• Unit testing—writing automated and repeatable tests for program code.

• Version control—keeping track of changes in program code and sharing program code among multiple developers.

• Documentation—helping other developers, and your future self, understand your program code.

• User-centred design—ensuring your software is as easy as possible for the intended audience to use.

• Alternative programming languages—some people would argue that Perl is not the best language for developing complex multi-developer projects. We therefore provide an introduction to other languages used in bioinformatics.

Arming yourself with this information, and getting a little practice, should open the door for you to work on larger, more complex projects and even help you understand how to make effective contributions to open-source projects, where you can improve your software engineering expertise in the virtual company of like-minded people.

6.1　Unit testing

It is obvious that anything newly designed needs to be tested before being released to users, be it a new piece of software, a new pharmaceutical, or a new toaster. In software development, the definition of testing can vary enormously. In the worst case it is limited to the developer having a quick play around with the software to see if it behaves as anticipated. This clearly lacks rigour, especially when the developer is not the intended user—any fundamental problems, such as references to non-existent libraries or syntax errors, will generate error messages, but more subtle mistakes in program logic may not be apparent. The field of software engineering offers us a number of more rigorous testing protocols, of which *unit testing* is one of the most popular. Indeed, many programmers would say that unit testing is essential even for the simplest single-developer projects.

A simple definition of unit testing could be 'the practice of creating tests that exercise small units of program code to ensure that they are working as expected'. In practice, this essentially means that we write programs that test small sections (or *units*) of the program that we are developing. What we mean by a unit can vary, but typically it would be a defined portion of logic in programming code, such as a module, class, method, or subroutine. The tests can be used to confirm that the unit functions as intended when it is first written, but can also be re-run as the program is further developed to ensure that nothing has been broken along the way. It is particularly useful when optimizing code for performance—this can involve extensive re-writing of code, so demonstrating that a unit still works after such major changes is important. Similarly, it is useful to test after performing maintenance, for example when code has had to be modified due to changes in the underlying language or libraries used. As a program gets larger, with more units of code, the number of unit tests that need to be applied will similarly increase, but since these tests are automated no significant developer time is needed to do this once the tests have been written. If done properly, it can therefore be both a thorough and efficient way of testing software.

If well written, the tests also fulfil a very useful secondary role, which is to formally define what exactly a given unit is supposed to do. The ultimate manifestation of this is *test-driven development*, where instead of starting off with a textual specification of what a piece of program code should do, with all the vagueness that can entail, it is possible to specify exactly how a piece of program code should behave by writing a test—if it passes the test it meets the specification. Just from this brief description, you have probably already thought of many difficulties associated with this approach. Most obviously, the quality of software developed in this way will be dependent on the quality of the tests, and how can we be sure of writing good tests that cover all eventualities? It turns out that this is a serious challenge, especially when user interfaces or database integration are involved. As a result there are whole books dedicated to this field, such as *Test-Driven Development: By Example* (Beck, 2003), which are well worth a look. In our opinion, a pragmatic view would be that unit testing is an extremely valuable tool but it does sometimes need to be supplemented by other testing methods.

6.1.1 Unit testing in practice

To understand what unit testing means in practice, let us consider an example of testing a module written in Perl. In the following code snippet we have the beginnings of a module called `EnsemblTools.pm`. The module is intended to be a collection of simple tools for working with data and identifiers from Ensembl.

```
package EnsemblTools;

use MooseX::Declare;

class IdParser {
  has 'id' => (is => 'ro', isa => 'Str', required => 1);

  method is_valid_id {
    if ($self->id =~ /^ENS\D*[G|T]\d+$/) {
      return 1;
    } else {
      return 0;
    }
  }
}
```

```
method species {
  if ($self->is_valid_id) {
    if ($self->id =~ /^ENS(\D*)[G|T]\d+$/) {
      if ($1 eq '') {
        return 'Human';
      } elsif ($1 eq 'MUS') {
        return 'Mouse';
      } else {
        return 'Unknown';
      }
    }
  }
}
}
1;
```

This object oriented code (using `MooseX::Declare`, which we covered towards the end of Chapter 3) is currently capable of performing only two tasks—checking that Ensembl IDs are correctly formed, and determining the species from Ensembl IDs.

Thus far we have been manually testing Perl programs like this, by running them and seeing if they work correctly. Then, after we make a change we have to run the program again to make sure it still works. Even for a simple program this is not only tedious, it is also hopelessly inadequate—we do not formally define what we mean by the program 'working correctly' and we usually cannot hope to verify the program's output for even a fraction of possible scenarios. As a program gets larger and more complex, this ad hoc method of testing only gets more painful and less effective—hence the need for unit testing.

We could just go ahead and start writing a Perl program to do unit testing on this module—automatically applying the classes to Ensembl IDs and checking that the correct results come back. However, this would require not just writing the tests but also developing program code around those tests, for example, to report the outcome of the tests in a consistent way. We can produce test code much more effectively by utilizing test tools that do all the mundane work for us. Some languages have unit testing functionality built-in as a fundamental part of the language, but in Perl this functionality is provided by libraries, of which there are several to choose from. The testing package that we are going to use in this example is `Test::More`; this is usually included as part of the standard Perl installation.

`Test::More` gives us a simple framework for writing tests on our code. It does this by providing methods to check the output of our code units, and by keeping track of which tests pass or fail. Before we get onto writing tests for our `EnsemblTools` module, let's introduce some of the basic functions of `Test::More` in an example:

```
#! /usr/bin/env per l

use strict;
use warnings;
use Test::More;

# The ok() function simply checks if the evaluated code returns
# true - if it does it passes, if not it fails.
ok(1 + 1 == 2); # pass
ok(1 + 2 == 4); # fail

# k() also allows you to label your tests, this is good for
# spotting which tests pass/fail in the program output.

ok(1 + 1 == 2, 'one plus one');

# The is() function lets us declare that something is supposed
# to be the same as something else.

my $gene_seq = 'ACTG';

is($gene_seq, 'ACTG'); # pass
is($gene_seq, 'AAAA'); # fail

# Again, you can label your tests

is($gene_seq, 'ACTG', 'gene_seq is as expected');

done_testing;
```

Here we introduce three key functions in `Test::More`.

- `ok`—This function simply checks that the evaluated code returns `true`. If it does, the test passes; if not, it fails.
- `is`—This checks that the two values passed to it are equal. If they are equal, the test passes. If not, it fails.
- `done_testing` With `Test::More` you must call the method `done_testing` at the end of the program so that it knows all of your tests have been run. If execution halts before `done_testing` is called, `Test::More` will indicate that the test suite did not complete.

If you save the above code into a file and run it as a normal Perl program it will generate the following output:

```
ok 1
not ok 2
```

```
# Failed test at test_more_basics.pl line 11.
ok 3 - one plus one
ok 4
not ok 5
#   Failed test at test_more_basics.pl line 24.
#          got: 'ACTG'
#     expected: 'AAAA'
ok 6 - gene_seq is as expected
1..6
# Looks like you failed 2 tests of 6.
```

This indicates that we ran six tests, and that two of them failed. As you can see, writing basic unit tests is really very simple when using `Test::More`. Some other useful test functions in `Test::More` (when working with modules) are:

- `use_ok`—This simply checks that the module we are testing compiles correctly and can be used (via the `use` declaration) within this script.

- `require_ok`—This is like `use_ok`, but it calls `require` on the module (and confirms the `require` was successful), so you can use the functions/classes in the test without prefixing everything with the namespace of the module under test.

Now that we have gone through the basics of `Test::More`, let's have a look at the test suite we have built for the `EnsemblTools.pm` example.

```perl
#! /usr/bin/env perl

use strict;
use warnings;
use Test::More;

use_ok('EnsemblTools');
require_ok('EnsemblTools');

# create some parser objects
my $bad_obj         = IdParser->new(id => 'qwerty');
my $human_gene_obj  = IdParser->new(id => 'ENSG00000139618');
my $human_tran_obj  = IdParser->new(id => 'ENST00000296271');
my $mouse_gene_obj  = IdParser->new(id => 'ENSMUSG00000018666');
my $mouse_tran_obj  = IdParser->new(id => 'ENSMUST00000093943');

# test id checking
ok($bad_obj->is_valid_id == 0, 'identifies bad ids');
ok($human_gene_obj->is_valid_id == 1, 'human gene ids');
ok($human_tran_obj->is_valid_id == 1, 'human transcript ids');
ok($mouse_gene_obj->is_valid_id == 1, 'mouse gene ids');
ok($mouse_tran_obj->is_valid_id == 1, 'mouse transcript ids');
```

```
# test species extraction
is($human_gene_obj->species, 'Human', 'human (gene id)');
is($human_tran_obj->species, 'Human', 'human (transcript id)');
is($mouse_gene_obj->species, 'Mouse', 'mouse (gene id)');
is($mouse_tran_obj->species, 'Mouse', 'mouse (transcript id)');
done_testing;
```

This code should be saved (call it `test_ensembl_tools.pl`) in the same directory as the module we are testing. All being well, executing the code should result in the following output:

```
ok 1 - use EnsemblTools;
ok 2 - require EnsemblTools;
ok 3 - identifies bad ids
ok 4 - human gene ids
ok 5 - human transcript ids
ok 6 - mouse gene ids
ok 7 - mouse transcript ids
ok 8 - identifies human (gene id)
ok 9 - identifies human (transcript id)
ok 10 - identifies mouse (gene id)
ok 11 - identifies mouse (transcript id)
1..11
```

This signifies that all 11 of the tests that we defined have passed, and the module is working as intended. When it comes to future development on the module, we only need to re-run this script to make sure we haven't broken any existing code. Of course, if new functionality is added to the module it will be necessary to add one or more additional tests, but it will still be important to run the original tests because adding new features is a great way to break existing functionality. To see what happens when a test fails, insert a trivial error into the module by changing `Human` to `Numan` in the `species` method. When the test program is run on this modified code, it will fail the two tests related to identifying human IDs (tests 8 and 9).

We should mention that, in this example, although we have tested every code path through the current `EnsemblTools` module, this is not really sufficient for production-quality code as we have only used a single variant of each type of ID considered. To feel more comfortable we would suggest having multiple variations on each type of ID, covering the widest possible range of IDs that we would expect to see.

Going further

Testing software is unlikely to ever have the same kudos as actually developing new software functionality, but we hope this brief introduction has convinced you that unit testing, in particular, can be a valuable part of building bioinformatics solutions and that it is not actually too much of an overhead if you use the right tools. Once you get used to writing tests, they don't take long, and if you get into the spirit of test-driven development they can even be quite fun. Exactly what you test, and how you test, comes with experience and will vary from project to project. In some development teams, unit testing is compulsory.

For more about unit testing Perl code, we recommend the `Test::Tutorial` on CPAN (`search.cpan.org/dist/Test-Simple/lib/Test/Tutorial.pod`), and the online documentation for `Test::More` (`perldoc.perl.org/Test/More.html`). Most other languages have unit testing either built-in or supported by third-party add-ons. For example, R has the `RUnit` and `testthat` packages (both available via CRAN).

6.2   Version control

Whether you are working on your own, or as part of a team, if you are writing program code it quickly becomes obvious that some form of version control is essential. In this section we introduce one of the more common version control tools used in bioinformatics today, but first we will give you an overview of what version control is and why you should be doing it.

Version control is the practice of tracking changes within files (in particular programming source code, but it can easily be used to track any other text-based file) so that you can see what has changed in your files over time, who has made the changes and (ideally) why the changes were made. This may sound like a tedious and unnecessary overhead on the already time consuming business of software development, but it turns out to be incredibly useful and modern version control software takes care of most of the tedium for us.

Most people when they start off use the 'versioned file' method—we have all done this—where you put a date or a made-up version number within a filename, and when it comes to doing a major edit you just save a copy with a new filename containing the updated version or date. Getting this to work long term requires great discipline as it can easily be forgotten, and when you start to share code files with others and work as part of a team you are then relying on others to be equally vigilant when it comes to saving new versions of the file. You will quickly find that this approach falls down and just does not work long term; this is where formal version control software can help.

As with all things in the software world, there are many different approaches to solving this problem, and many advocates and critics of each approach. Before we get into explaining any specific implementations in detail, we need to take a look at the basic concepts of version control itself, and introduce the two main types of version control that you are likely to come across.

6.2.1   The basics of version control

There are a number of basic conceptual items that are common to all forms of version control:

- *Repository:* The database storing the versioned files.
- *Server:* The computer hosting the repository.

- *Client:* A computer connecting to the repository, for example the development computer on which you are working with the files.

- *Working copy:* Your local directory of the files, where you make changes.

- *Revision:* A particular version of a file.

- *Trunk (or master):* The primary location/branch for code in the repository. Think of a version control repository as a tree—*branches* representing different versions split off from the *trunk* (we will look at branches in more detail shortly).

- *Codebase:* A term commonly used to describe the collection of program files associated with a particular project or version control repository.

In addition to these basic items, there are also fundamental actions that are common to all forms of version control:

- *Add:* Add a file into a repository so it can be tracked with version control.

- *Check out:* Download a file (or group of files) from the repository for editing.

- *Check in/commit:* Upload one or more files to the repository. If the file has changed, the file gets a new revision, and now other users can *check out* this new revision.

- *Diff:* Locate and show the differences between two revisions of a file.

- *Update:* Synchronize your files with the latest from the repository by updating to the latest version.

- *Revert:* Discard local changes to your file(s) and reload the latest version from the repository.

These actions are elaborated further in Figs. 6.1 and 6.2.

With a little thought, it becomes obvious that more complex scenarios can emerge from this version control process when multiple developers are working on the same codebase simultaneously. This gives rise to some more advanced concepts that we must become familiar with:

- *Branching:* The act of creating a copy of the file(s) under version control for a specific purpose (bug fix, adding a new feature, etc.). Again, think of the code like a tree, with branches stemming from the trunk.

- *Merging:* The act of applying the changes from one version of a file into another; that is merging the changed code from a branch into another (or into the trunk/master branch).

- *Conflict:* A conflict is what happens when pending changes from a merge operation contradict each other.

- *Resolve:* The act of fixing the differences that contradict each other (arising from a conflict) and checking in the fixed/correct revision.

Branching and merging are shown schematically in Fig. 6.3. We will leave further elaboration of conflicts and conflict resolution until a little later with a practical demonstration.

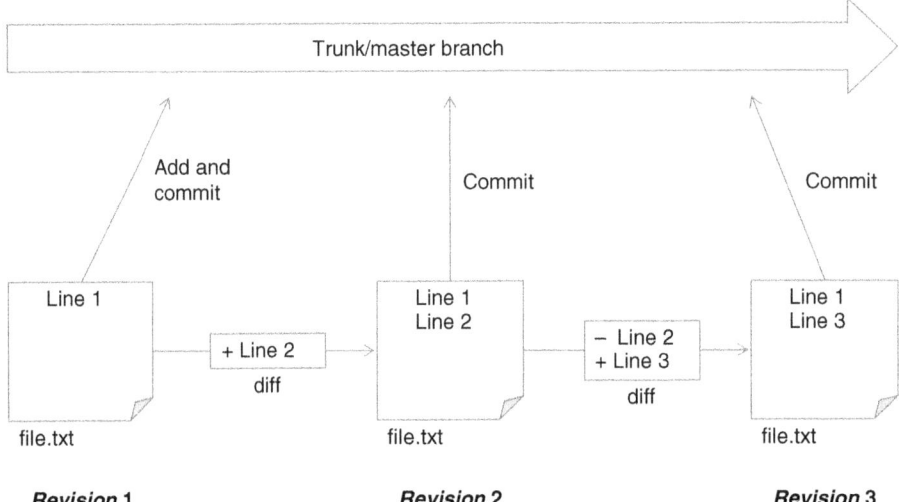

Fig. 6.1 The simplest scenario of version control. A user has a file (named `file.txt`) with some text in it; this is then added (and committed) to our repository's master branch. This file is then edited a subsequent two times with the changes made committed to the repository after each change, thus creating three individual revisions of our file within the repository (together with a record of the differences between each revision).

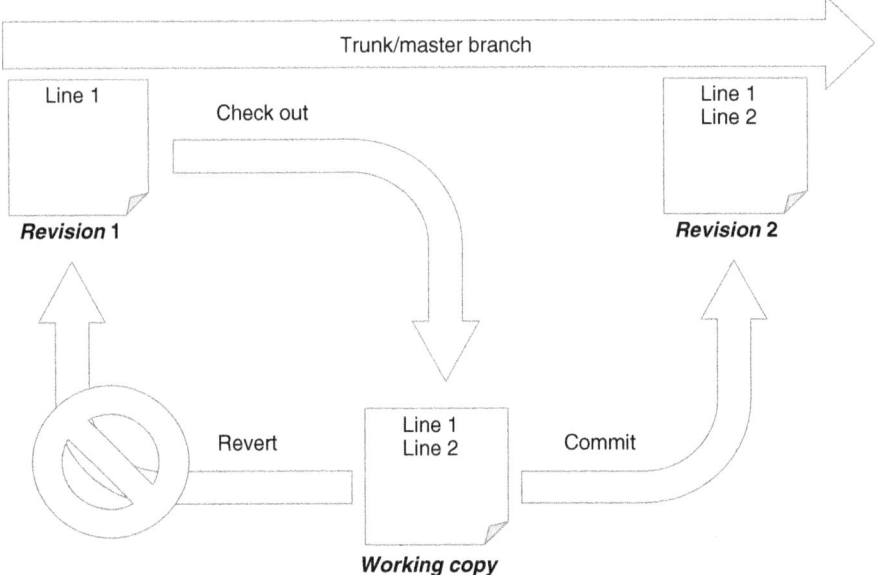

Fig. 6.2 An overview of how a developer would normally work with version control. The developer checks out a copy of the codebase from the master branch in the repository, makes some edits on their local (working) copy, and when they are finished they have the choice of either reverting or committing their changes.

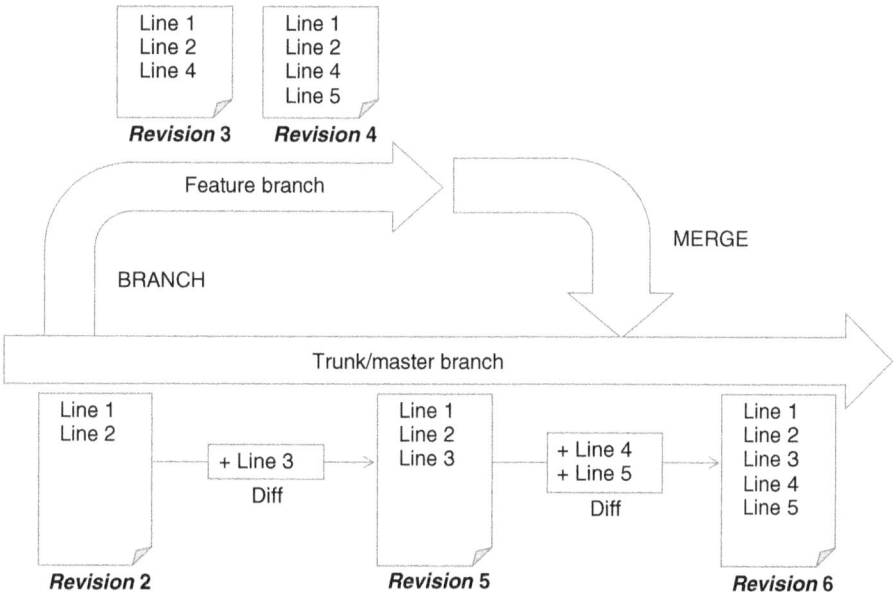

Fig. 6.3 Branching and merging. Branching allows a developer to work privately on the codebase without interrupting the master branch until his or her changes are ready to be added in. Here we see that a feature branch is created with a number of changes made, whilst at the same time an edit is made on the master branch. The final step is to merge this branch back into the master branch – thus combining the new edits with the current changes in the master branch.

6.2.2   Centralized versus distributed version control

There are two types of version control system in use today: *centralized* and *distributed*. In centralized version control systems there is a single server that acts as the host for the code repository, and all developers (using their computers as clients) need to connect to the server in order to check out, update, and commit code. A diagram of this model can be seen in Fig. 6.4.

Centralized version control systems tend to be slightly older and more traditional. Distributed version control systems, on the other hand, have gained popularity in the last decade and by comparison are relatively young. The main difference with a distributed version control system is that there is no requirement for a centralized server. Each developer has a full copy of the repository and they can push and pull a set of changes (often known as a *changeset*) between them. This is shown in pictorial form in Fig. 6.5, but it is worth noting that although this model of working with distributed version control systems is possible, it is almost never used like this in practice because it would quickly become impossible to track the changes from a large team of developers. All teams usually still have a central version control repository that developers push/pull changes to/from.

The obvious question then is: what is the point of distributed version control? Well, even though the vast majority of users do not use distributed version

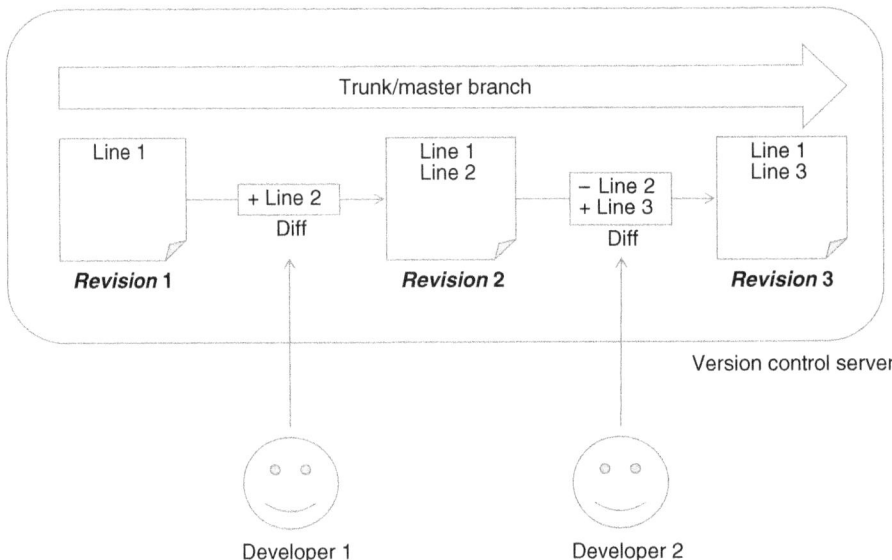

Fig. 6.4 The centralised version control model. With a centralised version control system all developers deal with the same version control server and commit all code changes there.

control in a truly distributed manner, this approach offers a number of very important benefits. Because every developer has a full local copy of the repository, distributed version control systems are fast, especially for branching and merging. Furthermore, commits and changes can be made offline, and backing up the repository is easy as every client has a complete copy of the codebase.

6.2.3 Git

Of the many version control systems available, we are going to focus here on a relatively new distributed system called *Git*. Git was originally released in 2005 to support the development of the Linux kernel. Following on from its use in the development of the Linux kernel, Git has become extremely popular in programming circles for its speed and ease of use, as well as its advanced feature set. Clearly, a system that is able to cope with something as complex as the development of the Linux kernel should be capable of supporting even the most ambitious bioinformatics project.

Getting started—installing Git

The first thing we must do before getting started with Git is to install it. Downloads are available for all major operating systems on the Git website, but we do not recommend using these downloads unless you are on Windows.

- Windows users, head to `git-scm.com/downloads`, download the Windows installer and install Git from there. Towards the end of the installation process select the 'Run Git from with Windows Command Prompt' option so that Git is added to the path and can be called from the command line in the following examples. All other options can be left at their default settings.

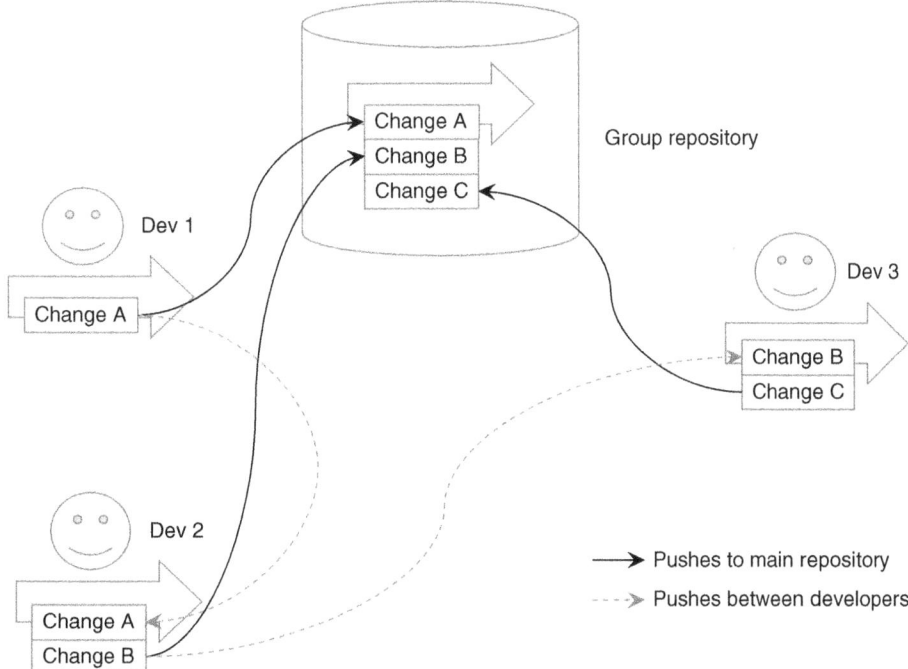

Fig. 6.5 The distributed version control model. In distributed version control, each copy of the repository (on a server somewhere, or on a developer's computer), has a full copy of the version control history, and changes can be pushed and pulled between all of them. In this example we have three developer's and a group repository, where the developers push/pull changes from each other (dotted arrows) as well as the main group repository (solid arrows). Whilst this activity is theoretically possible with distributed version control, the vast majority of groups only ever push/pull changes from a centralised server (solid arrows only).

- Linux users, install Git via your package manager.
- Mac users, our recommendation is to install Git via homebrew (as we did with MySQL in Chapter 2). In your terminal, simply run the command: `brew install git`.

The only further configuration necessary is to tell Git who you are, so that it can record who is making changes when you commit them. This can be done by setting your name and e-mail address by opening a command window and issuing commands like those below:

```
git config --global user.email f.sobotka@example.com
git config --global user.name "Frank"
```

Creating a Git repository

To start working with our fresh install of Git, we first need to create an example project to manage with it. Create a directory somewhere on your computer and call it `my_project`. Within that directory, create an empty text file called `file1.txt`. Now enter the following commands at the command line (substitute the first command to navigate to the location of your `my_project` directory).

```
cd my_project
git init .
git add .
git commit -m "Initial commit."
```

The three Git commands used above do the following:

- `init`—Initialize an empty Git repository within your project directory.
- `add`—Add all the current changes within the directory to the Git staging area.
- `commit`—Commit the changes in the staging area into the repository.

These are the three commands you will always use to start working with Git in a new (or even existing) project directory.

Git's staging/committing model

In the previous step you may have picked up on us using the term *staging area*—this is a concept common to many modern version control systems. The role of the staging area, and its relationship with the working directory and the Git repository, is explained in Fig. 6.6.

Viewing/committing new changes

Now that we have initialized the Git repository, we can start staging and committing changes to our project. First, open the text file `file1.txt` and add some lines of text to it, then save the changes. Now execute the following command:

```
git status
```

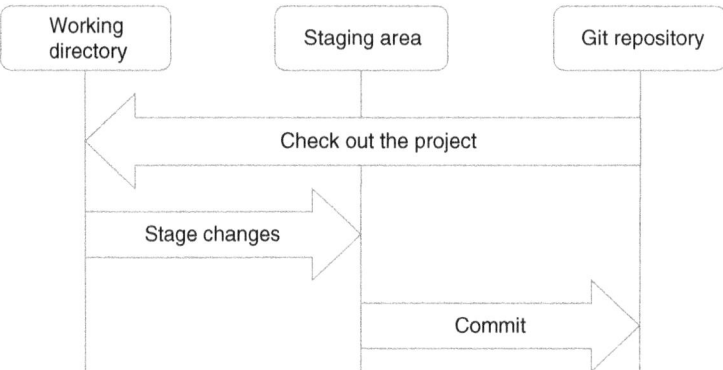

Fig. 6.6 The Git model. The Git repository is where Git stores the version control database for your project. The working directory is a single checkout of one version of the project. These files are pulled out of the git repository and placed on disk for you to modify. The staging area is a simple file, generally contained within your Git directory, which stores information about what will go into your next commit.

This should inform you that modifications have been detected within your file, but that no changes have been added to commit (within the staging area). At this point, it can be useful to see the changes in our files. This can be achieved with the following command:

```
git diff
```

This command will give you a readout of your changed file(s), whilst highlighting removed lines and text in red (also putting a – in the left gutter) and added lines and text in green (also with a + in the left gutter), thus showing exactly what has changed. For files that are too long to be shown in one go, the changes will be displayed in a paged file buffer—you can scroll up and down the document using the arrow keys, and when you are finished press the Q key to exit the buffer and return to the command prompt.

If we are happy with these changes, it is now time to stage and commit them into the Git repository. This can be achieved with the following two commands:

```
git add file1.txt
git commit -m "I made some changes."
```

The first command adds the named file to the Git staging area, whilst the second command commits the contents of the Git staging area to the Git repository using the commit message (denoted by the -m argument) "I made some changes.". This has now recorded our changes permanently within the Git repository.

Now, what if you want to see the recent activity in your Git repository? Run the following command:

```
git log
```

This lists all of the commits to your repository; showing the commit ID, the author, the date and time of the commit, and the commit message. These are listed in chronological order, with the most recent at the top. Here is an example of the output from git log for two commits:

```
commit a89c9c56d3ca33199df659118965ef5cf0e92b9a
Author: Frank <f.sobotka@example.com>
Date: Wed May 29 14:30:56 2013 +0100

    I made some changes.

commit 7a6270bda82711bbe78f485aa6d6e3951cb901bd
Author: Frank <f.sobotka@example.com>
Date: Wed May 29 14:29:53 2013 +0100

    Initial commit.
```

Branching and merging

When you want to make large changes to a codebase (or even small additions in many cases), a good strategy is to do all of your work in a branch as this leaves the master branch open for other changes and bug fixes. It also means that the master branch of your repository is always the latest stable version of your code, and doesn't have to contain any half-finished or experimental code (because you would not merge your changes until they are complete).

Before creating new branches and making changes, it would be good to see what branches we already have in our repository. We can do this with the following command:

```
git branch
```

This should only list one branch for our example so far: `master`. Master is the default branch in a Git repository and it is what we have been interacting with thus far. Other version control systems call this default branch *trunk* or *main*. You will also notice an asterisk next to the word *master*—this indicates that this is your currently selected branch. Now let's create a new branch in which to do some work:

```
git branch feature_branch
git checkout feature_branch
```

These commands create a new branch, called `feature_branch` and then switches our working copy (by using `checkout`) to be the new branch. In order to confirm that we have switched branches, you can run either `git branch` or `git status` and these will inform you of your current branch. Now, make some changes to `file1.txt` then add and commit them to the repository as before:

```
git add file1.txt
git commit -m "Updated content."
```

Now let's switch back to working on the master branch:

```
git checkout master
```

After running this command, take a look at the contents of the file that you just changed. You should now notice that your most recent changes have disappeared. You now have two concurrent versions of the file—`master` and `feature_branch`—that you can switch between. Let's now merge the changes made to the new branch back into the master branch with the following command:

```
git merge feature_branch
```

Once again, if you check the contents of the file you have edited, you will now see your most recent changes have been added in. The key thing to take from this example is that working in branches happens in the following steps:

1 Create a new feature branch to work in and check it out.

2 Make edits and commit in new branch.

3 Switch to target branch (for merging changes into).

4 Merge feature branch.

A final, optional but recommended, step is to delete the feature branch as it has now been merged back into master and is no longer needed. This will keep your list of branches small and manageable as your codebase grows. The branch can be deleted with the following command:

```
git branch -d feature_branch
```

Conflict resolution

Having started with branching and merging, we need to prepare ourselves for the inevitable: managing conflicts. Conflicts occur in a number of ways, but by far the most common is that the same line of code (or text in our example case) is edited in a feature branch, and on the master branch prior to merging, and Git (which knows nothing about programming or what you are trying to achieve) cannot determine which is the preferred version of the code to use. This is when a conflict is raised and a member of the team has to get involved to fix the mess.

   We can simulate this exact scenario to show what happens when a conflict occurs, and what you need to do to resolve it and move on with your work. First, ensure you are still working within your current Git repository, and that you are on the `master` branch. Now add a file called `conflict.txt` and enter the following text:

```
This is
some test
text
```

   Now add the file to the Git repository and commit it:

```
git add conflict.txt
git commit -m "Add a file to test conflicts on."
```

Then create and checkout a new branch:

```
git checkout -b conflict_test_branch
```

   Now we need to edit our file—let's just change the last line so that the file looks like this:

```
This is
some test
text for conflicts
```

Now add and commit these changes in our branch:

```
git commit -a -m "Add some text for testing conflicts."
```

Note the new trick on this commit line: the `-a` option. This combines the add and commit step into one command, by adding ALL changes made within the working copy to the staging area prior to committing. For brevity, you can combine the two options and simply use `-am` for the same effect, which is what we will do from now on. Now switch back to the `master` branch:

```
git checkout master
```

Opening the file will reveal that it has reverted to the original version without the 'for conflicts' addition. This time, we will modify the file in a different way so that it looks like this:

```
This is
some test
text to play with
```

Then commit these changes:

```
git commit -am "Correct the last line."
```

This is the final step needed to create the environment for a conflict. Let us quickly recap how this happened. First we created a text file with some content, then we created a branch, edited the file, and committed our changes within the branch. Finally we returned to the master branch, and edited the same line of text, and committed our changes in the `master` branch.

What we have done is to create two versions of the same file that are incompatible because they have two different versions of the third line. We will therefore generate a conflict if we try to merge the feature branch back into the master branch:

```
git merge conflict_test_branch
```

Executing this command will result in the following output telling you that a conflict has occurred:

```
Auto-merging conflict.txt
CONFLICT (content): Merge conflict in conflict.txt
Automatic merge failed; fix conflicts and then commit the
result.
```

If you now open `conflict.txt` you should see the following content:

```
This is
some test
<<<<<<< HEAD
text to play with
=======
text for conflicts
>>>>>>> conflict_test_branch
```

This is how Git lets us deal with conflicts in files: between the <<< and === characters you find the content from the 'left-hand' side of the merge (in our case, the master branch, that we were trying to merge into), and between the === and >>> you will find the content from the 'right-hand' side of the merge (the feature branch that we were trying to merge from). It is now up to us to sort this out.

Obviously, in this example the decision as to what is the correct commit to pick is only a matter of taste, but when this occurs in your program code, it will be much clearer what the intentions of both commits were and you will be able to correct the code easily. For the purpose of this example, let's just merge the two statements, so now modify the text in the file to look like this and save it:

```
This is
some test
text to play with conflicts
```

This has resolved the conflicts, but we now need to tell Git that we have fixed the conflict. To get a clue as to what we need to do next, run git status—this will give you the following output (or something similar depending on your version of Git):

```
# On branch master
# You have unmerged paths.
#    (fix conflicts and run "git commit")
#
# Unmerged paths:
#    (use "git add <file>..." to mark resolution)
#
#       both modified: conflict.txt
#
no changes added to commit (use "git add" and/or "git commit -a")
```

So, in order to register our conflict as resolved, we need to add our file to the staging area:

```
git add conflict.txt
```

Finally, re-commit to complete the merge (note that we do not supply a commit message here—Git will auto-generate one for us and open the message up in a text editor):

```
git commit
```

This should now present you with the following commit message:

```
Merge branch 'conflict_test_branch'
* conflict_test_branch:
  Alter some text.
Conflicts:
    conflict.txt
```

If you are happy with this description of the merge, simply save the text and exit the editor that Git opened up, and your merge will be completed and the conflict resolved. The text editor that Git launches in these situations is usually Vim, which is simple to use with as long as you know three basic commands:

◆ Press `i` to enter *insert* mode—this allows you to enter text into the editor.

◆ Press the escape key to leave *insert* mode (and return to *normal* mode).

◆ Type `:wq` in *normal* mode in order to save and quit the editor.

Although this was a contrived example of a conflict, it serves the purpose of demonstrating all the steps needed to resolve the conflicts that typically occur when using Git.

Tags

When you have hit a distinct milestone in your code, for example your first stable release to users, it is usually prudent to *tag* this version of your code so that, if needed, you can once again check it out. In `git` this is performed with the `tag` command. In your current repository you can enter the following command:

```
git tag v1.0.0
```

This command tags the most recent commit in the repository with the label 'v1.0.0'. However, if you want to tag a commit other than the most recent, you can simply pass a commit ID to the tag command to tag a specific commit, for example:

```
git tag v1.0.0 e6e2eb456725cb10006fb15f6771489f8c4c9ab9
```

If you tag a commit in error, removing a tag from your history is simply a case of using the `-d` command option (just like with branches). The following command will delete the tag 'v1.0.0' from the Git history:

```
git tag -d v1.0.0
```

If you would like to see your tags listed in the `git log` viewer (next to the related commit), simply add the `--decorate` command-line option when requesting the log. This can, however, be a touch easy to miss (especially when you have many branches and a lot of commits) so we would suggest the following options when running `git log` as this makes the view far more compact and also shows a visual representation of your branches.

```
git log --decorate --oneline --graph
```

Checking out old revisions or specific tags

As a project progresses you might find you need to be able to check out an old version of your code. This is a very simple operation in Git; you simply pass the commit ID or tag name to the `checkout` command:

```
git checkout e6e2eb456725cb10006fb15f6771489f8c4c9ab9
git checkout v1.0.0
```

Reverting unwanted commits

Sometimes things do not go to plan, and even the best programmers commit some code that is wrong or just plain broken. In these instances it is most useful to be able to revoke the offending changes and roll back to code that worked. With Git this is an easy and automated process, initiated by the `git revert` command. The command is simply `git revert` followed by the commit ID you wish to roll back the changes from, that is:

```
git revert e6e2eb456725cb10006fb15f6771489f8c4c9ab9
```

After you have run this command, your editor will be opened with a commit message pre-entered, simply save this and close the editor in order to complete the process—your commit has now been reverted.

Using remote Git repositories to share code

Thus far, all of our work has been done entirely locally on our computer—this is great for learning about Git and managing your own projects, but the real power of version control is the ability to work concurrently with other developers on a single codebase by using a remote Git repository. The basic concept behind this approach is that different branches of the same project can exist in different places. Let us imagine that a remote repository has been set up, containing the master codebase of a project. This code can be pulled down from a remote repository onto your local machine, and then modified in a newly created local branch. When you are ready to merge your changes back into the master branch, any recent changes to the master (on the repository) should first be pulled into your local copy—this helps reveal potential conflicts early. Your branch can then be merged into the updated master, and changes pushed up to the remote repository.

We are not able to work through an example of this type of workflow here, mainly due to the difficulty of maintaining a remote repository that every reader can commit code to. However, you can discover a lot about the practicalities of using Git with remote repositories by investigating the following Git commands:

- `clone`—Copy a remote repository into a new local directory.
- `fetch`—Download data, tags, and branches from a remote repository.
- `pull`—Fetch from and merge with another repository or a local branch.
- `push`—Upload data, tags, and branches to a remote repository.
- `remote`—View and manage a set of remote repositories.

GUI tools and IDE integration

The command-line interface to Git is fast and easy to use once you get familiar with it, but some developers prefer to use a graphical client to interact with their repositories. Git GUI is installed as part of the Windows installation, but there are a number of others to choose from including SourceTree (`sourcetreeapp.com`) if you are using Windows or Mac OS, or Giggle (`wiki.gnome.org/Apps/giggle`) or

QGit (`goo.gl/b87Cy`) if you are a Linux user. Another alternative for Windows is TortoiseGit (`code.google.com/p/tortoisegit`)—this integrates Git functionality right into the Windows explorer (as does Git GUI). In addition to these standalone GUIs, some text editors and many IDEs (integrated development environments) have integrated support for Git. Check the documentation for your tool of choice, as this may be available as a plugin.

Going further with Git

Thus far we have covered all of the basic tasks you would ever want to do with Git—armed with these commands alone (and a knowledge of remote repositories) you have more than enough information to be very effective at using Git to manage your code. However, there are many more advanced tricks possible with Git. Here are a few examples we would recommend looking up when you have time:

- `reset`—Reset your staging area or working directory to another point.
- `cherry-pick`—Pull a single commit from one branch into another.
- `rebase`—Re-apply a series of patches in one branch onto another and re-write history.
- `bisect`—Find by binary search the change that introduced a bug.
- `grep`—Print files with lines matching a pattern in your codebase.

There is a wealth of information and guidance online for working with Git, but the single best resource we can recommend is the Git website itself and the *Pro Git* book (Chacon, 2009) that in available there free of charge (`git-scm.com/book`).

6.2.4   Alternatives to Git

As we mentioned in the beginning of this section on version control, there are many alternatives to Git. Essentially they all do the same thing—track changes to files—but others may be more supported by your editor or IDE of choice, or your company or team may even require their use. When we last checked there were at least 30 different version control software packages in active use, here we give a quick overview of three of the more common alternatives to Git.

- *Concurrent Versions System (CVS):* Dating back to 1990, this is one of the oldest version control systems that is still widely used today. It is a centralized version control system (thus requiring a connection to a dedicated CVS server). We would not suggest using it if you have the choice, as compared to more modern equivalents it is slow, complicated to use, and has only a very basic feature set.

- *Subversion (SVN):* SVN, released in 2000 was developed as a direct successor to CVS and as such is almost identical in day-to-day use. SVN also fixed a lot of the shortcomings found in CVS, thus leading to it being one of the most used version control systems in the world. It is fast, can handle branching and merging quite well, and has good support in many editors and IDEs. If you choose not to use a distributed version control system, we would recommend Subversion.

- *Mercurial (hg):* Mercurial, like Git, is a distributed version control system. It was also developed and released at the same time as Git, in 2005. In fact, they had the same intended purpose—managing the Linux kernel project—but Git was chosen for this task. As such Mercurial has an almost identical feature set (and day-to-day usage pattern) to Git so almost everything you have learnt thus far with Git is applicable to Mercurial, which is good news if you come to need to use it.

6.2.5 Hosting and sharing your code on the Internet

Recent years have seen a proliferation of online code repositories that allow code to be stored, backed up and shared online. The benefits are obvious. Simply having your code stored somewhere else means that you are protected against loss of valuable work if your local storage dies. More interestingly, a well-managed repository can act as a central point to share code with co-developers and with the world at large—this is the very essence of open source.

Here we introduce four of the most popular options for hosting a project online. As well as the core functionality of the version control system(s) that they support, each of these repositories also provides additional features to help support development, and it is often the quality of these extra features that persuades people to use one system over another. These extra features typically include issue tracking, wikis for creating documentation, pull requests and code review tools, hosting of downloadable builds (i.e. software that end users can install and run), and automated e-mail notification of significant code changes such as new commits. Issue trackers allow users and developers to post bug reports and feature requests that can then be allocated to individual developers and subsequently updated if and when they are addressed in the code—this can be a very effective basis for managing a complex multi-developer project. Pull requests and code review tools allow other developers to easily contribute code changes/fixes to a project and give an interface to show the changes being proposed and a way for other developers to comment on the changes before merging into a project.

GitHub (github.com)

When it comes to Git hosting, GitHub is the biggest, and many would say the best, service for hosting source code and projects online. If your project is open source, the hosting is free. Private repository hosting is available for a monthly fee. In our opinion, GitHub has the most intuitive and friendly user interface of all code hosting sites. To see it in action, take a look at `github.com/dazoakley/bbs-v2`, our repository containing the main code examples from this book.

Bitbucket (bitbucket.org)

Bitbucket, from a company called Atlassian, is a direct competitor to GitHub and offers many of the same services as well as a few others that GitHub does not provide. In particular, Bitbucket supports Mercurial as well as Git. Bitbucket's issue tracking features are integrated with Atlassian's project management tool, Jira. There are both free and paid accounts available depending on your needs, with unlimited private repository hosting that is free of charge for projects with five or fewer collaborators.

SourceForge (sourceforge.net)

SourceForge is the home of many prominent open-source projects, and offers developers of open-source software useful tools to manage their project free of charge. SourceForge supports Git, Subversion, and Mercurial. As well as the typical extra features, SourceForge also provides a project homepage and discussion forums.

Google Code (code.google.com)

Alongside GitHub and SourceForge, Google Code is one of the more popular code hosting services among bioinformatics developers, and once again offers full functionality for open-source projects free of charge. Git, Subversion, and Mercurial are all supported. An example of a Google Code project is mzq-lib (`code.google.com/p/mzq-lib`), a Java library built around the mzQuantML file format for working with the results of quantitative proteomics experiments.

6.2.6   Running your own code repository

Thus far we have only talked about third-party code-hosting solutions. These are fine for most of the bioinformatics projects that are run out of academic institutions, where open source is usually a must, but depositing valuable unreleased code on other people's servers can be less palatable in commercial environments. In these cases it can make more sense to keep your code and project management completely private on your own servers. There are many tools out there to help you do this, but we cannot justify covering the complexities of this here. If this type of use case is important to you, we recommend that you investigate Redmine (`www.redmine.org`) and Gitlab (`gitlab.org`).

6.3   Creating useful documentation

Writing good documentation is sadly very often overlooked, especially by people starting out in programming. It is easy to understand why—it takes time and effort but it is nowhere near as interesting as building your software or adding new features to it. However, if you want your software to be useful to end users, or want other developers to help maintain your code (perhaps via one of the code repositories introduced in the previous section), or incorporate it into their own programs, it needs good documentation. There are three types of documentation to consider:

+ Standalone documentation for end users (user manuals).

+ In-program documentation for end users (e.g. at the command line).

+ Documentation about your code for other developers.

We have all seen examples of the first type of documentation, typically in the form of an instruction manual for a piece of software. Writing such documentation requires considerable skill and patience, and benefits massively from being trialled on typical users. However, it is the latter two types of documentation that we focus on here, and in particular the technical underpinnings of these.

6.3.1   Documenting command-line applications

If you have used many command-line applications in the past you will know that the best of them have good help text letting you know the various options and parameters that they support. If you are not sure what we are talking about, try typing git --help in your terminal. This command will produce a detailed list of subcommands or options that can be passed to Git, to make it perform different tasks. Providing this kind of help functionality within our Perl programs would obviously be a great boon for our users. The good news is that support for this functionality is readily available in Perl.

Getopt::Long is a standard part of Perl that allows your programs to take command-line options in exactly the same way as Git. This is a great library that allows you to build up powerful command line utilities, but the one thing it doesn't do is automatically generate help text. This is where the module Getopt::Long::Descriptive (available from CPAN in the usual way) comes in. This is modelled on Getopt::Long, but is slightly simpler to use, and has the added benefit of automatically supplying help text generated from the command-line options you put in your code. Here is a short example script to demonstrate how Getopt::Long::Descriptive works:

```perl
#! /usr/bin/env perl
use strict;
use warnings;
use Getopt::Long::Descriptive;

my ($opt, $usage) = describe_options(
   'my-program %o <some-arg>',
   [ 'server|s=s', "the server to connect to" ],
   [ 'port|p=i', "the port to connect to", { default => 3306 } ],
   [ 'verbose|v', "print every detail" ],
   [ 'help', "print usage message and exit" ],
);
print($usage->text), exit if $opt->help;

# connect to our server... for example...

print "Connected to server!\n" if $opt->verbose;
```

As can be seen, Getopt::Long::Descriptive provides the describe_ options method in which we can describe the options available in our program. The output of this method is two variables: $opt, the passed options from the user, and $usage, a representation of the help text, automatically generated.

With this example, and the more detailed documentation on CPAN (bit. ly/14OF8B7), it should be easy to see how this library can be used.

Now if you save the above code into a file (we'll call it opts_test.pl), and run it, but adding --help after the program name in the command line, you will see the automatically generated help text. For example:

```
$ perl opts_test.pl --help

my-program [-psv] [long options...] <some-arg>
      -s --server      the server to connect to
      -p --port        the port to connect to
      -v --verbose     print extra stuff
      --help           print usage message and exit
```

As you can see, this directly maps to what was entered in the code. This is a great way of getting useful documentation to your users whilst making your programs work in a familiar way.

6.3.2 Documenting Perl code

Code documentation in Perl is in a format known as POD, which stands for Plain Old Documentation (see perldoc.perl.org/perlpod.html for the detailed specification) and refers to the format of documentation that Perl programmers write within their code. POD documentation allows you to document your programs, methods, functions, and classes to help other developers to understand your software, or even to remind yourself if you have been away from it for a while, and to guide other programmers in the development of programs that interact with your classes and functions.

The best way to illustrate the key POD concepts is to go through an example Perl program to which POD code has been added, and show how you can subsequently extract the documentation out of this. For this example we return to the code from Section 6.1 (unit testing), where we started putting together a small module for dealing with data from Ensembl. Here is the code again, except this time with some POD added:

```perl
package EnsemblTools;

use MooseX::Declare;

class IdParser {
  has 'id' => (is => 'ro', isa => 'Str', required => 1);

  method is_valid_id {
    if ($self->id =~ /^ENS\D*[G|T]\d+$/) {
      return 1;
    } else {
      return 0;
```

```
      }
    }
method species {
  if ($self->is_valid_id) {
    if ($self->id =~ /^ENS(\D*)[G|T]\d+$/) {
      if ($1 eq '') {
        return 'Human';
      } elsif ($1 eq 'MUS') {
      return 'Mouse';
    } else {
      return 'Unknown';
    }
    }
    }
  }
}
1;


__END__

=head1 EnsemblTools

EnsemblTools is a small helper library for interacting with
content from Ensembl (http://www.ensembl.org).

=head2 IdParser

EnsemblTools::IdParser - a helper object for parsing Ensembl ID's.

=over 1

=item new()

Create a new instance of EnsemblTools::IdParser. This method
requires an id to be passed in.

  my $id_parser = IdParser->new(id => 'ENSG00000139618');

=item is_valid_id()

Checks the validity of the id. Returns 1 (valid), 0 (non-valid).

=item species()

Extracts the species encoded in the id.

=back

=cut
```

As can be seen in the above, this is exactly the same code, except for the extra content at the end of the file that makes up the POD documentation. The POD documentation does not have to come at the end of the file—it can be placed anywhere within the code, but for an example as small as this, it seems like the best place to put it (you will soon decide which approach you personally prefer once you start writing—and maintaining—some POD documentation of your own).

Before we move on to showing how to extract this documentation out of your code, we should first explain the one extra bit of syntax we have added in this example that you would have not seen before – the `__END__` literal. This is one of Perl's special literals (`perldoc.perl.org/perldata.html#Special-Literals`) and is there simply to tell the Perl compiler that this is the end of the code—no need to parse anything more from here. The reason we do this is that in large modules/codebases, the documentation could be huge—if we let the compiler run over all of the documentation, it would not cause any errors, but it would be unnecessary processing that would make your programs slower to start. With this directive, Perl knows to miss the final portion of the file and move on to whatever it needs to do next.

Now that we have all of this documentation in our code, what can we do with it? As we have seen, in this example it is quite easy to read the documentation as it is, but we can make it even better via the use of the `pod2text`, `pod2html`, `pod2man`, and `pod2latex` command-line utilities that come with Perl.

Each one of these tools simply takes the name of the Perl file(s) to process as a command-line argument and then extracts, converts, and formats any POD documentation found in the file(s) into the selected format. For example, this is what `pod2text` produces when we run it on the example program file:

```
$ pod2text EnsemblTools.pm
EnsemblTools
    EnsemblTools is a small helper library for interacting
    with content from Ensembl (http://www.ensembl.org).

  IdParser
    EnsemblTools::IdParser - a helper object for parsing
    Ensembl ID's.

   new()
    Create a new instance of EnsemblTools::IdParser. This
    method requires an id to be passed in.

    my $id_parser = IdParser->new(id => 'ENSG00000139618');

   is_valid_id()
    Checks the validity of the id. Returns 1 (valid), 0
    (non-valid).
```

```
species()
  Extracts the species encoded in the id.
```

As you can see, POD documentation is a great resource for other developers who want to use our module as it gives them clear and concise instructions of how to use the functions contained within. Having the code and documentation together in a single file makes it easy to update the documentation as the program changes, and means that changes to the documentation can be tracked by a version control system along with the changes to the code. Most modern programming languages provide something similar to POD, so when using another language we recommend spending a little time finding out about that language's in-code documentation method before starting on any serious projects.

6.4   User-centred software design

If you are writing bioinformatics software that will be used by other people, it is worth thinking about who those people are before designing and building it. Historically, many software developers working in bioinformatics have conceived and designed software (its functionality, look and feel, display options, etc.) in isolation from the intended users. Some of the resulting applications were not user-friendly, being too complex and difficult for biologists to use.

It sounds obvious, but in many applications the developer is not the user. The person you are designing the software for may not have the same needs, background, skills, and interests as you. Hence, taking a user-centred design (UCD) approach can help you focus on getting the design right for the target audience (Pavelin *et al.*, 2012). Established UCD techniques can help you to explore the problem your software will solve, characterize the people who will use it, and inform the design of the user interface. Thus, using a UCD philosophy, you can improve the likelihood of the research community benefiting from your efforts. Such is the importance of UCD that it has become a discipline in itself, with companies and individuals dedicated to maximizing the usability of their clients' software. Specific examples of techniques you could employ to improve the user experience include:

- **Interview potential users**: aim to understand what your users need, how your solution will fit into their existing activities, and the language they use to describe it. Use open questions such as 'tell me about . . .'; 'last time you tried to do this, what happened?'

- **Derive user profiles**: create two or three specific user types to describe why and how people will use your software. Include their motivations for using the application (such as their research questions), and their journeys through the application (so-called 'user stories').

- **Card sort**: get users to organize items/specific data into piles and give each stack a name. This is to define the information architecture; for example, to build intuitive menus and labels in your application.

+ **Create mockups of your software:** these can be paper or digital (such as using a wire-framing tool like Balsamiq Mockups, available from `balsamiq.com`). The 'paper prototyping' approach is a cheap and quick way to test the flow of steps through a website or application, and can show whether displays are intuitive. For testing interactions, you can even use a PowerPoint slideshow of interface designs.

+ **Test the software with users:** try testing initial software designs with users— this can be at the paper or interactive prototype stage. Using specific scenarios ('Imagine you have just done an experiment . . .') and tasks ('Find the information about . . .'; 'Produce a sequence alignment . . .'), you will see the application in action, and spot issues with the design. Ask the user to think out loud, and as observer remember to refrain from explaining or showing the participant what to do.

In summary, the key to creating good user experiences is to apply evidence of user behaviour to the design of your software. Do not rely on your own assumptions. For a full account of applying UCD to a bioinformatics project see de Matos *et al.* (2013).

A final thing to mention is that UCD does take resources, time, and effort, but it is worthwhile when developing any software that requires extensive user interaction because it can make that software more usable. Your software may have a novel algorithm or provide access to high quality data, but if users cannot understand how to use it, your work will go to waste. Indeed, observing just a small number of people interacting with a prototype can make a huge difference, and if issues are spotted early, they are easier to fix.

6.5   Alternatives to Perl

As stated earlier, Perl is just one of many programming languages that can be used in bioinformatics. In fact, any programming language could be used for bioinformatics tasks, some are just more common than others due to technical benefits or their established user base. Every language has strengths and weaknesses, so when it comes to deciding which one to use for a given project you need to carefully consider project requirements and the experience of your development team. In the following few sections we will introduce some alternatives to Perl, namely Python, Ruby, and Java. These are the more common languages used in bioinformatics, and most programmers would agree that they are all in some way better than Perl, so having an awareness of them is a good thing. Clearly, we do not have space to teach you how to program in these three languages, but we can show you how to get started in each language and explain the main differences among them. Finally, we briefly cover a potential alternative to programming altogether—the Galaxy platform.

6.5.1   Python

Python is general-purpose programming language, first introduced in 1991. Its syntax is similar to Perl and it could be used in any situation where Perl would be considered useful. Given this, Python has become a very popular language in the bioinformatics, scientific, and engineering fields over the last few years. As a result, there is a large number of very advanced third-party libraries for complex data processing activities that were previously the preserve of dedicated data analysis packages such as R or Matlab. Indeed, there are libraries for natural language processing, machine learning, image analysis, and statistical computation, among others. This makes Python a very compelling language to look into when dealing with complex data, as there may already be a library there to get you started very quickly. Python is also more suited than Perl to object-oriented programming.

Downloading and installing Python

Getting Python installed and set up on your computer is a fairly easy process. If you have a Mac, it is already there as it is part of the default operating system install, and if you are a Linux user this is most likely the case too. Simply type the following at the command line to find out which (if any) version of Python you have installed:

```
python --version
```

If you get an error indicating that Python is not installed, simply install it via your package manager. If you do have Python already installed, but it is an old version (older than 2.7), please look into upgrading your python install to the most recent version of the 2.7 series.

Windows users should head to the Python website (`python.org/download`) and download and run the Windows installer for the most recent build of Python 2.7 (2.7.4 at the time of writing). You will probably have to manually add the location of Python to the Windows path. You can do this by executing the following command (substituting `C:\Python27\` with whatever is the location of Python on your computer).

```
set PATH=%PATH%;C:\Python27\
```

At the time of writing there are two current stable releases of Python: 3.3.1 and 2.7.4. Python 3.x introduced changes that broke backwards-compatibility with 2.7.x and resulted in many widely-used third-party libraries becoming unsupported. While there are efforts to support 3.x, much of the community still relies on the 2.7.x series, so at this time, we would recommend you stick with the 2.7 series.

Something else to be wary of when getting into Python is that, unlike in most other languages, whitespace is significant. When we get on to code examples in a short while we shall explain this some more, but it is worth noting up front that blocks in Python (if-else, while, for etc.) are not delimited by brackets or braces but by the indentation of the code that follows them. This can come as a surprise to Python newcomers. We shall discuss this more a little later, but for now let us look at some simple Python examples.

Hello, world

The following is a basic Hello World program written in Python:

```
#! /usr/bin/env python
print "Hello, world!"
```

Like Perl, we open our program with a similar looking shebang line, this time linking to the Python interpreter (but as with Perl, not strictly necessary in Windows as we explained in Chapter 3). We then have a single-line program that is essentially the same as its Perl equivalent save for two small differences. The first is no semi-colons at the end of the line—you can use semi-colons if you want, but they are strongly discouraged as such behaviour is not considered *Pythonic*. The second difference is that there is no newline character in the string; this is not needed as Python's print command automatically appends one onto your string unless a comma character follows it, as in the example below.

```
print "hello",
```

The last thing to note is that Python programs typically have the file extension .py. Just to check you have Python installed and working correctly, save the above code into a file called hello.py and try running it from the terminal:

```
python hello.py
```

Variables and data structures

The basic data structures and variables should be familiar from Perl:

```
#! /usr/bin/env python

# this is a comment
a = 'ACBCDEFG'           # string
b = 12                   # integer
c = [ 1, 2, 3, 4, 5 ]    # a list (array) of integers
d = { 'a': 1, 'b': 2 }   # a dictionary (similar to a hash)
e = dict(a=1, b=2)       # a dictionary identical to d, with the
                         # exception that keys do not require
                         # quoting
# accessing list and dictionary elements
c[0]                     # This returns 1
d['a']                   # This also returns 1
```

The syntax for creating data structures in Python is almost identical to Perl references, but with the two main differences. The first is that variable names do not need to begin with a $, /@ or % symbol; and the second is that hashes (or *dictionaries* as they are known in the Python world) have a slightly different syntax. Accessing the elements of both arrays and hashes should also be very similar. The above code includes a few comments—these are also formatted the same as in Perl.

Let's look at some more complex data structures for which we used references in Perl, this should not come as too much of a surprise:

```
# a multi-dimensional list
a = [ [1,2,3], [1, 'a'] ]

# a dictionary containing other dictionaries and lists...
b = { 'an_array': [1,2,3], 'a_dictionary': {1: 'a', 2: 'b'} }

# accessing the elements
a[0][1]                 # This returns 2
b['a_ dictionary'][1]   # This returns 'a'
```

String interpolation

If we wanted to output a variable as part of a string in Perl, we might use the following code:

```
my $user = 'Dave';
say "Hello $user";
```

Here is the equivalent in Python:

```
user = 'Dave'
print "Hello {0}".format(user)
```

This is quite different to what we got used to in Perl. In Python, you cannot simply embed variables in strings and have them returned correctly; instead you must use placeholders (in this example the {0}) and call the format method on the string, passing the variables to be used in the placeholders—this is analogous to how we prepared database statements in Perl using DBI—create a statement using placeholders, and then pass the variables into the placeholders. When there is more than one variable, we need more than one placeholder, for example:

```
user = 'Dave'
family = 'Brent'
print "This is a message for {0} {1}".format(user,family)
```

Control structures, loops, and logic operators

Like most other programming languages, Python has the usual selection of conditionals, loops, and control structures. Here are some examples of the Python syntax for the common constructs that we saw previously in Perl:

```
# IF-ELIF-ELSE blocks
if False:
    print 'this code will not run'
elif True:
    print 'this code will run'
```

```
else:
    print 'it will not get this far'

# IF NOT blocks (UNLESS equivalent)
x = False
if x is not True:
    print 'x is {0}'.format(x)

# FOR loops
a = range(0,10)    # create a list containing the range 0 to 9
for i in a:
    print "element: {0}".format(i)  # print the variable in
                                     # the string

# WHILE loops
i = 0
while i < 10:
    print "i = {0}".format(i)
    i += 1    # increment i by 1, Python does not have i++
```

The above code examples are mostly self-explanatory, but please take note of our earlier warning: there are no brackets or braces to indicate these blocks are opening or closing. This is due to the significance of the leading whitespace within the blocks—this tells Python when the blocks start and end. If you remove some of the indentation from the above code before running it, you will likely get an `IndentationError` message telling you that this is invalid syntax. The recommended standard is to use four spaces each time you indent code—never tabs.

The final thing from the above code that is unfamiliar (compared to Perl) is the use of the keywords `True` and `False`. In Python, `True` and `False` are built-in Boolean types.

Regular expressions

Regular expressions in Python are quite different to those found in Perl. Let us start with a short example:

```
import re

gid = 'ENSG000041_A'
matcher = re.compile('^ENSG\d+')

if matcher.search(gid):
    print "String is an Ensembl Gene ID."
```

This is a port from Perl to Python of one of the regular expression examples in Chapter 3, demonstrating how you can search for a basic pattern such as that of an Ensembl Gene ID. The first thing to note in this example is the `import` statement at the top of the code; this is Python's equivalent to the Perl `use` statement—what we are importing here is Python's built-in regular expression module 're' as this is not loaded into all Python programs by default. We then compile a regular expression using `re.compile` (note that the regular expression itself is just a string in Python, it does not have to be enclosed in `//` characters) and then use the `.search` method on the compiled regular expression, passing in a string to search—this will return either a `MatchObject` (an object representing the matched regular expression) or `None` (a built-in Python type) depending whether the regular expression matches or not.

Let us expand upon this example to show how we would use matched variables in Python:

```
import re
string  = 'The Human Gene ID is ENSG000041 revision 1'
matcher = re.compile('The (.*) Gene.*(ENSG[0-9]+).*')
match   = matcher.search(string)

if match is not None:
    print "Our gene ID is: {0}".format(match.group(2))
    print "Our species is: {0}".format(match.group(1))
```

The main difference to take home in this example is that we now interact with the returned `MatchObject` from the `.search` method, by calling the `.group(x)` method on it to retrieve the matched variables from the regular expression. This is equivalent to the `$1`, `$2`, and so on variables in Perl's regular expressions.

File operations

File operations in Python are best illustrated with a few well-commented code examples:

```
# Read the whole contents of a file into a string
contents = open("BRCA1.fasta","r").read()

# Read the whole contents of a file into an array in memory
contents = open("BRCA1.fasta","r").readlines()

# Iterate over the lines in a file
filehandle = open("BRCA1.fasta","r")
for line in filehandle:
    print line.rstrip()

# Note for above, .rstrip is similar to chop() in Perl

# Write to a file
output_file = open("output.txt","w")
```

```
output_file.write("This is some text\n")
output_file.close()

# Write to a file using a context manager
# Note - this will automatically close the file for you
with open("output.txt","w") as file:
    file.write("This is some text\n")

# Append data to the end of a file.
with open("output.txt","a") as file:
    file.write("This is some more text\n")
```

Error handling

Error handling in Python is very similar to the `eval` method of error handling in Perl, as it is performed via the use of `try` and `except` blocks. Here are some examples of how to handle unexpected errors, and raise your own using `raise` (Python's equivalent of Perl's `die`). First, this is the syntax for raising your own errors:

```
if age > 45:
    # deliberatly raise an error.
    raise Exception("age must be under 45.")
```

Second, this is how to catch an unexpected error:

```
try:
    file = open("input.txt","r")

    # ...
    # some code with potential for error
    # ...

except IOError as e:
    # Handle a specific error class
    print "I/O error({0}): {1}".format(e.errno, e.strerror)
except:
    # Rescue all other errors.
    print "Unexpected error"
    raise
```

Object-oriented programming

Python embraces a lot of object-oriented programming (OOP) concepts, so it should come as no surprise that writing OOP code in Python is straightforward. However, unlike some more modern alternatives, Python is not a totally object-oriented language. Here we provide a quick tour of what is possible, using some familiar examples. The first example creates a basic class and performs some simple interactions with it.

```python
class Dog(object):
    def __init__(self, name=None, breed=None, age=None):
        self.name = name
        self.breed = breed
        self.age = age

def __str__(self):
    str = "{0} is a {1} and is {2} years old."
    out = str.format(self.name, self.breed, self.age)
    return out

# now interact with Dog...
sam = Dog()
sam.name = "Sam"
sam.breed = "Terrier"
sam.age = 5

max = Dog("Max", "Hound", 8)

print sam
print max
```

This snippet of code demonstrates the basics of building a class in Python, making instances of that class and some other useful tricks. In the first line we create a class called `Dog` (that inherits from `object`—a built in Python class—signified by it being in parenthesis after the class name), we then define two methods on the class (note the use of the command `def` here, this is Python's equivalent to Perl's `sub`).

The first method (`__init__`) is a constructor method that is invoked when the object is created, and specifies that the object can take up to three arguments (`name`, `breed`, and `age`), but note that these are all optional as they have default values assigned to them (`None` in this case). These arguments are then stored in the object's attributes with the same name. The second method (`__str__`) is automatically called when an instance of a `Dog` class is called in a string context, for example when we try to print it. Finally, we create two instances of `Dog`. For the first we create an instance with no properties and add them in subsequent lines, in the second we populate all of the attributes as we create the instance, just to demonstrate the ways in which objects can be built.

One thing to note from this code example is that we did not have to declare object properties in advance like we did with Moose in Perl (using the `has` method). This can be a good thing as you don't have a lot of setup code defining properties/interfaces, but also a bad thing in that it is not clear anywhere (except in the rest of the code that interacts with the `Dog` class—which you would have to read) what the complete set of properties for the given class are.

Inheritance

As with every other object-oriented language, Python supports class inheritance, in fact we have already seen this as our classes in the previous section all inherited their properties from `object`. Here is a simple example of how to use inheritance with your own classes. To allow direct comparison with the Moose code, let's use the same example, starting with an `Animal` class followed by a `Dog` and `Cat` class that both inherit from `Animal`.

```
class Animal(object):
    def __init__(self, name=None, breed=None, age=None):
        self.name = name
        self.breed = breed
        self.age = age
        self.human_years_multiplier = 1

    def human_age(self):
        return self.age * self.human_years_multiplier

    def __str__(self):
        str = "{0} is a {1} and is {2} years old."
        out = str.format(self.name, self.breed, self.human_age())
        return out

class Dog(Animal):
    def __init__(self, name=None, breed=None, age=None):
        parent_proxy = super(Dog, self)
        parent_proxy.__init__(name, breed, age)
        self.human_years_multiplier = 7

class Cat(Animal):
    def __init__(self, name=None, breed=None, age=None):
        parent_proxy = super(Cat, self)
        parent_proxy.__init__(name, breed, age)
        self.human_years_multiplier = 5
```

In this example we introduce a few more new Python concepts. The first is the use of parenthesis after the name when referring to `self.human_age` within the `str.format` statement above; this is required as we are calling a method, not retrieving a property (as is the case with `name` and `breed`).

The second concept that we have introduced above is the use of the function `super`. This is used when you need to call methods from a parent class that have been overridden in a child class. To use `super` you must pass the type of the current class (e.g. `Dog` in the `Dog` class), and an instance of the class in question (`self`); this then gives you a proxy to the methods on the parent class (we call this `parent_proxy` above) which can be used to invoke the parent's version of the methods that have been overridden. In the example above we execute the constructor method on the Animal class (`__init__`) within our child classes (`Dog` and `Cat`) so that the properties (`name`, `breed`, and `age`) can be set without us duplicating the code needed to do this, then we override the `human_years_multiplier` property with the desired value from the child class.

Multiple inheritance

The final OOP concept we would like to cover with Python is that of multiple inheritance. Multiple inheritance means that a class is permitted to inherit behaviour from multiple parent classes. Here is an example of that in practice:

```
class Animal(object):
    def eat(self):
        print "i eat"

class Mammal(Animal):
    def breathe(self):
        print "i breathe"

class WingedAnimal(Animal):
    def flap(self):
        print "i flap my wings"

class Bat(Mammal, WingedAnimal):
    def __init__(self):
        print "i am a bat"

bat = Bat()
bat.eat()
bat.breathe()
bat.flap()
```

In this simple example we have a `Bat` class, which inherits from all three of the other classes defined (`Animal`, `Mammal`, and `WingedAnimal`) to build up its behaviour; this is done simply by listing more parent classes in the parenthesis following the class name.

Multiple inheritance is a controversial subject in programming circles. Some people believe it allows modelling all aspects of the world cleanly and clearly, whereas opponents believe it can introduce too much complexity into a system. As always, the best way to make a judgement is to experiment with multiple inheritance in a given application on which you are working to see whether or not it brings benefits.

It is worth noting that Python is not the only language that supports multiple inheritance; there are many others that allow you to use this programming style, in fact Moose allows you to use multiple inheritance in Perl. The other two programming languages covered in this book (Ruby and Java) do not implement multiple inheritance, but similar functionality can be achieved using Ruby *mixins* (covered later) and Java *interfaces*.

The Python ecosystem

Python's equivalent of CPAN is the Python Package Index (`pypi.python.org/pypi`), a collection of over 29,000 user-contributed libraries. These packages can be installed via a command line tool called Pip (`www.pip-installer.org`).

Python is a popular language for creating web applications. Some of the more popular web frameworks for this task are Django (`www.djangoproject.com`), Flask (`flask.pocoo.org`), and Bottle (`bottlepy.org`) but there are many other alternatives.

For scientific and engineering applications, including bioinformatics, Python is extremely popular and well catered for in terms of supporting libraries. We list here some that we believe to be the most useful for bioinformatics work:

- BioPython (`biopython.org`)—a set of tools for biological computation in the same vein as BioPerl.

- matplotlib (`matplotlib.org`)—a 2D plotting/graphing library for creating publication-quality figures.

- SciPy (`www.scipy.org`)—a comprehensive library for mathematics, science, and engineering.

- scikit-learn (`scikit-learn.org`)—a suite of tools/libraries for machine learning tasks.

- scikit-image (`scikit-image.org`)—tools for image processing.

- IPython (`ipython.org`)—this provides an interactive front end for Python, much like R Studio provides a front end for R.

- Pandas (`pandas.pydata.org`)—a library providing high-performance, easy-to-use data analysis tools and data structures.

Connecting to a MySQL database

Python has many different libraries for talking to databases, but the one that we would recommend at this point is `MySQLdb` (sometimes also referred to as 'MySQL-python'). This is not a standard part of the Python distribution, so to get started with it you will first need to install it. Linux users, you will find it in your package manager, Windows and Mac OS users will need to install it via Pip:

```
pip install MySQL-python
```

If you have any problems installing this, `pip` usually guides you with some very helpful instructions, but if you are still having trouble, head over to the forum at `www.bixsolutions.net` for assistance.

Now that `MySQLdb` is set up, let's write a basic program to interact with the database we made earlier in Chapter 2:

```
#! /usr/bin/env python

import MySQLdb
```

```
# connect to the database
con = MySQLdb.connect('localhost', 'user', 'password',
      'database');

# get a 'cursor' for interacting with the db
# 'MySQLdb.cursors.DictCursor' allows us to use column names
# in resultsets
cur = con.cursor(MySQLdb.cursors.DictCursor)

# run a query
cur.execute("SELECT * FROM Scientist")

# fetch all the rows of data
for r in cur.fetchall():
    print "{0} {1}".format(r["title"], r["family_name"])

# close the database connection
con.close()
```

The MySQLdb interface is quite similar to Perl DBI, you connect to your database and get a database handle (or *cursor* as it's more commonly known), then run a query (via the `.execute` method on the cursor), and finally retrieve the results.

Going further with Python

Our recommended reference for Python programming is the Python documentation website (`docs.python.org`). All the core features and the standard library are well documented there. The site also hosts a good tutorial (`docs.python.org/2/tutorial`) which would be a great way to follow up what we have discussed in this chapter. For a more in-depth look at Python we recommend *Think Python* (Downey, 2012)—this book is available to purchase as a paperback but also freely available in electronic form (`www.greenteapress.com/thinkpython`).

6.5.2 Ruby

Ruby is another scripting language (like Perl and Python) that has become popular in the bioinformatics community. Introduced in 1995, it is the youngest language that we cover in this book. Although Ruby does not quite have the depth and breadth of third-party libraries available to it for complex data analysis tasks that Python does, it has a very strong set of tools and culture for making web applications and unit testing. Ruby is also a fully object-oriented programming language, with a simple and concise syntax that allows for the construction of full-featured, complex applications with very little—but very clean and readable—code.

Downloading and installing Ruby

There are various ways to install the latest version of Ruby, as detailed on the Ruby website's download page (`www.ruby-lang.org/en/downloads`).

For Windows users, we would recommend using the suggested Ruby Installer—available at `rubyinstaller.org`. During the installation, ensure the 'add Ruby executables to Windows path' is checked so that you get effortless access to Ruby from the command line.

Linux and Mac users may already have Ruby installed, but it is a rapidly developing language so we would recommend using the RVM (the Ruby Version Manager), available from `rvm.io`, to ensure you have the latest version (2.0.0-p195 at the time of writing).

Hello, world

Our first demonstration of Ruby code is the familiar Hello World example:

```
#! /usr/bin/env ruby
puts 'Hello, world!'
```

Like Perl and Python, the program begins with a shebang line (pointing to the Ruby interpreter) and then we have a single line program. The only difference here you may notice (compared to Perl and Python) is the command `puts`; this is a Ruby function to print a line of text that is automatically followed by a newline character.[1] Ruby also has the `print` function (and this works identically to the Perl `print` function), but the `puts` function is more commonly used in the Ruby community.

The other main difference you will notice with this small example—there are no semi-colons at the end of the lines. Like Python, these are also optional in Ruby; and again like Python, you can use semi-colons if you like (out of habit), but this is not encouraged.

Ruby programs typically have the file extension `.rb`. To check you have Ruby installed and working correctly, save the above code into a file called `hello_world.rb` and try running it from the terminal with the following command:

```
ruby hello_world.rb
```

Variables and data structures

Ruby takes many of its roots from Perl, so it should come as no surprise that variables are very similar between the two languages. Some examples are shown below.

```
#! /usr/bin/env ruby

# this is a comment
a = 'ACBCDEFG'             # string
b = 12                     # integer
c = [ 1, 2, 3, 4, 5 ]      # an array of integers
d = { 'a' => 1, 'b' => 2 } # a hash
```

---

[1] In fact, Perl does have an equivalent function to `puts`, called `say`, but it is not widely used.

```
# accessing array and hash elements
c[0]                          # This returns 1
d['a']                        # This also returns 1
```

Like Python, the syntax for creating data structures in Ruby is essentially identical to Perl references. Comments also use the same (#) character. If we look at some examples of more complex data structures, that we would have used references for in Perl, you will see that the Ruby code is very clean, compact, and easy to read, reducing the possibility of error:

```
# a multi-dimensional array
a = [ [1,2,3], [1, 'a'] ]

# a hash containing other hashes and arrays...
b = { 'array' => [1,2,3], 'hash' => {1 => 'a', 2 => 'b'} }

# accessing the elements
a[0][1]                       # This returns 2
b['hash'][1]                  # This returns 'a'
```

String interpolation

If we wanted to output a variable as part of a string in Perl, we might use the following code:

```
my $user = 'Dave';
;say "Hello $user";
```

The equivalent in Ruby is:

```
user = 'Dave'
puts "Hello #{user}"
```

In Ruby, it is necessary to surround any code that you want to be evaluated within the string by the #{} construct. This may seem like extra work for no extra benefit in this simple example, but this does allow for *any* Ruby code to be executed within the context of the string. For example:

```
require 'date'
puts "It is #{DateTime.now} exactly..."
```

This code creates a new `DateTime` object for the current time (`now`), and outputs it as a string (within the context of our main string). As with Python, there is a special method that is called when we want to represent an object as a string (`__str__` in Python) - this is `to_s` (meaning *'to string'*) in Ruby. So, `DateTime.now.to_s` is essentially what we called above.

This small snippet of code also introduces another keyword of Ruby: `require`—this is analogous to the Perl `use` keyword and is used for including library code in your programs.

Conditionals, loops, and logic operators

Like all the other languages in this chapter, Ruby has the usual selection of conditionals and loops. Examples of Ruby syntax for the most common constructs are shown below.

```ruby
# IF/ELSE blocks
if false
  puts 'this code will not run'
elsif true
  puts 'this code will run'
else
  puts 'it will not get this far'
end


# UNLESS blocks
unless 1 == 10
  puts '1 will never equal 10'
end


# FOR loops
a = (1..10)                 # create an array containing the
                            # range 1 to 10
for i in a
  puts "element: #{i}"    # print the variable in the string
end


# another way of doing FOR loops - an EACH block
# this will have the same result as above...
a.each do |i|
  puts "element: #{i}"
end


# WHILE loops
a = 0
while a < 10
  puts "a = #{a}"
  a += 1                    # increment a by 1, Ruby does not
                           # have a++
end
```

As you can see, Ruby syntax is similar to Perl and even more similar to Python. One novelty here is the part that begins with `a.each`. This (specifically the section of code between the `do` and `end`) is what is known as a *block* in Ruby. A block can be thought of as an anonymous function that (in this example above) is passed each element in the array as the variable i. The `do`/`end` form of blocks is most common in Ruby, but you can also use curly brackets to form blocks:

```
# This is equivalent to the above.
each { |i|
  puts "element #{i}"
}

# As is this, but compacted onto a single line
a.each { |i| puts "element #{i}" }
```

Regular expressions

Regular expressions in Ruby are almost identical to those found in Perl. Here is the familiar Ensembl ID example written in Ruby:

```
id = 'ENSG000041_A'

if id =~ /^ENSG[0-9]+/
  puts "String is an Ensembl Gene ID."
end

if id.match("^ENSG[0-9]+")
  puts "String is an Ensembl Gene ID."
end
```

As you can see, Ruby supports the `=~` operator that will be familiar from Perl, but it also has a `.match` method for string objects that can be passed regular expressions. The example below further illustrates Ruby regular expressions with matched variables, again showing both the use of the `=~` operator and the `.match` method:

```
string = 'The Human Gene ID is ENSG000041 revision 1'

if string =~ /^The (.*) Gene.*(ENSG[0-9]+).*$/
  puts "Our gene ID is: #{$2}"
  puts "Our species is: #{$1}"
end

if match = string.match("^The (.*) Gene.*(ENSG[0-9]+).*$")
  puts "Our gene ID is: #{match[2]}"
  puts "Our species is: #{match[1]}"
end
```

This example takes regular expressions a little further via the use of matched groups; as you can see the `=~` method is identical to the Perl equivalent, as the matched groups are captured in the $1, $2 ... $n variables, whereas the `.match` approach is more like Python, in that a `MatchData` object is returned (and stored in the match variable above), and you can then call the matched groups found from it.

File operations

File operations in Ruby are easy, much like in Perl and Python. Here are some examples:

```
# Read the whole contents of a file into a string
file = File.open("BRCA1.fasta","r").read

# Read the whole contents of a file into an array
# (this would also be the approach to use if you wanted to
# read a file line by line)
file = []
File.open("BRCA1.fasta","r").each_line do |line|
  file.push(line.chomp)
end


# Write to a file
file = File.open("output.txt","w")
file.puts "This is some text"
file.close

# Write to a file using a block
File.open("output.txt","w") do |file|
  file.puts "This is some text"
end

# Append data to the end of a file
File.open("output.txt","a") do |file|
  file.puts "This is some more text"
end
```

Error handling

Error handling in Ruby, like Python, is similar to the `eval` method of error handling in Perl, and is performed via the use of `begin` and `rescue` blocks. Here are some examples of how to handle unexpected errors, and raise your own using `raise`, as in Python but with different syntax. Here is an example of raising an error:

```
if age > 45
  # deliberatly raise an error.
  raise "age must be under 45."
end
```

This is how we can catch unexpected errors:

```
begin
  file = File.new("input.txt","r")

  # ...
  # some code with potential for error
  # ...

rescue Errno::ENOENT => error
  # Handle a specific error class
  # - in this case, "file not found"

  warn "File, 'input.txt' not found!"
  raise
rescue => error
  # Rescue all other erorrs.

  warn "error: '#{error}' was raised..."
  raise
end
```

Object-oriented programming

Ruby is an inherently object-oriented programming language. Indeed, everything in the language is an object. Even high-level Ruby functions (e.g. `puts`) belong to a special object called `Kernel`. So everything we have done thus far with Ruby has been creating and interacting with objects—when we have used the dot operator in a statement like `something.method_name` we have been calling methods on objects. The way in which classes are defined and objects created is similar to Python, but with different syntax, as shown in the example below.

```
class Dog
  def name
    @name
  end

  def name=(name)
    @name = name
  end

  def breed
    @breed
  end

  def breed=(breed)
    @breed = breed
  end
```

```ruby
  def age
    @age
  end

  def age=(age)
    @age = age
  end

  def to_s
    "#{@name} is a #{@breed} and is #{@age} years old."
  end
end

# now interact with Dog...
sam        = Dog.new
sam.name   = "Sam"
sam.breed = "Terrier"
sam.age    = 5
puts sam.to_s
```

This code is fairly verbose, but illustrates the basics of building a class in Ruby and making an instance of it. In the first line we create a class called `Dog`, and define a number of methods on the class—all of which (bar one called `to_s`) are *getters* and *setters* for instance variables within the class. These work in exactly the same way as the `has` attribute declarations in Perl's Moose, but without the built-in type checking. It is also worth noting the use of the `@` character in Ruby—this defines an instance variable (not an array). An instance variable is a variable that is available to all other methods within an instance of the same class. The final method, `to_s`, in the class returns information about the instance of a class when it is referenced as a string, just like `__str__` in Python. While the above example is useful to illustrate Ruby's OOP concepts, it is not typical Ruby code. Here is the same example, written using a more common coding approach:

```ruby
class Dog
  attr_accessor :name, :breed, :age

  def initialize(args={})
    @name  = args["name"]
    @breed = args["breed"]
    @age   = args["age"]
  end

  def to_s
    "#{name} is a #{breed} and is #{age} years old."
  end
end
```

```
sam = Dog.new({ "name" => "Sam", "breed" => "Terrier", "age"
    => 5 })

max       = Dog.new
max.name  = "Max"
max.breed = "Hound"
max.age   = 8
puts sam.to_s
puts max.to_s
```

   This version of the code creates exactly the same Dog class, but does it with much less code and allows for two ways of instantiating an instance of a Dog object. The main reason for the brevity of this code is the removal of all the *getter* and *setter* methods, which have been replaced with a single line of attr_accessor declarations; attr_accessor is a built-in Ruby method that builds the getter and setter methods for you—just like the has method from Perl's Moose. The other major change in this version of the class is the inclusion of the initialize method in the class; this is what is known as a constructor method and is what is called when you call Dog.new() within the rest of the code (equivalent to __init__ in Python). In our example we have an initialize method with an optional argument hash that can be passed to it (called args)—this is optional due to the '={}' following its name (without it the calling code would have to pass something into the new method); this means that if it is not passed by the caller, an empty hash is used as its value. The value of this initialize method setup is shown at the end of the example in the two different ways in which you can instantiate instances of the Dog class—with a one line command or with multiple commands.

Inheritance

A short example of how inheritance is performed in Ruby is shown below. This re-uses the same example of object inheritance that we used in both Moose and Python—we start with an Animal class followed by a Dog and Cat class that both inherit from Animal.

```
class Animal
  attr_accessor :name, :breed, :age

  def initialize(args={})
    @name = args[" name"]
    @breed = args["breed"]
    @age = args["age"]

    @human_years_multiplier = 1
  end

  def to_s
    #{@name} is a #{@breed} and is #{human_age} years old.
```

```
  end
  def human_age
    @age * @human_years_multiplier
  end
end

class Dog < Animal
  def initialize(args={})
    super
    @human_years_multiplier = 7
  end
end

class Cat < Animal
  def initialize(args={})
    super
    @human_years_multiplier = 5
  end
end
```

Two new pieces of Ruby syntax have been introduced here. These are the use of the < character to indicate that one class inherits from another, and the `super` function. The `super` function in Ruby works in a slightly different way to Python, in that you do not need to build or access a proxy object for the parent class—we simply call `super`, and that will run the same method in the parent class (with the same arguments that were passed to the current method) and then return to, and carry on the method in, the child class. So, in our example above, calling `super` causes the parent class's `initialize` method to run, we then carry on running the `initialize` method in the child class, where we can override any of the variables setup in the parent class—in our case we update the @human_ years_multiplier variable to the appropriate number for our given animal.

Modules and mixins

The final OOP concept we are going to look at in Ruby is that of modules and mixins. The term *module* in Ruby refers to a collection of functions stored together within a single namespace, much like the Perl modules we have been using previously; and the term *mixin* refers to the way in which the modules are used; they are mixed-in to other classes and objects to extend functionality. Mixins are very similar in concept to the Moose roles and traits that we looked at in Chapter 3, in that they can provide additional functionality to a class or object, but the main difference is that they do not perform the 'interface' function that is common with traits—that is, you cannot declare that a consuming class must provide certain functions. This is best explained with an example:

```
module A
  def method_a_1
```

```ruby
    puts "I'm calling Module A : method_a_1"
  end

  def method_a_2
    puts "I'm calling Module A : method_a_2"
  end
end

module B
  def method_b_1
    puts "I'm calling Module B : method_b_1"
  end

  def method_b_2
    puts "I'm calling Module B : method_b_2"
  end
end

class Sample
  include A
  include B

  def method_s_1
    puts "This is the method_s_1 in Sample"
  end
end

samp = Sample.new
samp.method_a_1
samp.method_a_2
samp.method_b_1
samp.method_b_2
samp.method_s_1
```

Here we create two modules, A and B, and then a class called `Sample`, which can then consume (or `include` as is the common term, and keyword, in Ruby) the two previous modules. We then instantiate an instance of `Sample` and demonstrate how we can call all of the functions provided in both A and B via a `Sample` object.

The Ruby ecosystem

Ruby's equivalent of the downloadable Perl modules (available from CPAN) are known as *gems*. The main repository of gems—the equivalent of CPAN—can be found at `rubygems.org`. From here you can browse tens of thousands of gems that have been made for Ruby. Gems can be installed automatically via the command line client `gem`, which comes with Ruby. For example, to install the popular `biomart` gem for interfacing with Biomart servers, we simply type:

```
gem install biomart
```

Ruby is a particularly popular language for developing web applications thanks to web frameworks such as Ruby on Rails (`rubyonrails.org`) and Sinatra (`www.sinatrarb.com`), but it is equally useful for any general programming task that you might have used Perl for. The main reason why people may prefer Perl or Python over Ruby is that Ruby's bioinformatics ecosystem is not as mature as either Perl or Python, but there are projects trying to address this. The most notable of these is BioRuby (`bioruby.org`)—a suite of bioinformatics tools for Ruby in the same vein as BioPerl and BioPython.

Connecting to MySQL

Ruby has many different libraries (gems) available for connecting to and interacting with databases, it even has a `DBI` interface analogous to the Perl variant, but the one that we would recommend in standalone programs[2] is Sequel (`sequel.rubyforge.org`). Sequel is a third-party gem; it is not part of the standard Ruby library so you will first need to install it (and the mysql2 gem that provides the drivers for talking to MySQL):

```
gem install sequel
gem install mysql2
```

Windows users will need to install the Ruby Installer 'Development Kit' from `rubyinstaller.org` before you can install and use these gems.

Once these are installed, using Sequel is quite a simple process. Here is an example program interacting with the MySQL database we built in Chapter 2:

```
#! /usr/bin/env ruby

require "sequel"

# connect to the database
DB = Sequel.connect(
  :adapter => "mysql2",
  :database => "my_database",
  :user => "me",
  :password => "my_password"
)

# iterate over the Scientist table
DB["Scientist"].each do |s|
  puts [s[:title], s[:given_name], s[:family_name]].join(" ")
end
```

The code above needs little explanation—we simply connect to a database, and then use a simple Sequel *data set* function (`[]`) to iterate over every row in the

---

[2] Ruby web applications are usually written using frameworks such as Ruby on Rails, which tend to come with their own preferred way of talking to databases.

Scientist table. Within the block you will notice that the row of data from the table is returned as a hash. This is the simplest way of interacting with a database via Sequel. Here is a second example, showing how to use custom SQL queries:

```ruby
#! /usr/bin/env ruby

require "sequel"

# connect to the database
DB = Sequel.connect(
  :adapter => "mysql2",
  :host => "ensembldb.sanger.ac.uk",
  :database => "homo_sapiens_core_47_36i",
  :user => "anonymous"
)

# save an SQL statement into a string (using a heredoc in Ruby)
query = <<"SQL"
  SELECT es.synonym, sr.name, g.seq_region_start, g.seq_region_end
  FROM seq_region sr, gene g, external_synonym es
  WHERE es.xref_id = g.display_xref_id
  AND sr.seq_region_id = g.seq_region_id
  AND es.synonym IS NOT NULL
  ORDER BY g.seq_region_start ASC
  LIMIT 500
SQL

# use .fetch to run the query and iterate over the results
DB.fetch(query) do |row|
  puts "synonym: #{row[:synonym]}, name: #{row[:name]}"
end
```

In this example we once again connect to a database (this time the public Ensembl mirror at the Sanger institute), then we prepare a SQL query in a string, and finally run this query on the database (using the .fetch method on our DB connection object) and iterate over the results.

Going further with Ruby

If you want to learn more about Ruby, we would recommend the book *Programming Ruby: The Pragmatic Programmers Guide* (Thomas *et al.*, 2013). This is the standard text for any new Ruby programmer and will get you up to speed very quickly. A highly recommended alternative is *The Ruby Way* (Fulton, 2006)—this book takes more of a cookbook approach to learning Ruby, by showing you specific examples of how you would solve a given problem. Finally, as with Python, Ruby has comprehensive documentation on the Internet (hosted at ruby-doc.org), which serves as a great reference and is of course likely to be more up to date than printed books.

6.5.3 Java

Java is a ubiquitous programming language, running not only on the desktop computers and servers that we use for bioinformatics but also on smartphones (Android apps are written in Java), Blu-ray players, smart cards, and a host of other devices. This wide reach is due to the fact the Java was conceived from the very beginning as a platform-independent language. In Java, programs are not written to run on a specific operating system, but on something called the Java virtual machine (JVM). JVMs are provided for all manner of different platforms, including Windows, Linux, and Mac OS. Java programs interact exclusively with the JVM, which in turn interacts with the operating system and underlying hardware. Because the virtual machine is designed to look identical regardless of the platform on which it is running, you can write a program in Java and run it unmodified on any platform for which a JVM exists.[3] Another benefit of the virtualization approach is increased security—because every Java program is separated from the operating system by the JVM there is a limit to the damage a malicious or poorly written program can do. Java also has potential performance advantages, because the source code is compiled to optimized *bytecode* prior to execution.

In terms of programming, Java is an inherently object-oriented language that has a lot in common with the other languages that we have looked at—there are variables of familiar types (strings, integers, arrays, etc.), conditional statements, loops, comments, a documentation generator (called Javadoc), a unit testing framework (JUnit), and mechanisms for dealing with files and errors. However, Java does not share the scripting roots of Perl, Python and Ruby. It more closely resembles the C++ language, which means that program code is generally more formal, consistent, and verbose. Such languages tend to be favoured by computer scientists, who consider them to lead to more robust, efficient, and readable code. They are generally less popular among infrequent coders, such as biologists, who find the learning curve off-putting. However, Java is widely used in bioinformatics, particularly in the proteome informatics community, for web development, and for applications where a platform-independent client-side user interface is required.

It is worth noting that, from the software development perspective, Java comes in a range of distinct *editions*, each tailored to a specific type of application development. In this chapter we will exclusively use the Standard Edition (Java SE), which is intended primarily for producing computer-based applications that run locally. The other edition of interest for bioinformatics work is the Enterprise Edition (Java EE), for developing web-based applications.

Downloading and installing Java

It is possible to develop Java software as we have done for the scripting languages, by typing programs in a text editor, and compiling and running them at the

---

[3] Of course, programs written to take advantage of platform-specific hardware will not be fully functional on devices that do not have that hardware.

command line. In practice, this is rarely done because of the relative complexity of Java program code. Instead, many developers make use of a freely available cross-platform IDE called NetBeans, which is highly optimized for Java. NetBeans helps us edit and manage our code, even to the extent of providing automatic code generation for routine tasks. For this tour of Java, we will be exclusively using NetBeans.

Before installing NetBeans, we need to ensure that key Java components are in place. Firstly, for the reasons explained earlier, you must have a JVM (often referred to as the *Java Runtime Environment*, *JRE*, or simply *Java*) installed on your computer before any Java program will run. You probably already have it, but you can check and install it if necessary by following the simple instructions at `java.com`.

The JVM only contains the basics needed to run Java programs, not the tools to develop and compile them. Before we can start developing Java programs, we need to install the Java software development kit, often called the *Java Development Kit* (*JDK*) or *Java Platform*. The Java SE version of this can be downloaded via `goo.gl/UIPvF`. Here you have the option to download the JDK by itself, or together with NetBeans. Since we will be using NetBeans, we recommend the combined download. After downloading, the installation is straightforward and default installation options can be used.

Hello, world

To get started writing Java code, we first need to launch NetBeans. This will bring up a welcome page with links to various tutorials and information. The functionality of NetBeans is almost overwhelming at first, with all this information available and lots of menus across the top, each containing a large number of options. It is worth spending considerable time exploring these things, but for the moment we will go straight into writing our familiar example program that writes out 'Hello, world'. The first step is to create a new NetBeans project. From the File menu, select New Project. In the window that appears, click Next to start a new Java Application. This will bring up a second window in which you can specify the name of the project and the location of the program files. Make changes there if you want, then click Finish to create the project. You will then be taken to the code editor, where you will see that NetBeans has already created the bare bones of your first Java program.

As in Ruby, everything in Java is object-oriented, so your application is a class, with a method (called `main()`) that is called automatically when you execute your program. Note that `main()` has a string parameter called `args`—this will contain any command-line parameters passed to the program. For a simple example, we just need to add a line of code between the curly brackets ({ and }) that denote the code block belonging to the `main()` method so that the program looks like the example below. Note that for reasons of space we have omitted the white space and comments (in Java any text surrounded by /* and */) that NetBeans inserts.

```
package javaapplication1;
public class JavaApplication1 {
    public static void main(String[] args) {
        System.out.println("Hello, world!");
    }
}
```

We can compile and run this within NetBeans by clicking the Run Project button (the green arrowhead) on the toolbar at the top of the NetBeans window. The command-line output, together with some status information from NetBeans, will appear in a pane within the NetBeans window, below the program code.

The line that we added uses the pre-wrapped `System.out` object, which represents the standard output steam—the command line in our case. This field supports a number of methods, including `println()` which is used to output text. As in Perl, lines must be terminated by a semi-colon. Compared to the other languages in which we have implemented this example, there is no doubt that Java is the most verbose by far. The program has at least twice as many lines as the equivalent Perl code, and the command to print the welcome message is considerably more complicated. This is one of the things that dissuades people from using Java. However, in fairness, the coding overhead of Java is highly exaggerated in a trivial example such as this. It is in larger, more complex, applications where the formality of Java really pays dividends. As one of our colleagues succinctly put it 'for complex projects you need a complex language'.

The NetBeans IDE does a lot to mitigate Java's complexities. For example, you may have noticed that as you typed the new line within NetBeans, hints appeared to help you complete it, including a list of fields supported by the `System` class, a list of methods supported by `out`, and the variable types accepted by `println()`. Right clicking on an item within the program code will bring up a menu, from which you can access Javadoc documentation for that particular item. It is worth taking a look around to see what else NetBeans offers as there really is a wealth of functionality there, including advanced code navigation, integrated testing, debugging, profiling (finding out which parts of your program take up the most compute time), and version control (Git is one of several supported systems).

Running Java programs from the command line

Running programs within the NetBeans IDE, as we do in this chapter, is very convenient but does hide important steps that take place behind the scenes. What actually happens when you click the Run Project button is that the classes get compiled from the human readable `.java` text files into JVM readable `.class` bytecode files. NetBeans then uses the locally installed JVM to execute the classes, and captures the default output stream so that it can be displayed within the NetBeans window.

Because Java applications usually contain multiple classes, all the classes for the application are typically packaged into a single Java archive (`.jar`) file for distribution. This can be executed from the command line simply by passing the name

of the .jar file to the Java VM using a command like the one below. Java is therefore just as suitable at Perl, Python, or Ruby for developing command-line tools.

```
java -jar JavaApplication1.jar
```

To see this in action for our example program, click on Clean and Build Project
from the Run menu in NetBeans. This builds the project into a .jar file and will
show you, via the output window, the command needed to execute it.

Variables and data structures

In Java, all variables must be declared and their type specified prior to use, or
when they are first assigned a value. Variables can only exist in methods and
classes, so must be defined within those code blocks in which they are to be
used. In the example program below some variables are declared and used within
main():

```
package javaapplication1;
public class JavaApplication1 {
  public static void main(String[] args) {
    int x;                      /* integer declared but not set */
    String a = "ACBCDEFG";      /* sting variable set and declared */
    int b = 12;                 /* integer declared and set */
    double p = 3.14;            /* fractional number */
    int [] c = {1,2,3,4,5};     /* an array of integers */
    x = b + c[0];               /* referencing first element of c
                                   array */

  }
}
```

In Java, all but the most primitive data types are actually classes, so the instan-
tiations of these are objects rather than variables. In the above example, a is an
object because all strings are objects, and c is an object because it is an array. Like
any objects, a and c have properties and methods that can be used to interrogate
and manipulate them. All other more complex data types are also implemented
as classes.

String interpolation

The easiest way to output a variable as part of a string is to use the format
method, which is supported by System.out. Like Python, this requires placeholders to be included in the string and variables to be provided for each of these
placeholders. The placeholders are then instantiated with the values of the variables at runtime. Here are some examples that could be inserted towards the end
of the code block in our simple program, above to report the values of the variables that were declared.

```
System.out.format("a = %s%n",a);
System.out.format("The value of p is %f%n",p);
```

```
System.out.format("The value of p to one decimal place is
                  %.1f%n",p);
System.out.format("The value of x is %d + %d = %d%n",b,
                  c[0],x);
```

Note that the placeholders must indicate the type of variable to be interpolated: `%s` for a string, `%d` for an integer and `%f` for a fractional number. There are other converters for date and time variables, and `%n` which inserts a newline character. Detailed formatting information can be included in the string, such as the number of decimal places to report for p in the third line.

Control structures, loops, and logic operators

The Java syntax for the key programming constructs is not dissimilar to Perl. In some cases, Java offers multiple ways of writing the same thing, but these are the most common:

```
/* single line conditional statement */
if (false) System.out.println("this code will not run");

/* multi-line conditional statement with else code */
/* brackets only needed for multi-line code blocks */
if (false) {
    System.out.println("this code will not run");
}
else {
    System.out.println("this code will run");
}

/* ten iteration for loop */
for (int a = 0; a < 10; a++) {
    System.out.format("i = %d%n",a);
}

/* ten iteration while loop */
int i = 0;
while (i < 10) {
    System.out.format("i = %d%n",i);
    i++;
}
```

The `for` loop is a typical example of the formality, or thoroughness, of Java. When initializing the loop we must declare the loop variable and specify its type, then explicitly state the condition that must be satisfied for the loop to continue and how the loop variable will be altered each time around the loop. While this may seem verbose, it allows for the creation of a wide range of different loops while remaining eminently readable.

If you enter the `if` examples in NetBeans you will notice a small yellow warning icon appear next to some of the lines. Mousing over this icon reveals the hint 'the branch is never used'. This is an example of the real-time code checking that NetBeans does for us—it uses its knowledge of the Java language to flag up typos, syntax errors, and even logic errors like this one, where it has discovered that those particular lines of code will never get called.

Object orientation

As already mentioned, Java was created with object-oriented programming in mind from the outset, and we have had to use classes and objects to achieve even the most fundamental things like writing text to the command line. As you might expect, creating and working with your own objects in Java is both elegant and efficient. Here is a version of our familiar pet example:

```
package javaapplication1;
public class JavaApplication1 {
  public static void main(String[] args) {

    /* define the Dog class */
    class Dog {
      String name;
      String breed;
      int age;

      String getDescription() {
        String out = String.format("%s is a %s and is %d
                    years old.", name, breed, age);
        return out;
      }
    }

    /* use the Dog class to create a dog */
    Dog sam;
    sam = new Dog();

    /* set some of the Dog's properties */
    sam.name = "Sam"
    sam.breed = "Terrier"
    sam.age = 5;

    /* tell us something about the dog */
    System.out.println(sam.getDescription());
  }
}
```

This clearly follows the same pattern as the similar examples for other languages, with arguably more readable syntax. All of the other standard functionality, such as inheritance, that makes OOP so efficient, is supported also.

Building a graphical user interface

Thanks to its object-oriented philosophy and a well-established library called Swing, Java is particularly well suited to producing standalone software tools with graphical user interfaces. Such tools are having something of resurgence in the bioinformatics community, as the size of some data sets makes web-based tools impractical for certain applications. The SeqMonk sequence data visualization tool (www.bioinformatics.babraham.ac.uk/projects/seqmonk) is an excellent example of the sophisticated interfaces that can be created by an experienced Java developer.

To illustrate the use of the Swing library, we are going to create a much simpler application. Following the same process as before, create a new NetBeans project called GUIverse. After creating the project, NetBeans will take you to the code editor with the familiar automatically generated code to get us started. This time, rather than start writing code, add a new file by selecting New File from the File menu. In the dialogue box, select Swing GUI Forms from the left list and then JFrame Form from the right list. Click Next, then Finish to create the form with the default name NewJFrame. NetBeans will then take you to a newly created tab in the editor, displaying a rather boring grey rectangle—this is our *frame*, or *window* as we might more commonly call it. On the right of the NetBeans window is a palette of GUI objects that can be added to this window. For the purpose of this tutorial, just drag across one standard button and one label—it doesn't matter exactly where you place these in the frame. Clicking once on one of the objects in the frame will cause its properties to be displayed in a panel to the right of the NetBeans window. These properties can be edited, for example you can change the text property of the button to OK. The button will change in the frame to reflect this edit.

To see our new frame in action, we need to add the code below to the main() method of GUIverse.java to create and display a frame like the one we created (use the tabs at the top of the code panel to navigate between the different files). The code for creating the frame object should be familiar from the earlier dog example. The setVisible() method is used to bring the newly created frame into view.

```
NewJFrame mainwindow;
mainwindow = new NewJFrame();
mainwindow.setVisible(true);
```

Running this code will result in a window like the one you designed appearing on the screen. It can be moved around and re-sized, but otherwise it does nothing. To remove the window and quit the application, click the window's close icon.

If you clicked the OK button that we created, you will notice that it does nothing. To make it perform some action we need to attach program code to it. To do this, go back to the NewJFrame.java frame designer and double click the button in the form. This will take you to the program code that underpins the form, all of which has so far been auto-generated. In particular, it will take you

to a new method called `jButton1ActionPerformed()`. This is the event handling method that is automatically called when the user clicks the button. To add
some basic functionality to this button, insert the following command into this
method:

```
jLabel1.setText("The button changed me!");
```

Now, if you run the program and click on the OK button you will see that the
label changes. It is from this simple concept that even the most complex GUIs
can be built. Each GUI object has a range of events associated with it, and by writing code to deal with these events we can build up a fully functional application.
To take a more practical example, we might want to build an interface to scroll
through a visual representation of some data. A scrollbar would be added to the
frame and code written for the scrollbar's `adjustmentValueChanged` event to
update the view of the data according to the position of the scrollbar whenever
the user moves it.

A major attraction of using Java to develop applications with GUIs is that the
GUIs are, like Java itself, platform independent. This is achieved by some clever
technologies, particularly layout managers that dynamically reorganize GUI
elements to fit the resolution of the display on which it is running. The details
of how GUI elements are displayed—the so-called *look and feel*—also changes to
match the operating system on which the application is being run, or you can set
it manually within the application.

The Java ecosystem

There are thousands of Java libraries available, and thanks to Java's fundamental
strengths in terms of consistent syntax, object-oriented philosophy, and documentation generation, they are generally easy to use. The most important library
to be aware of is the Java API, which comes with the JDK that we already downloaded. The Java API is a collection of themed packages, each of which contains
a number of ready-baked classes that together cover most generic programming
tasks. At the time of writing, the current API is version 7, which contains literally thousands of classes, all of which are documented in detail at `docs.or-
acle.com/javase/7/docs/api`. The Swing package that we used for our GUI
example is part of the Java API. JDBC, for connecting Java programs to relational
databases is another very important part of the Java API.

For bioinformatics work, a collection of classes for common sequence analysis and structural biology tasks is provided by the BioJava project (`biojava.
org`). There are plenty of other relevant Java libraries out there but (unlike Perl,
Python, and Ruby) there is no single accepted repository for these, so finding
what you need is a case of searching the web or scanning the literature.

Going further with Java

While this section has discussed some of the distinguishing features of Java, and
demonstrated how to start programming in Java with NetBeans, it is probably fair
to say that becoming proficient in Java will take longer than it would in the other

languages that we have covered. Java makes available a lot of powerful functionality, but experience is needed to take full advantage of it. Our recommended book for beginning Java is *Learning Java* (Niemeyer and Leuck, 2013). *Head First Java* (Sierra and Bates, 2005) is another introductory Java book, which has become popular thanks to its unusual but effective style of explaining complex concepts.

Official Java tutorials and the very latest documentation for the Java language and the current Java API can be found on the Oracle website (`www.oracle.com/technetwork/java/javase/documentation` for the Java SE edition). As we have already seen, the NetBeans welcome page has links to a range of information about Java programming and NetBeans. More information about NetBeans itself can be found at `netbeans.org`.

6.5.4   Using Galaxy

As discussed in Chapter 1, recent years have seen the development of a number of software platforms for joining bioinformatics tools together into more complex pipelines, ostensibly removing the need to write bespoke software that connects different tools. Such frameworks are not a total replacement for any of the programming languages that we have been looking at in this chapter, because it is still necessary to write new tools for new biological applications. However, using these frameworks can save us from a lot of the mundane programming work needed to integrate tools and interact with users.

At the time of writing, the pre-eminent workflow platform in bioinformatics is Galaxy. Galaxy is accessed via a web browser, with information displayed in the three main panes—the leftmost pane contains a list of tools available in the Galaxy instance being used, the middle pane is used primarily to configure tools and display results, and the rightmost pane records a history of files that have been uploaded, analyses that have been done, and the results generated. Tools can be executed individually, one after the other, or can be linked together in workflows for automated analysis, either by schematically creating a workflow or by creating one automatically from a recent analysis history.

The best way to get a feel for Galaxy is to visit the publicly available Galaxy server at `usegalaxy.org`. For more serious work, Galaxy can be installed locally on Linux or Mac OS by following the simple instructions at `getgalaxy.org`. There is no official support for Windows, so if you want to run Galaxy under Windows you will need to create a Linux virtual machine as explained in Appendix C (note that the Bio-Linux distribution has Galaxy pre-installed).

The simple concept underlying Galaxy, and most other workflow frameworks, is that most bioinformatics tools can be considered as a piece of programmatically accessible software (e.g. a command-line tool or web service) that takes some input data, usually together with some user-specified parameters, processes that data, and produces some output. For a tool to be useable within Galaxy it must have a *wrapper* that defines the types of input and output needed, the parameters required from the user, and the structure of the command needed to execute the tool. Galaxy wrappers must be written in a standard XML format (documented at `wiki.galaxyproject.org/Admin/Tools/ToolConfigSyntax`). For each tool, Galaxy is able to use the information contained in the wrapper to render a simple user interface for specifying the input data and parameters, and can determine how to deal with the output produced when the tool is executed. It can also work out which tools can be connected together in a workflow, by checking they have compatible inputs and outputs.

If you are producing a command-line bioinformatics tool, there are compelling benefits to writing a Galaxy wrapper for it: the web interface that Galaxy generates for the tool will be a lot easier for most biologists to use than the command line, and integrating your tool with others in a pipeline should be effortless. However, writing wrappers can become complicated for some tools and debugging is not easy because a lot is happening out of sight, behind the scenes.

6.6   Summary

When starting out in programming, it is easy to dismiss the software engineering practices introduced in this chapter as an unnecessary overhead, especially in bioinformatics where a piece of software may initially have few users and there is pressure to get results out for an imminent publication or deadline. However, once you have a few programming projects under your belt, the benefits become clear: good software engineering helps us create programs that are more reliable, easier to use, easier to understand, and easier to extend and maintain. Perhaps most importantly, it provides an effective framework for the development of software by multi-developer teams. Thanks to various extensions, we can adopt all the major software engineering practices in Perl. However, more modern languages have been designed from the ground up with these practices in mind, so they are seriously worth considering if you are beginning a new project. Indeed, we have witnessed an increasing uptake of Python and Ruby in the bioinformatics community. For the moment though, due to the well-developed ecosystem and army of experienced developers, Perl still dominates.

Regardless of the specific languages, libraries, and tools used, the core principles of data storage, programming, and data analysis covered in this book have been established for many years, and are likely to remain at the core of bioinformatics for decades to come. If you have grasped the concepts in this book, you should be well prepared for the future. We wish you every success in building your own bioinformatics solutions.

References

Beck, K. (2003). *Test-Driven Development: By Example*. Addison-Wesley: Boston, USA.

Chacon, S. (2009). *Pro Git: Everything You Need to Know about the Git Distributed Source Control Tool*. Apress: New York, USA.

Downey, A.B. (2012). *Think Python: How to Think Like a Computer Scientist*. O'Reilly: Sebastopol, California, USA.

Thomas, D., Fowler, C. & Hunt, A. (2013). *Programming Ruby 1.9 & 2.0 (4th edition): The Pragmatic Programmers' Guide*. The Pragmatic Programmers: Texas, USA.

Fulton, H. (2006). *The Ruby Way: Solutions and Techniques in Ruby Programming*. Addison-Wesley: Boston, USA.

de Matos P., Cham J.A., Cao H., Alcántara R., RowlandF., Lopez R., & Steinbeck C. (2013). Enzyme Portal: a case study in applying user-centred design methods in bioinformatics, *BMC Bioinformatics*, 14: 103.

Niemeyer. P. & Leuck, D. (2013). *Learning Java (4th edition)*. O'Reilly: Sebastopol, California, USA.

Pavelin K., ChamJ.A., de Matos P., Brooksbank C., Cameron G. & Steinbeck C. (2012). Bioinformatics meets user-centred design: a perspective, *PLoS Computational Biology*, 8(7): e1002554.

Sierra, K. & Bates, B. (2005). *Headfirst Java*. O'Reilly: Sebastopol, California, USA.

Sommerville, I. (2010). *Software Engineering*. Addison-Wesley: Boston, USA.

Appendix A: Using command-line interfaces

This appendix provides a brief introduction to the use of command-line interfaces, for the benefit of those readers who do not have experience of working with these. In this book, we use command lines to interact with three distinct programs: MySQL, R, and the operating system (which may be Windows, Linux, or Mac OS). Although specific commands vary between programs, all command-line interfaces work in a similar way. In what follows, we concentrate on the operating system as an example. Specific information about the command lines of MySQL and R can be found throughout Chapters 2 and 4.

A.1 Getting to the operating system command line

Depending on your computer's operating system, the command line can have several names. On Windows PCs it is known as the *command prompt*, on Mac OS and most recent Linux systems it is known as the *terminal* (older Unix variants, and people experienced with such systems, may refer to it as the *shell*). The way the command line is accessed varies according to the operating system.

Opening the Windows command prompt

The following method should work on all recent versions of Windows:

* Press the Windows key on the keyboard. This will bring up the Start screen (or the Start menu depending on your version of Windows).

* Start typing 'command prompt' (without the quotes) until the command prompt appears at the top of the list of apps (or programs in older versions of Windows).

* Hit the Enter key.

This should open a new window that by default has a black background with white text on it. In that window you will see the command prompt, which will look something like:

```
C:\Users\Conrad>
```

Building Bioinformatics Solutions. Second Edition. Conrad Bessant, Darren Oakley and Ian Shadforth.
© Conrad Bessant, Darren Oakley, and Ian Shadforth 2014. Published 2014 by Oxford University Press.

The prompt tells you that the operating system is ready for you to type a command. It also shows you (in the section before then > symbol) which *directory* you are currently in. A directory is exactly the same as a Windows folder, but the term directory is more typically used in programming languages and at the command line. For example, you can list the contents of a directory by typing dir (short for directory) and hitting return. In the list of items that is returned, subdirectories are flagged with the label <DIR>.

You can move to another directory using cd (short for change directory). For example, to move into a subdirectory called Desktop, we would use the command:

```
cd Desktop
```

Most folders contain a special folder denoted by two dots (..). This is actually a shortcut to the parent folder. So, issuing the command below will move you up one folder:

```
cd ..
```

The above examples move you to directories relative to the directory you are currently in. If you want to move directly to a directory regardless of your current location, you need to specify the new location from the disk drive downwards, for example:

```
cd c:\Progam Files\R
```

This would take you to the R directory, within the Program Files directory on your C: drive. In cases like this, where a file or folder name contains a space, it is good practice to enclose the name in quotation marks to avoid confusion.

```
cd "c:\Progam Files\R"
```

To change to another drive, just type the letter of that drive followed by a colon. So, to change to the E: drive just type e: and hit the Enter key.

Opening a Linux terminal

Depending on how you have Linux configured, you may not have a graphical interface and will therefore already be at the command line. If not, you first need to open up a *terminal* (or *shell*) window. To do this, look for an icon that looks either like a picture of small computer monitor in black with white characters on it, or, in some instances, an icon that looks like a sea shell. You may need to access a list of applications to find this, for example in recent versions of Ubuntu you need to click on the Ubuntu icon (Dash home), type the word 'terminal' (without the quotes) and the Terminal icon should appear for you to click.

Any of these methods should open up a window that will have a solid background with contrasting text, often white on black. Within this window you should see the Linux command prompt. Exactly how this looks will depend on your particular flavour of Linux, but it will likely be similar to one of these:

```
$ /home>
username@computername:/home$
username@computername[/home]
```

This shows you your current directory—in this case the top level `home` directory. You can see the contents of the current directory by typing `ls` followed by hitting the Enter key. As in Windows, you can change directory using `cd`. For example:

```
cd BBS
```

takes you to a folder called BBS.

```
cd ..
```

takes you up one directory.

```
cd /usr/local/
```

will take you to the directory `local` within the `usr` directory. Note that the slashes delineating the different directory levels are forward-slashes, whereas Windows uses backslashes.

Opening a terminal in Mac OS

Mac OS has a terminal application very similar to the ones found in Linux. You can find this in the Applications > Utilities directory. Once you have opened up the Mac OS terminal (also sometimes referred to as *Terminal.app*), the situation is very much the same as Linux, and the instructions described above apply.

A.2   General command-line concepts

Working directory and path

In operating systems (and in R) the concept of the *working directory* is important when using the command line. The working directory is the directory you are in at any given time. If you try to execute a file by typing its name, this is the first place the operating system will look for the file. If the file is not in the working directory it will not be found, unless its location is included in the *path*. The path is a list of directories that you want the operating system to search every time you try to access a program. Depending on the operating system you are using, there are various ways of viewing or modifying the current path.

In Windows you can view the current path by typing `PATH` at the Command Prompt. The path can be edited via Windows Control Panel (`PATH` is a system environment variable, so search for System Environment Variables within Control Panel), but a simpler way to temporarily add a directory to the path is by modifying the path at the command line using a command like the one below. This example will add the `C:\Python27` directory to the path for the duration of the current Command Prompt session.

```
PATH=%PATH%;C:\Python27\
```

In Linux and Mac OS the current path can be viewed by typing `echo $PATH`. The path can be edited using the `export` command. For example, the command below temporarily appends `/home/user/bin` to your existing path for the duration of the current terminal session.

```
export PATH=$PATH:/home/user/bin
```

Parameters, arguments, options, and switches

Some commands, such as `ls` and `dir` are very simple and can just be typed by themselves. Many other commands require one or more parameters to be specified. This is known as *passing* parameters (or *passing arguments*) and is done by typing these parameters after the command, on the same line. One example is the `cd` command shown earlier, which requires the name of the target directory to be specified after the command.

On a similar theme, some commands have options or *switches*, which are used to modify their behaviour. A good example of this is the Perl version switch (`-v`) demonstrated at the start of Chapter 3. If this switch is included as part of the Perl command, as shown below, Perl will just print its version number to the screen, instead of actually executing any code.

```
perl -v
```

Through a combination of command-line arguments and switches, we are able to control the behaviour of even very complex programs. A good example is the blastn program (for more information see `www.ncbi.nlm.nih.gov/books/NBK1763`) used to search for a sequence in a database. An example of blastn in use is shown below.

```
blastn -db nt -query seq1.fasta -out results.out
```

Here, the combination of switches and parameters defines precisely how the search is conducted. Specifically, the nt sequence database is selected with the `-db` switch and the file containing the query sequence (the sequence we want to look for) is specified as `seq1.fasta` by the `-query` switch. Similarly, the file to which results will be written is specified using the `-out` switch. Looking at the documentation at NCBI we can see that `blastn` supports many other possible options, but where we are happy with an option being set to a default setting we can simply omit these options.

Although this way of interacting with programs might seem arcane, and typing such long commands may be prone to error, it actually becomes much more efficient than using a graphical interface if the same—or similar—operations need to be completed repeatedly. In particular, it is easy to automatically generate and execute these commands in Perl using its `system()` function, as described in Chapter 3.

A.3 Command-line tips

There are various shortcuts that make life in the command line a little easier. The details of these will vary according to which particular program or operating system you are using, but something that is common to most is the ability to recall previous commands using the up and down arrow keys. This is particularly useful if you want to repeat a command or type a command very similar to one that you entered earlier (for example, if you made a typing error the first time, or if just want to change one of the switches or parameters).

On recent versions of Windows and most Linux systems, the Tab key may also be used to attempt to automatically complete directory, file, and command names before you have finished typing them. This can speed up typing considerably as you only need to enter the first few letters of a directory or file then press Tab; however, the letters you enter should uniquely point to a single instance, or you may either be given a list of possibilities or, on some systems, one of the options will be selected for you. It is therefore wise to check any automatically completed text. Automatic Tab completion for table and field names is also enabled in some versions of MySQL.

Another handy feature is the path, described earlier. Having key tools, such as Perl and MySQL, in your path makes life a lot easier because you don't have to spend so much time switching between directories. Similarly, file *associations* can be useful. A file association is a link—known to the operating system—between a specified file type extension, and a program that deals with such files. An example of such an association would be to associate `.pl` files with the Perl interpreter program, `perl.exe` (on a Windows system—it's just called `perl` on Linux and Mac). If this has been done, then a Perl program can be executed simply by typing its name at the command prompt, so typing `hello_world.pl` on a Windows system, or `./hello_world.pl` on Linux or Mac OS would run the program with that name directly, without you having to specify that Perl is needed to run it.

Finally, we should note that, throughout this book, when we show examples of commands to be entered at the command line we do not show the command prompt. The only exception to this is where we illustrate a sample command-line session, and in those cases the command prompt is used to differentiate between text that is typed in by the user and text returned to the screen by the computer. This is most frequently used to demonstrate interactions with R in Chapter 4, and occasionally with MySQL in Chapter 2.

Appendix B: Getting started with Apache HTTP Server

To do any serious web development work, you will need access to your own fully fledged web server. As mentioned in Chapter 5, Apache HTTP Server (commonly just referred to as *Apache*) is the server software of choice for most applications in bioinformatics and beyond. There are several ways to start using Apache:

- *Use a server provided by your organization.* If you are working for a company, or in an educational establishment, your organization may have server hardware on which Apache is already installed. The administrator of this server would have to give you a user account with permission to upload files to this server. You can then copy files and web applications over to the server, and access them by pointing your web browser to that server.

- *Buy a hosting package.* There are many companies around the world offering space on their servers for hosting websites, often at very competitive prices. However, it is important to realize that not all web hosting is created equal. In particular, to get a server capable of running applications that you have written yourself, particularly incorporating database functionality, can be expensive.

- *Use a spare computer.* If you own more than one computer, you can install web server software such as Apache on one computer and use that as a web server. You can then use this second computer to host your web pages and applications. One benefit of this approach is that the operating system on the server (Linux would be a good choice) can be independent of the operating system on your main computer (e.g. a Windows PC).

- *Use your own computer.* This is the cheapest and simplest option, and is perfect for development work. In this appendix we explain how to install Apache on your computer—this is referred to as running the server *locally* (as opposed to over a network). You can then connect to the Apache server from a browser on that same computer, just as if you were accessing it over the Internet. This is not a practical way of hosting a publicly available website, as it requires your computer to be permanently on and accessible via the Internet, and you will be

Building Bioinformatics Solutions. Second Edition. Conrad Bessant, Darren Oakley and Ian Shadforth.
© Conrad Bessant, Darren Oakley, and Ian Shadforth 2014. Published 2014 by Oxford University Press.

sharing your computer's processor and memory with visitors to your website. However, it is a perfect approach for learning about developing web resources.

♦ *Use a virtual machine (VM) on your own computer.* This is a combination of the previous two options, giving you the opportunity to run a server on a different operating system than the one you use day to day. To do this, you would set up a VM (see Appendix C), install the server software within that, and then connect to it either from a browser within the VM or a browser in the host operating system. This is useful if you need to build operating system specific functionality into your web application.

B.1 Installing Apache

The process for installing and configuring your own instance of Apache depends on your operating system.[1] We recommend making sure that all network related programs (e.g. browsers, e-mail, Skype, and any other chat software) are closed prior to installing Apache as these can interfere with the installation process. Also, we do not recommend running more than one web server on a single computer.

♦ Windows users will find all the necessary instructions within the Apache documentation (`httpd.apache.org/docs`). Click on the documentation item for the latest version of Apache and then you will find specific instructions for Windows installation in the Platform Specific Notes section. As with software downloaded in other chapters, it is best to look for the MSI file for easy installation. These files are not hosted on the Apache website itself, but on a number of mirror sites that are listed at `www.apache.org/dyn/closer.cgi/httpd/binaries/win32`. Click through to one of the mirror sites, then look through the list of downloads for the MSI file for the most recent no-SSL version of Apache. It will be called something like `httpd-2.2.22-win32-x86-no_ssl.msi`. Click on the filename to download it, and open the file to begin installation of Apache. During installation, you will be prompted to enter a domain name, server name, and e-mail address. Depending on how your computer is set up, these fields may already be filled in, but if not you can enter `localhost` for both domain and server, and your normal e-mail address.

♦ If you are a Linux user, consult the documentation specific to your distribution— Apache is a central part of the success of the Linux operating system and is easy to install on all distributions.

♦ Macintosh users already have Apache installed as part of the operating system, and all web documents (HTML files) are stored in the Documents directory found in `/Library/WebServer/`. All you need to do to start Apache is to activate 'web sharing' in the system preferences, or type `apachectl start` in the terminal.

---

[1] At the time of writing, there is a choice of Apache installations—one with additional security (SSL) and one without. For simplicity we recommend using the version without SSL while you are learning, but suggest investigating SSL in future if you need secure connections to your web pages.

The way to test an Apache installation is to point your browser to your server—the URL will be `localhost` if you are running Apache locally. If the installation is successful you should see a simple 'It works!' message. At this point, you might begin to wonder who else can see this page, and which of your other files they can see. The answer is that only files placed in a specific location on your computer can be accessed via Apache (see the next section), and these can only be seen from remote computers if you do not have a functioning firewall, or you have allowed incoming access to port 80 through your firewall. If you do not want your web server accessible from other machines, make sure that your firewall is turned on and is blocking incoming connections; if you do want to allow access to your web server, consult the documentation for your firewall (most likely found with the documentation for your operating system) and look for information on how you can open up connections on specific ports to outside machines.

Once Apache is installed, using it to serve up your documents is simple. The only things that you need to know are where Apache expects certain types of documents to be stored, and the permissions that need to be applied to these documents.

B.2 Apache fundamentals

To ensure that users can access and view web content on your server, it is important to understand how to organise that content.

HTML documents

HTML documents are text-based files that get served up as static web pages by your web server. In order for these to be served by Apache, they must be stored in a specific directory on the server—this is often known as the DocumentRoot or `htdocs` and its location should be detailed within the server installation documentation. Basically, if you store all of your HTML files here (or in subdirectories within this directory), you will then be able to access these files via a web browser by pointing your web browser at the server. On Windows, your DocumentRoot will be something like `C:\Program Files (x86)\Apache Software Foundation\ Apache2.2\htdocs`, depending on where you installed Apache and which version you have (in this case 2.2); on Linux it is usually located at `/var/www/`, on Mac OS it is located at `/Library/WebServer/Documents`.

Unix file ownership and permissions

Understanding the basics of file ownership and permissions is critical to working successfully with Apache on Linux and Mac OS, so we provide here an overview of the key concepts. Windows users can skip most of this section as file permissions in Windows are much less strict, so there is no need to consider permissions and ownership. In Windows you only need to ensure that you can modify the content of Apache's `htdocs` folder.[2]

---

[2] If you find that you are unable to modify the content of `htdocs` in Windows 8, right click on the folder, select Properties, then in the Security tab of the window that appears select Users, and then click Full control and OK.

All files and directories on a Unix based system are owned by a user and a group, and the user, the group, and everyone else can be given specific privileges to use, or indeed not use, a file or directory. To find out what user and group a file belongs to, open up a terminal and run the following command in a directory of your choice: `ls -l`. You will see output similar to that shown below:

```
-rw-rw-r--  1 daz staff 4.5K 2007-02-28 14:40 image.gif
drwxr-xr-x 11 daz staff  374 2007-09-10 11:13 directory1/
drwxr-xr-x  8 daz staff  272 2007-12-01 15:31 directory2/
drwxr-xr-x 16 daz staff  544 2007-03-29 21:48 directory3/
-rw-rw-r--  1 daz staff 3.9K 2007-02-28 14:40 index.html
```

In this example, all the files belong to the user `daz` and the group `staff`, indicated by the names to the left of the file listing. We also need to take note of the permissions granted to the files and directories—this is the matrix of d, r x, and w symbols on the left. This basically tells us if the line refers to a directory (signified by a d), and then which users can read (r)/write (w)/execute (x) a file. The layout of this matrix is shown below.

It is basically three groups of the letters r, w, and x for each possible type of user that could access a file. If a letter is present, it means that the appropriate permission is granted to that type of user.

The reason that we need to have an understanding of this is that the Apache web server runs as if it were being run from a user account—not from the `root` (administrator) account—in order to protect the server from being compromised should a security flaw be found in Apache. Therefore, any document that you create for Apache to use must have appropriate permissions set so that Apache can both access and handle the file accordingly. As such, from here we shall assume that all documents that you create and place in your DocumentRoot directories will be owned by the `root` user and also belong to the `root` user group (as if they were created by the administrator of your system). If you decide to create files in your own directories and then move/copy them into the Apache served directories, you can change the ownership of the files (over to the `root` user) using the chown command, for which the generic syntax is:

```
chown new_owner:new_group file_name
```

An example of this would be:

```
chown root:root index.html
```

Then, finally, you will need to ensure that all of your files (i.e. HTML, CSS, JavaScript, etc.) are readable by the server process, so we should ensure that the files are readable by all users, but not writable or executable. These permissions can be altered using the `chmod` command, the syntax for which is:

```
chmod modification_code file_name
```

The modification code refers to a code that defines the permissions that will be applied to the file after using this command. The way these codes are constructed is complicated, but there is only really one that you are likely to need whilst working with Apache and plain HTML: 744. Code 744 sets the permissions of a file to be readable/writable/executable by its owner, but only readable by everyone else, (`rwxr--r--` if viewed via `ls -l`). For example:

```
chmod 744 index.html
```

Appendix C: Setting up a Linux virtual machine in Windows

This appendix explains how to set up a virtual Linux machine within Windows, using Oracle's VirtualBox software package. As a Windows user, if you follow these instructions you can have a fully functioning instance of Ubuntu Linux running within a window on your Windows desktop. There are many motivations for doing this: to get hands-on experience of Linux without affecting your Windows installation; to use software that is not currently available on Windows; and to test that software you develop in Windows is compatible with Linux. There are many virtualization platforms to choose from. We suggest Oracle's VirtualBox because it is free and, in our experience, has proved to be reliable and easy to configure and use.

We are going to use VirtualBox with Windows as the *host OS* and Ubuntu Linux as the *guest OS*—so we will install VirtualBox on Windows and then set up a Virtual machine (VM) running Ubuntu within VirtualBox. VirtualBox can also be installed on Linux and Mac OS, and a whole range of different operating systems can be installed within VirtualBox, so there are many more possible host/guest combinations. For example, a Linux user may use VirtualBox to run an instance of Windows, or even a different version of Linux. The latter option may sound pointless, but it is a great way to test software compatibility on different Linux distributions, or to evaluate different Linux distributions without the upheaval of changing the OS that you use day to day.

C.1 Installing VirtualBox and configuring a virtual machine

To get a Linux VM running within Windows, it is necessary to complete three main tasks. First, we must install the VirtualBox application on Windows. We must then create a virtual machine within VirtualBox. Finally, we need to install a Linux operating system (in this case Ubuntu) onto the virtual machine. The following instructions lead you through the process step by step.

Building Bioinformatics Solutions. Second Edition. Conrad Bessant, Darren Oakley and Ian Shadforth.
© Conrad Bessant, Darren Oakley, and Ian Shadforth 2014. Published 2014 by Oxford University Press.

Step 1: Obtain the latest Ubuntu disk image

Before we start the main tasks, we need to get hold of an operating system to install on the VM when it is finally set up. Appropriately for a virtual machine, operating systems can be obtained as disk images. Although a disk image is just a single file (with the extension .iso), to VMs within VirtualBox it looks like a physical disk full of files, just like a CD that you would use to install an operating system on a real computer.

We are going to use the Ubuntu disk image, which is available from www.ubuntu.com. When you navigate to the download page you will see that there are a few different flavours of Ubuntu to choose from. The most appropriate for our purposes is Ubuntu Desktop. There are differently numbered versions of this—most noticeably a well-established version that has long term support (LTS) and a newer version with cutting-edge features. Our recommendation would be to select the LTS version (there is nothing stopping you setting up a separate VM with the newer version some time later). You will also need to choose between a 32-bit or 64-bit version of Ubuntu. This distinction can be important for some software, but if you do not have a particular reason to do otherwise we suggest you download the 64-bit version unless you are running a 32-bit version of Windows (you can find out by checking your system properties within Windows).

Step 2: Install VirtualBox

From the VirtualBox downloads page (www.virtualbox.org/wiki/Downloads), download the installer for the latest version of the VirtualBox platform package for Windows hosts. Execute the installer and follow the installation process—there should be no need to deviate from the default installation settings.

Step 3: Create and configure a virtual machine

When you launch the VirtualBox application for the first time you will be presented with an essentially empty window, because there are no VMs to display. To begin creating your first VM, click the New icon in the toolbar. You will be prompted to give the machine a name (choose whatever you want), a type (choose Linux) and a version (choose Ubuntu 64bit, or simply Ubuntu if you downloaded the 32-bit version).

Clicking Next will take you to the next step, where you specify how much of your computer's memory (RAM) to allocate to this particular VM. The amount of RAM to allocate will depend on several factors, particularly how much RAM your PC has, what you intend to use the VM for, and what else Windows will be doing while the VM is running. A convenient way to think about this is to remember that, for all intents and purposes, the VM is a self-contained computer, so how much memory would you want installed in the computer? Two gigabytes is probably an absolute minimum, but if your host PC only has 2GB that will clearly be impossible so a compromise will be needed. There is no need to agonize too much over this as RAM allocation can easily be changed later.

The next steps in setting up the VM deal with the creation of a virtual hard drive for the machine to use. You will be invited to select various options during this process. The default settings should be acceptable. The only thing we would say is that if you have a lot of free disk space it can be a good idea to create a fixed size hard drive (rather than the dynamically allocated default) because dynamically allocating disk space has an impact on performance and can confuse software that checks free disk space (e.g. installers). Note that you can select the location of virtual hard drive—it does not have to be the same hard disk on which VirtualBox was installed. Regarding disk size, as with RAM, you need to think about what would be sensible for the work you are proposing to do.

On completion of the VM creation process, the VM will be added to the previously empty list of VMs in the left part of VirtualBox's main window, and the rest of the window will show the specification of the VM (see Fig. C.1 for an example).

Step 4: Installing Ubuntu
You can now double click the VM in the list to start it. Dialogue boxes may pop up while the VM is starting telling you about keyboard and mouse capture—it is safe to dismiss these by clicking OK. Because there is no operating system installed on the VM, VirtualBox will then prompt you to select a startup disk. You need to

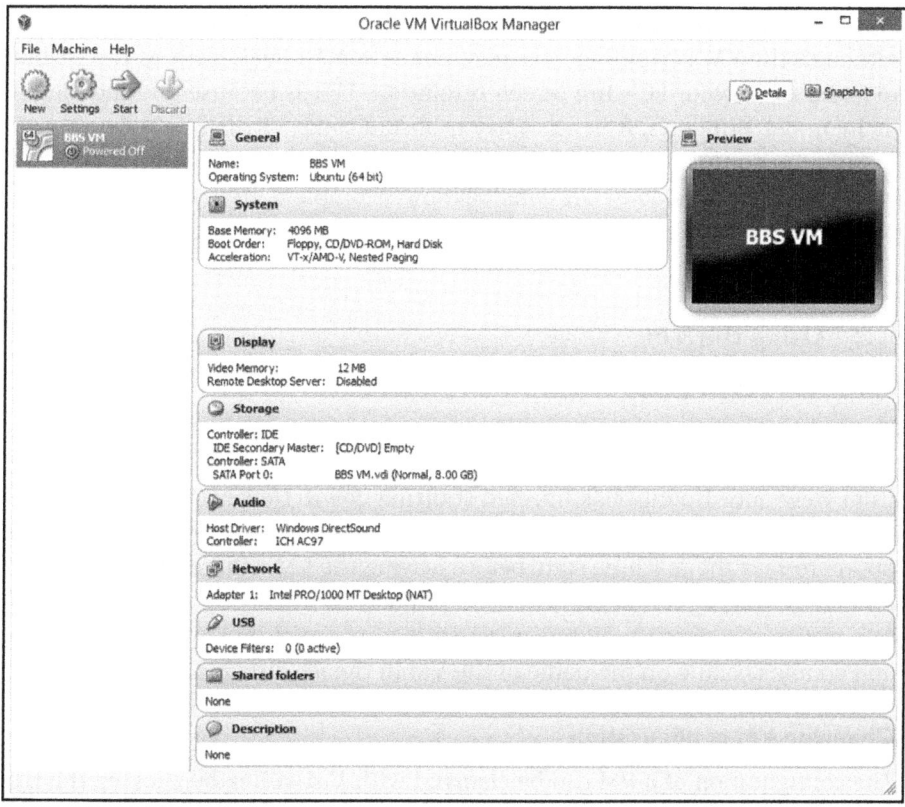

Fig. C.1 The main window of VirtualBox, with a single VM called "BBS VM" shown in the list of VMs in the pane on the left of the window, and the configuration details of that VM on the right.

select the Ubuntu ISO disk image that you downloaded back in Step 1. Having selected this, the Ubuntu installation process will start within the VM and you will be given the option to select your language and either Try Ubuntu or Install Ubuntu. Since we are starting with a new (virtual) machine, we need to select the latter option.

As with any software installation, you will be given various options during the process. It is safe to go with the default options throughout, even though some sound very alarming, for example at one point you need to confirm that you are going to 'erase disk and install Ubuntu'. It is important to remember that the disk Ubuntu is talking about is the virtual hard drive (this is the only disk it can see at this point)—not your actual hard disk—so your Windows installation and any local files are safe. Similarly, when you are prompted to restart the computer at the end of the installation it will be the VM that re-starts, not your whole computer.

When the VM re-starts after installation, you should have a fully functioning VM running the latest version of Ubuntu. You will see that VirtualBox seamlessly routes the VM's hardware interactions through Windows and out to your PC, so you can access all the essential hardware on your PC including keyboard, mouse, networking, and audio right away.

Step 5: Install guest add-ons

Although the VM is running, you may notice some issues, such as the inability to make use of your PC's full screen resolution. This is because, to function optimally under Windows, it is necessary to install some VirtualBox add-ons within Ubuntu. To do this, go to the Devices menu at the top of the VM's window and click Install Guest Addons. Once installation is complete, re-start Ubuntu. Ubuntu should now adapt its resolution to whatever size you make the VM window.

C.2   Using the VM

Starting and stopping a VM

If you shut down the VM within Ubuntu (e.g. by clicking the cog icon in top right and selecting Shut Down), the VM will automatically 'power off' and disappear from view once Ubuntu has finished shutting down. However, if you simply try to close the VirtualBox window in which the VM is running, you will be given the option to save the machine state before leaving the VM—this allows you to carry on exactly where you left off next time you use the VM.

However you close the VM, you can start it in future by launching VirtualBox and double clicking on the name of the VM in the VM list.

Changing VM configuration

The configuration of a VM can be changed with VirtualBox by clicking the titles of the configuration panes for the VM in the main VirtualBox window, so if you come to regret any of the selections made when setting up the VM you can change them here. Note that, just like a real computer, this can only be done when the VM is 'powered off'—not when it is in use or in a saved state.

Getting data to and from the VM

There are various options for moving data between Windows and the VM. VirtualBox's Shared Clipboard feature can be used to transfer small amounts of data, such as commands or blocks of text. This can be enabled via the Shared Clipboard item in the Devices menu when a VM is running. Enabling this in bidirectional mode will all you to copy something from a Windows window and paste it directly into a window in Ubuntu, and vice versa. For whole files and folders the Drag'n'Drop functionality, which can also be enabled via the Devices menu, makes it possible to drag files into Ubuntu from the Windows desktop.

A more elegant and programmatically accessible way of sharing data is to set up shared storage. This could be a shared hard drive accessed over a network, or a cloud storage solution such as Dropbox. It is also possible to set up a folder that can be shared between Ubuntu and Windows on the same computer. These folders can be created and configured by clicking on Shared Folders in a VM's configuration panel in the VirtualBox main window. When adding a folder you will need to specify the location of the folder on the host (Windows) machine and give the folder a name to be used within the VM. For ease of use, it is best to make the folder auto-mount and permanent. When you restart the VM, you will find the folder in the `\media` directory – it will have `sf_` prefixing your chosen name to indicate that it is a shared folder. To be able to access the folder from within Ubuntu, you need to add yourself to a group called `vboxsf`, by executing the following command at the command line (where `username` is your username), and then rebooting the VM.

```
sudo usermod -a -g vboxsf username
```

Running a server within a VM

All of the server software that we may want to use—including Apache, Morbo, and Galaxy—should run faultlessly under Ubuntu with a VirtualBox VM. However, by default the server will only be accessible from inside the VM (i.e. as `localhost`). To access from outside (e.g. from the host OS) you will need to set the VM's Network Adapter (accessed via the Devices menu) to Bridged Adapter rather than the default NAT. You can then find the IP address of the VM by typing `ifconfig` at the command line within the VM (the IP address is shown as the `inet addr`) and use this address to access the server from your Windows browser.

C.3   Other uses of virtual machines

Creating a second VM running Bio-Linux

VirtualBox does not limit you to having a single VM configured at one time. A new VM can be added to the list in the VirtualBox by clicking the New icon and following the process explained earlier. This allows you to experiment with different distributions of Linux, or indeed different operating systems altogether. One distribution of particular note is Bio-Linux, which provides a version of Ubuntu pre-loaded with hundreds of popular bioinformatics tools and documentation, including a fully functioning Galaxy installation. To get a Bio-Linux VM up and running, you just need to download the disk image from `nebc.nerc.ac.uk/nebc/tools/bio-linux` and follow the instructions explained earlier in steps 3, 4, and 5 of Section C.1.

Distributing virtual machines

As with most virtualization platforms, VirtualBox provides the facility to export a VM to a file. This file can then be passed to other people who can import it into their own installation of VirtualBox and use that VM, complete with the OS and all other software installed within it. This is particularly valuable for those of us building bioinformatics solutions, because we can get our software into people's hands without them having to worry about any complex installation procedures or dependencies (e.g. a particular version of a Perl may be required). This is not really suitable for ultimate deployment of new bioinformatics tools, but it is an efficient way to share prototype software with colleagues and can even be used to make available all the software and data generated with a particular research project. For example, a VM from the ENCODE project can be downloaded from `scofield.bx.psu.edu/~dannon/encodevm`.

Index

Lightning Source UK Ltd.
Milton Keynes UK
UKOW05f0637100216

268081UK00002B/3/P